William Fawcett
Alfred Barton Rendle

Flora of Jamaica Containing Descriptions of the Flowering Plants Known from the Island

Volume I

Elibron Classics
www.elibron.com

Elibron Classics series.

ISBN 0-543-91360-0 (paperback)
ISBN 0-543-91359-7 (hardcover)

This Elibron Classics Replica Edition is an unabridged facsimile of the edition published in 1910 by the British Museum, London.

FLORA OF JAMAICA

VOL. I.

ORCHIDS OF JAMAICA

FLORA OF JAMAICA

CONTAINING DESCRIPTIONS OF THE FLOWERING PLANTS KNOWN FROM THE ISLAND

BY

WILLIAM FAWCETT, B.Sc., F.L.S.

FORMERLY ASSISTANT, DEPARTMENT OF BOTANY, BRITISH MUSEUM (NATURAL HISTORY)
LATE DIRECTOR OF PUBLIC GARDENS AND PLANTATIONS, JAMAICA,

AND

ALFRED BARTON RENDLE, M.A., D.Sc., F.R.S., F.L.S.

KEEPER OF THE DEPARTMENT OF BOTANY, BRITISH MUSEUM (NATURAL HISTORY)

WITH ILLUSTRATIONS

VOL. I.

ORCHIDACEÆ

WITH THIRTY-TWO PLATES

LONDON:
PRINTED BY ORDER OF THE TRUSTEES OF THE BRITISH MUSEUM

AND SOLD BY

LONGMANS & CO., 39, PATERNOSTER ROW, E.C.; B. QUARITCH, 11, GRAFTON STREET, NEW BOND STREET, W.; DULAU & CO., LTD., 37, SOHO SQUARE, W.; AND AT THE BRITISH MUSEUM (NATURAL HISTORY), CROMWELL ROAD, S.W.

1910

LONDON:
PRINTED BY WILLIAM CLOWES AND SONS, LIMITED,
DUKE STREET, STAMFORD STREET, S.E., AND GREAT WINDMILL STREET, W.

PREFACE

DURING his residence in Jamaica for twenty-one years as Director of Public Gardens and Plantations, Mr. William Fawcett gave attention to the botanical exploration of the Island. He was specially interested in the Orchids, and, with the assistance of Mr. William Harris, Superintendent of the Gardens and an excellent collector, a very fine collection illustrating the family was got together. A beautiful series of drawings from the living plants was also made under Mr. Fawcett's direction by Miss Helen Wood. During one of Mr. Fawcett's visits to England we agreed to collaborate in a monograph of the Jamaican Orchids, and in 1904 an account of the genus *Lepanthes* was published in the Transactions of the Linnean Society. The work, however, proceeded slowly until 1908, when Mr. Fawcett retired and came to England. The permission of the Trustees having been obtained for the publication of the work as a British Museum Catalogue, Mr. Fawcett was fortunately able to devote himself to the task, and I have given what time I could spare in the intervals of other work. Though Mr. Fawcett has done the larger part of the work of preparation, we have both been carefully through each species. Our work has been greatly facilitated by the loan of the Jamaican Herbarium, for which we are much indebted to the Government of Jamaica.

As Mr. Fawcett is able to continue his work at the Museum, it is proposed that the present volume shall form the first part of a complete "Flora of Jamaica."

A. B. RENDLE.

DEPARTMENT OF BOTANY,
BRITISH MUSEUM (NATURAL HISTORY),
CROMWELL ROAD, LONDON, S.W.

October, 1910.

INTRODUCTION

The Island of Jamaica, the largest of the British West Indian Islands, is 144 miles long and 30 to 40 miles broad. It lies about 90 miles south of Cuba, and about 380 miles N.E. of Honduras, the nearest part of the mainland.

"Jamaica is remarkable for the limited area in which very many species are found, and for the wonderful contrasts in neighbouring districts in geological structure and soil, in elevation, in rainfall and in temperature.

"The Blue Mountains in the eastern end of the island attain the highest elevations, rising to 7,423 feet. The distance from the ridge to the coast on the north and south in a straight line is about 20 miles. On the Blue Mountains and north to the coast the north-east trade winds bring almost constant rain throughout the year, the rainfall varying from 150 to 200 inches annually. From the ridge to the south coast the rainfall decreases, and as the Blue Mountains run not directly west but inclined towards the north-west, and as the prevailing winds are from the north-east, the result is a decrease in the rainfall in travelling along the coast from Morant Point to Kingston; thus from Morant Point to Morant Bay there is a rainfall from 70 to 55 inches annually, then a belt 50 to 40 inches, and then for a considerable distance to Kingston and beyond, 35 to 30 inches. The slopes of the mountains are frequently precipitous, and there may be great differences in elevation, in moisture and temperature within short distances. For instance, one may pass from the southern slopes of the John Crow Peak of the Blue Mountains at about 5,500 feet, where xerophytic conditions sometimes occur and often cold nights with temperature down to 50° F., round the shoulder to the north, and straight down 2000 feet to the Mabess River, where there is almost constant rain, continual moisture from a perennial stream, and a warm temperature at nights. The geological structure of the district consists chiefly of brittle shale, easily decomposing, with occasional masses of coral limestone, marble, granite, &c., interspersed.

"There is a range of mountains close to the east end, running in a northerly and southerly direction, and rising to an elevation of 2,100 feet. The coast in the neighbourhood is rocky, and continues so for some distance to the west of Port Antonio. These mountains, known as the John Crow Mountains or Blake Mountains, are so wet, and the limestone rock is so difficult to travel over, that it is believed only one white man has ever crossed them. Nathaniel Wilson may have collected on the slopes of these mountains on the east side, or on the southerly slopes, some of the species that have not been found since. The limestone rock of these mountains occurs over the greater part of Jamaica, and in large districts it has disintegrated so as to form, as in the John Crow Mountains, such a series of precipitous cliffs that it is impossible to travel through it. The so-called 'Cockpit country' is of this nature, and it is in such places that new species of plants may naturally be looked for." (W. F. in Urban, Symbolæ Antillanæ, vi. 86.)

The Orchids of Jamaica are of special interest from the large number of endemic forms which they include. The total number of genera is 61, one of which, *Homalopetalum*, is endemic. The total number of species is 194, of which 73 are confined to the Island. There are also some endemic varieties.

The most striking affinity of the Orchid flora is with that of Cuba, 82 species, out of a total of 121, which are not endemic, occurring also in that island, and of these 14 are restricted to the two islands. The affinity with Hispaniola is apparently not so marked, there being only 29 species known from both islands, of which two, *Epidendrum repens* and *Broughtonia domingensis*, occur nowhere else. The difference is due no doubt to some extent to our greater knowledge of the botany of Cuba as compared with that of Hispaniola. With the much better explored island of Porto Rico, which lies further east beyond Hispaniola, the affinity is apparently greater, there being 40 species common to Jamaica and this island, four of which are known elsewhere only in Cuba. Ten species are common to Jamaica and the Bahamas, and 15 to Jamaica and Florida, the relation of which to the West Indies is very close. Passing further east and south from Porto Rico, 38 species are common to Jamaica and one or more of the Lesser Antilles. Forty species occur both in Jamaica and Trinidad, of which nine are not recorded from the Lesser Antilles, but six of these are known from Cuba, and three from Cuba and Porto Rico. *Epidendrum nutans* has a somewhat remarkable distribution, being known only from Trinidad, Venezuela and Brazil, in addition to Jamaica.

The relationship with Central America is well marked; 37 species are common to Jamaica and some part of Central America, and if we include Mexico the number is raised to 44. Of these 28 are also found in equatorial South America, of which sixteen represent an element widely distributed over the West Indian and tropical American area. Of the sixteen which do not occur in equatorial South America, four are confined to Jamaica and Central America and ten to Central America (with Mexico), Jamaica and Cuba.

The affinity with equatorial South America is also strongly marked, 48 species being known to occur in some portion of tropical South America and Jamaica. A number of these also represent a widely distributed tropical American element occurring more or less generally in the West Indies and Central America, some few being also known from Florida and the Bahamas. If we consider Trinidad as part of equatorial South America, six of these forty-eight species occur nowhere else in the West Indies except in Jamaica. These comprise three species of *Epidendrum* —*E. verrucosum*, known also only from Columbia, *E. nutans* from Trinidad, Venezuela and Brazil, and *E. imbricatum* from Central America and Brazil—and *Brassavola nodosa, Maxillaria rufescens* and *Cryptarrhena lunata* from Central America and more or less widely distributed in equatorial South America.

These six species which do not otherwise occur in the West Indies, together with the Central American group already referred to, probably represent an element of the Jamaican flora associated with two former extensions of the land surface between Cuba and Yucatan, and Jamaica and Honduras. The sea-bottom between Jamaica and Honduras comes within the 6000 fathom limit and contains large banks within the 600 fathom limit. If we include Jamaican species which occur also in the Greater Antilles, there are twenty-four common to these and equatorial South America which are not found in the Lesser Antilles and may also be associated with the extensions of land already referred to. A south-eastern line of association with equatorial South America through Hispaniola and Porto Rico is without doubt represented by the series of the Lesser Antilles and Trinidad; as we have mentioned above thirty-eight Jamaican species occur in the Lesser Antilles and forty in Trinidad.

These facts of distribution are in accord with geological evidence of a former linking up of Central America and Venezuela by means of the line of the West Indies.

" The Antilles are formed by the summits of a mountain chain which separates the Caribbean Sea from the Atlantic Ocean and the Gulf of Mexico. A second arc running from Honduras through Jamaica and the south-west of Haiti appears to join the principal chain. Even the great deeps which occur in certain places, as, for instance, between Virgin Gorda and Anguilla, do not interrupt the continuous course of the mountain chain."

" With the slopes of the Mexican plateau in the State of Oaxaca, says Seebach, the compact northern continent terminates. To the south and to the east of the isthmus of Tehuantepec, Central America commences, which already belongs to the island world of the Antilles. 'The mountain series of the Greater Antilles, which further to the east in Porto Rico and San Domingo—the eastern part of Haiti—forms a single main chain, divides in the middle of the latter island, giving rise to a southern branch which proceeds through the elongated peninsula of Jacmel towards Jamaica and Honduras, and a northern branch which extends beyond Cuba towards Yucatan . . . Is it merely a remarkable accident that the Sierra Maestra, consisting of crystalline schists and massive rocks and situated in the south-east of Cuba (where the Greater Antilles reach their greatest elevation of 2,338 meters above the sea), should run through the Cayman group, the bank of Misteriosa, the Viciosas, and Swan Island to the depths of the gulf of Honduras, and that from the edge of this mountain ridges similarly constituted should rise abruptly and proceed with constant strike into the interior.' "—(Suess, " The Face of the Earth," Engl. Transl. i. 543.)

Our knowledge of the flora of Jamaica starts with the work of Sir Hans Sloane, who collected in the Island during a stay of fifteen months in the years 1687–89, and published the results in his work on the Natural History of Jamaica (1707–25). Sloane's Jamaican plants occupy Vols. I. to VIII. of the Sloane Herbarium at the British Museum, and these volumes have been carefully gone through in connection with our work on the Orchids. The herbarium of Sir Joseph Banks, which became at his death the foundation of the Department of Botany, included collections from Jamaica by William Houstoun (1729–33), William Wright (1771–85), Roger Shakespear (1780–2), and Francis Masson (1781), and also a large number of specimens from Olof Swartz (1784–6), including types of many of the new species published in his "Prodromus" and elsewhere. Besides these and the later collections at the British Museum we have worked through the material in the Kew Herbarium, which includes the collections of James Macfadyen (1825–50), W. Purdie (1843, 4), Nathaniel Wilson (1846–58), R. C. Alexander Prior (1849, 50), W. T. March (1857, 8), and others.

We have to thank the Government of Jamaica for the loan of the whole of the material contained in the Jamaican Herbarium, the formation of which was begun by Sir Daniel Morris (1879–86) and was continued by one of us (1887–1908) with the very efficient help of Mr. William Harris as collector. In this connection special reference should be made to Mr. Harris's work as an indefatigable and successful plant-collector, to which is owed a considerable increase in knowledge of the flora of Jamaica. Our work has been greatly facilitated by the use of a fine series of drawings made at the Hope Gardens, Jamaica, from the living plants, under the supervision of one of us, by Miss Helen A. Wood; many of these have been reproduced in the plates at the end of the volume, for the large number of which we are grateful to the Trustees of the British Museum. We have also to thank Dr. N. L. Britton, Director of the New York Botanic Garden, for specimens and for the loan of a collection of Orchids made in Jamaica by Sir D. Morris. A number of private individuals resident in or visiting the Island have also contributed specimens; most of these are contained in the Jamaican Herbarium. By the kindness of Dr. Radlkofer of the Royal Botanical Museum, Munich, we have been able to work through the Orchids collected by H. R. Wullschlaegel (1847–9). As Wullschlaegel's specimens are cited by Grisebach in the "Flora of the British West Indies," it has been of interest to revise his determinations by comparison with the original collection. Prof. Ignatius Urban of Berlin has generously entrusted to us the volume of Swartz's "Icones Ineditæ," and has allowed us to cite the drawings. During the preparation of our work Prof. Urban has given a detailed account of the botanical exploration of Jamaica since the publication of Grisebach's "Flora" in his "Symbolæ Antillanæ" (vi., pp. 70–131, 1909). To Prof. Cogniaux, who is elaborating the Orchids of the West Indies for the "Symbolæ," we are indebted for the loan of specimens and helpful information.

Our descriptions of the genera have been drawn up with special reference to the species which occur in Jamaica. For measurements we have used the metric system, but for convenience of tourists we have added English measures for the height of the plant and the size of the flower. The figure below renders possible a ready comparison of the two systems, *e.g.* 2·5 cm. are equivalent to an inch, 2 mm. to $\frac{1}{12}$ inch, and 1 dm. to about 4 inches.

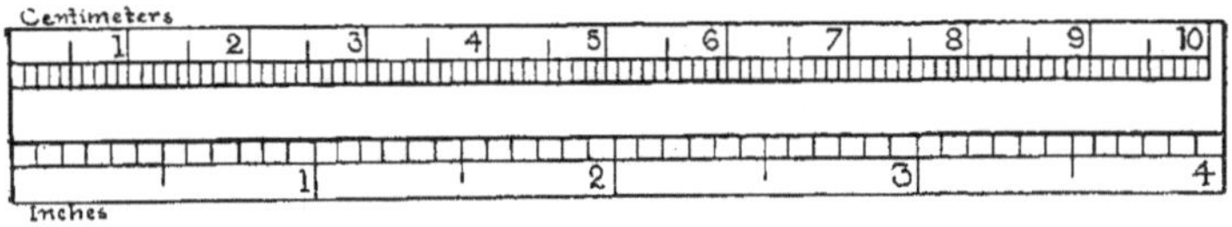

The J.P. (Jamaican Plants) numbers cited are those given by Sir D. Morris to the species in the Jamaican Herbarium; the numbers cited as Fl. Jam. (Flora Jamaicensis) are those attached to specimens, as collected, later.

The mark "!" indicates that we have seen a specimen.

In order to conform to the international rule, which requires that diagnoses of new species shall be published in Latin, descriptions of novelties have been inserted in the "Journal of Botany," and are quoted here from that work.

W. FAWCETT.
A. B. RENDLE.

LIST OF PLATES

WORKS REFERRED TO IN THE TEXT

(Where no date is given the work is in progress)

Ait. Hort. Kew.—Hortus Kewensis, ed. 2. W. T. Aiton. London. 1810-13.

Ames Orch.—Orchidaceæ, Illustrations and Studies of the Family. Oakes Ames. Boston and New York.

Ames Orch. S. Fla.—Contribution...Orchid Flora of S. Florida. O. Ames. Cambridge, Mass. 1904.

Andrews Bot. Rep.—Botanists Repository. H. C. Andrews. London. 1797-1816.

Ann. & Mag. Nat. Hist.—Annals and Magazine of Natural History. London.

Ann. Sc. Nat.—Annales des Sciences Naturelles. Paris.

Aubl. Pl. Guian.—Histoire des Plantes de la Guiane Française. J. B. C. F. Aublet. London and Paris. 1775.

Bateman Orch. Mex. & Guat.—The Orchidaceæ of Mexico and Guatemala. J. Bateman. London. 1837-43.

Bauer Ill. Orch.—Illustrations of Orchidaceous Plants. F. A. Bauer. London. 1830-38.

Beitr. Orch. Centr. Amer.—Beiträge zu einer Orchideenkunde Central Amerikas. H. G. Reichenbach fil. Hamburg. 1866.

Benth. & Hook. f. Gen. Pl.—Genera Plantarum. G. Bentham and J. D. Hooker. 1862-83.

Ber. Deutsch. Bot. Ges.—Berichte der Deutschen Botanischen Gesellschaft. Berlin.

Bijdr. Flor. Ned. Ind.—Bijdragen tot de Flora van Nederlandsch Indië, K. L. Blume. Batavia. 1825-26.

Biol. Centr. Amer. Bot.—Biologia Centrali-Americana. F. D. Godman and O. Salvin. London. Botany by W. B. Hemsley. 1879-88.

Bonpl.—Bonplandia. Ed. by W. E. G. Seemann and B. Seemann. Hanover. 1853-62.

Bot. Mag.—Botanical Magazine. Ed. by W. Curtis and others. London.

Bot. Reg.—The Botanical Register. London. 1815-47.

Bot. Zeit.—Botanische Zeitung. Berlin and Leipzig.

Browne Hist. Jam.—Civil and Natural History of Jamaica. P. Browne. London. 1756.

Bull. Sc. Acad. Pétersb.—Bulletin scientifique publié par l'Académie Impérial des Sciences de St. Pétersbourg. St. Petersburg. 1836-40.

Catesby Nat. Hist. Carol.—Natural History of Carolina, &c. M. Catesby. London. 1731-43.

Comp. Bot. Mag.—Companion to the Botanical Magazine. W. J. Hooker. London. 1835-37.

Duperr. Voy. Coq. Bot.—L. I. Duperrey. Voyage autour du Monde sur *La Coquille*, 1822-25. Botanique par A. T. Brongniart. Paris. 1829.

Engl. & Prantl Nat. Pflanzenf.—Die natürlichen Pflanzenfamilien. H. G. A. Engler and K. A. E. Prantl. Leipzig. Orchids by E. Pfitzer, (ii. pt. 6. 1889).

Fawc. Fl. Pl. Jam.—List of the Flowering Plants of Jamaica. W. Fawcett. Kingston. 1893.
Fedde Repert.—Repertorium novarum specierum regni vegetabilis. F. Fedde. Berlin.
Fl. Bras.—Flora Brasiliensis. K. F. P. von Martius and A. W. Eichler, completed by I. Urban. Munich. 1840–1906.
Fl. Cub.—Flora cubana, &c. R. de La Sagra. Phanerogams by A. Richard. Paris. 1853.
Fl. Cub. Ic.—Icones plantarum in Flora Cubana. R. de La Sagra. Phanerogams by A. Richard. Paris. 1863.
Flora—Flora. Regensburg.

Gard. Chron.—Gardeners' Chronicle. London.
Gard. Mag. Bot.—Gardeners' Magazine of Botany. T. Moore and W. P. Ayres. London. 1850–52.
Griseb. Cat. Cub.—Catalogus Plantarum Cubensium. A. H. R. Grisebach. Leipzig. 1866.
Griseb. Fl. Br. W. Ind.—Flora of the British West Indian Islands. A. H. R. Grisebach. London. 1859–64.

H. B. & K. Nov. Gen. & Sp.—Nova genera et species plantarum. Humboldt, Bonpland and Kunth. Paris. 1815–25.
Hamburg Gartenz.—Hamburger Garten und Blumenzeitung. Hamburg.
Herb. Sloane—Sloane's Herbarium in the Department of Botany, British Museum.
Herm. Parad.—Paradisus Batavus. P. Hermannus. Ed. W. Sherard. Leyden. 1698.
Hook. Exot. Fl.—Exotic Flora. W. J. Hooker. London. 1823–27.
Hook Ic. Pl.—Icones Plantarum. W. J. Hooker and others. London.
Hook. Journ. Bot.—Journal of Botany. W. J. Hooker. 1834–42.

Ic. Fl. Germ.—Icones Floræ Germanicæ et Helveticæ. H. G. L. Reichenbach and others. Leipzig. Orchids by H. G. Reichenbach fil. xiii & xiv, (1850, 1).

Jacq. Collect.—Collectanea. N. J. von Jacquin. Vienna. 1786–96.
Jacq. Enum. Pl. Carib.—Enumeratio Plantarum, quas in Insulis Caribæis, etc., detexit N. J. von Jacquin. Leyden. 1760.
Jacq. Ic. Pl. Rar.—Icones Plantarum rariorum. N. J. von Jacquin. Vienna. 1781–93.
Jacq. Sel. Stirp. Amer.—Selectarum Stirpium Americanarum Historia. N. J. von Jacquin. Mannheim. 1763. Ed. pict.—Edition with coloured plates. 1788.
Journ. Bot.—Journal of Botany. J. Britten. London.
Journ. Linn. Soc.—Journal of the Linnean Society of London.

Kew Bulletin—Bulletin of Miscellaneous Information, Royal Botanic Gardens, Kew. London.
Kuntze Rev. Gen. Pl.—Revisio Generum Plantarum. C. E. O. Kuntze. Leipzig. 1891–8.

L. Sp. Pl.—Species Plantarum. C. Linnæus. Stockholm. 1753.
L. Syst.—Systema Naturæ. C. Linnæus. (Ed. 10. Stockholm. 1759.)
Lam. Encycl. Méth.—Encyclopédie Méthodique. Paris, &c. Botany by J. B. Lamarck and others. 1789–1817.
L'Hérit. Sert. Angl.—Sertum Anglicum. C. L. L'Héritier. Paris. 1788–92.
Lindenia—Lindenia. Gand.
Lindl. Coll. Bot.—Collectanea Botanica. J. Lindley. London. 1821.
Lindl. Fol. Orch.—Folia Orchidacea. J. Lindley. London. 1852–59.

Lindl. Gen. & Sp. Orch.—Genera and Species of Orchidaceous Plants. J. Lindley. London. 1830–40.
Lindl. Orch. Scel.—Orchidearum Sceletos. T. Lindley. London. 1826.
Linnæa—Linnæa. Berlin. 1826–82.
Llave & Lex. Nov. Veg. Descr.—Novorum Vegetabilium Descriptiones. P. de La Llave and J. Lexarza. Mexico. 1824–25.
Lodd. Bot. Cab.—Botanical Cabinet. C. Loddiges. London. 1817–33.
Lour. Fl. Cochinch.—Flora Cochinchinensis. J. Loureiro. Lisbon. 1790.

Mém. Mus. Par.—Mémoires du Muséum d'Histoire Naturelle. Paris. 1815–32.
Mém. Soc. Phys. & Hist. Nat. Genève.—Mémoires de la Société de Physique et d'Histoire Naturelle de Genève. Geneva. 1821–76.
Merian Pl. Ind.—Recueil des Plantes des Indes. M. S. Merian. Paris. 1768.
Miller Gard. Dict.—Gardeners Dictionary. P. Miller. (Ed. 8. London. 1768.)
Mus. Bot. Lugd.-Bat.—Museum Botanicum Lugduno-Batavum. K. L. Blume. Leyden. 1849–56.

Nederl. Kruidk. Arch.—Nederlandsch Kruidkundig Archief. Leyden and Amsterdam.
Nov. Act. Upsal.—Nova Acta Regiæ Societatis Scientiarum Upsaliensis. Upsala.
Nutt. Gen. Amer.—Genera of North American Plants. T. Nuttall. Philadelphia. 1818.

Orch. Rev.—Orchid Review. London.

Paxt. Fl. Gard.—Paxton's Flower Garden. J. Lindley and J. Paxton. London. 1850–84.
Pl. Hartw.—Plantæ Hartwegianæ. G. Bentham. London. 1839–57.
Pluk. Alm.—Almagestum Botanicum. L. Plukenet. London. 1696.
Plum. Cat.—Catalogus Plantarum Americanarum. C. Plumier. Paris. 1703.
Plum. Ic. ined.—Plumier's original drawings in the Department of Botany, British Museum.
Plum. Nov. Pl. Amer. Gen.—Nova Plantarum Americanarum Genera. C. Plumier. Paris. 1703.
Plum. Pl. Amer. (Burm.).—Plantarum Americanarum. C. Plumier. Ed. J. Burmann. Amsterdam. 1755–60.
Poepp. & Endl. Nov. Gen. & Spec.—Nova genera ac species plantarum. E. F. Poeppig and S. L. Endlicher. Leipzig. 1835–45.
Proc. Biol. Soc. Wash.—Proceedings of the Biological Society of Washington. Washington, D.C.

Redouté Liliacées—Les Liliacées, peintes par P. J. Redouté. Text by A. P. De Candolle and others. Paris. 1802–16.
Reichb. Fl. Exot.—Flora Exotica. H. G. L. Reichenbach. Leipzig 1834–6.
Rep. Miss. Bot. Gard.—Report of the Missouri Botanical Garden. St. Louis.

Salisb. Ic. ined.—R. A. Salisbury's original Drawings and Mss. in bound volumes in the Department of Botany, British Museum.
Salisb. Prodr.—Prodromus stirpium in horto ad Chapel Allerton vigentium. R. A. Salisbury. London. 1796.
Saund. Ref. Bot.—Refugium Botanicum. W. W. Saunders. London. 1869–73.
Schrad. Journ.—Journal für die Botanik. H. A. Schrader. Gottingen. 1799–1803.

Schum. & Lauterb. Nachtr. Fl. Deutsch. Südsee.—Nachträge zur Flora der Deutschen Schutzgebiete in der Südsee. K. Schumann & K. Lauterbach. Leipzig. 1905.

Sert. Bot.—Sertum Botanicum. P. C. van Géel. Brussels. 1828-36.

Sloane Cat.—Catalogus plantarum quae in Insula Jamaica sponte proveniunt...H. Sloane. London. 1696.

Sloane Hist.—Voyage to ... Jamaica, with the Natural History. Hans Sloane. London. 1707-25.

Sm. Ic. Pict.—Icones pictæ plantarum rariorum. J. E. Smith. London. 1790-93.

Small Fl. S. E. Un. St.—Flora of the South Eastern United States. J. K. Small. New York. 1903.

Smith Spicil.—Spicilegium botanicum. J. E. Smith. London. 1791-92.

Spreng. Syst.—Systema Vegetabilium. C. Linnæus. Ed. 16, C. Sprengel. Gottingen. 1825-8.

Steud. Nom.—Nomenclator botanicus. E. G. Steudel. Ed. 2. Stuttgart and Tübingen. 1840-41.

Sw. Adnot. Bot.—Adnotationes Botanicæ. O. Swartz. Ed. by J. E. Wikström. Stockholm. 1829.

Sw. Fl. Ind. Occ.—Flora Indiæ Occidentalis. O. Swartz. Erlangen. 1797-1806.

Sw. Icon. ined.—Descriptiones et Icones Plantarum incognitarum quas in India occidentali detexit atque delineavit O. Swartz. Lent by Prof. I. Urban. Berlin.

Sw. Obs. Bot.—Observationes Botanicæ. O. Swartz. Erlangen. 1791.

Sw. Prodr.—Nova Genera et Species Plantarum. O. Swartz. Stockholm. 1788.

Symb. Ant.—Symbolæ Antillanæ. I. Urban. Berlin. Orchids by A. Cogniaux.

Trans. Linn. Soc.—Transactions of the Linnean Society of London.

Vet. Handl. Stockh.—Kongl. Vetenskaps Academiens nya Handlingar. Stockholm.

Walp. Ann.—Annales Botanices systematicæ. W. G. Walpers and others. Leipzig. 1848-68. Orchids by H. G. Reichenbach fil.

Willd. Sp. Pl.—Species Plantarum. C. Linnæus. Ed. 5. C. L. Willdenow. Berlin. 1797-1810.

Williams Orch. Alb.—The Orchid Album. R. Warner, B. S. Williams and others. London. 1882-96.

Xen. Orch.—Xenia Orchidacea. H. G. Reichenbach fil. Continued by F. Kraenzlin. Leipzig. 1858-1900.

NAMES OF THE MORE IMPORTANT COLLECTORS AND CONTRIBUTORS WITH DATE OF COLLECTION

BANCROFT, E. N. 1811–42.
BARRETT, Miss T. MOULTON, 1887–
BERTERO, C. 1821.
BLAKE, Sir HENRY and Lady, 1887–97.
BRITTON, Dr. N. L. and Mrs. 1906–
BROWNE, P. 1746–55.
CAMPBELL, E. 1894–6.
CRADWICK, W. 1888–
DISTIN, H. 1830–40.
EGGERS, H. VON. 1888.
FAWCETT, W. 1887–1908.
GOSSET, Miss B. 1909–
HARRIS, W. 1881–
HART, J. H. 1880–6.
HITCHCOCK, A. S. 1890–
HOUSTOUN, W. 1729–33.
JENMAN, G. S. 1873–9.
LONGSTAFF, Dr. and Mrs. 1907.
MACFAYDEN, J. 1825–50.
MARCH, W. T. 1857, 8.
MASSON, F. 1781–
MCNAB, G. 1838–59.
MOORE, A. 1883–
MORRIS, Sir DANIEL. 1879–86.
NICHOLLS, C. N. 1891.
NICHOLS, G. E. 1903, 6.
PRIOR, R. C. ALEXANDER, 1849, 50.
PURDIE, W. 1843, 4.
SHAKESPEAR, R. 1780–2.
SLOANE, Sir HANS. 1687–9.
SULLIVAN, C. 1883–90.
SWARTZ, O. 1784, 5, 6.
SYME, G. 1879–84.
THOMPSON, W. J. 1889–
THOMSON, R. 1862–79.
TOMLINSON, R. K. 1900–
WATERS, J. about 1826–30.
WATT, D. E. 1890–
WILES, J. 1793 1805.
WILSON, N. 1846–58.
WRIGHT, W. 1771–7, 1782–5.
WULLSCHLAEGEL, H. R. 1847–9.

ORCHIDS OF JAMAICA

Orchids are perennial herbs showing much variety in their habit. They are either terrestrial, growing in the soil, or more or less epiphytic, that is, growing on stems or branches of trees, on rocks, etc., often surrounded by a dense growth of moss. In the simplest form of growth, the stem continually produces new leaves at the apex, while the flowering shoots spring from the axils of older leaves. This method, known as the monopodial, occurs in some epiphytic orchids, as in *Dichæa*, where the internodes and leaves are both very short. In all terrestrial orchids and many epiphytic, growth of the main axis soon ceases, usually at the end of one season, while a lateral shoot continues the growth next season. The apparent main axis of the plant is then a sympodium, consisting of the basal scale-bearing portions of successive shoots, the upper parts of which are aërial and leafy. The shoots may end in an inflorescence, or flowers are borne on special lateral branches. The development of the sympodial rhizome governs the habit of the plant. If the basal portion of the yearly shoot is short, the successive aërial shoots are crowded and the habit is bushy or clustered, whereas if the basal portions are lengthened a creeping or climbing habit results.

In terrestrial orchids the basal portions of the annual shoots form a thin or fleshy root-bearing rhizome (as in *Spiranthes*), or a tuber is produced each year (as in *Habenaria*, *Bletia*). The aërial stem is slender, bearing one to many foliage-leaves and ending in a single flower or an inflorescence. The leaves are inserted along the stem (cauline) or are clustered at the base (radical).

The majority of epiphytic sympodial forms are pseudobulbous; that is, the stem is thickened to serve as a reservoir of food and water. The pseudobulb consists of only one internode and bears one or more leaves at its apex, or of several internodes and bears leaves throughout its length, or bears scales at the lower nodes and a terminal tuft of leaves. The pseudobulb is of very various shapes, sometimes globose, usually more or less elongated; its surface is generally smooth and shiny. In *Pleurothallis* and allied genera the shoots are not thickened to form a pseudobulb, and bear a single leaf which spreads at right angles to the axis or apparently forms a continuation of it.

The leaves are simple, generally sessile, and wither and decay on the stem, or, as in most epiphytic species, separate by a distinct joint. The margin is entire and the apex often denticulate or unequally cut; the venation is as a rule parallel. In some cases the blade passes gradually into a stalk which has generally a sheathing base.

At the basal nodes of the shoots leaf-scales are borne, which pass gradually into foliage leaves, or the latter may start abruptly. The leaves are generally arranged in two ranks, one each side of the stem.

Leaves of epiphytic forms are generally thick and fleshy; and by a thick cuticle they are adapted for storage of water. A main root is never present; its place is taken by adventitious roots which arise especially from the nodes and often show as regular an arrangement as the leaves. In terrestrial orchids the roots are sometimes thickened to form stores of reserve material. The air-roots of epiphytic orchids have a special development of the outer layers forming a sponge-like tissue (*velamen*), by which water is absorbed for the plant from the atmosphere. Three kinds of air-roots may be distinguished. Clinging roots which creep close to the substratum and are inseparable from it; absorptive roots, or branches of the former which grow into the humus which collects about the plant; and the fine aërial roots which hang down sometimes to a considerable length.

The flowers are hermaphrodite and irregular. The perianth consists of an outer whorl of three sepals and an inner whorl of three petals which alternate with the sepals. The odd or median sepal is normally on the lower side of the flower and furthest from the axis. The odd petal, which generally differs considerably in size and shape from the two lateral petals and is termed the *lip*, is normally on the upper side of the flower next the axis. This position is, however, generally reversed by a twist of the ovary so that the lip in the open flower is on the lower side. The sepals are usually smaller and less conspicuous than the petals, but in some genera (as *Pleurothallis*, *Lepanthes* and *Brassia*) they are much larger. They are free from each other or more or less coherent; in some genera, as *Pelexia* and *Stenorrhyncos*, the lateral sepals are united at the base to form a spur which in *Pelexia* encloses the elongated base of the lip. The petals show a great variety in form and colour; the lateral pair are generally smaller than the lip. The lip which is, as a rule, the most conspicuous feature of the flower, is often large and spreading, or has upwardly curving sides. It is simple (undivided) or two- three- or four-lobed and generally bears on the upper face thickened veins or swellings (*calli*), or fleshy warts, which are often, as in *Oncidium*, large and conspicuous. The lip is sometimes produced at the base into a sac or spur, which may arise partly from the axis of the flower.

The most characteristic feature of the flower is the development of the receptacle, which at an early stage becomes cup-shaped, and finally forms a hollow cylinder. The carpels (three in number) arise as outgrowths from the edge of the cup, on the interior of which the ovules are developed on three double placental lines alternating with the carpellary outgrowths. The perianth springs from the upper edge of the inferior ovary, where there is often also an expansion of the axis known as the *foot*. The lateral sepals are continued along the sides of the foot, from the tip of which the lip springs at a sharp angle, forming externally a chin, inside which at the base nectar is often secreted.

The stamens and stigmatic surfaces are raised above the perianth on an elongation of the floral axis known as the column; occasionally some of the other organs are raised up on the column, as in *Ponthieva*, where the petals and lip are carried up above the sepals. Comparative study of the orchid flowers among themselves, and especially with those of other families of *Monocotyledons*, suggests that the stamens are derived from two alternating whorls, each of three members, the members of the outer whorl alternating with the petals. The stamens are, however, much reduced, and are represented in the great majority of orchids by one anther, which is variously placed on or near the apex of the column, and represents the median stamen of the outer whorl, opposite the median sepal. There is generally no trace of the other members of this whorl, but they sometimes appear as small outgrowths on the side of the column; and the lateral pair of the inner whorl is also often similarly represented; in *Epidendrum Ottonis* they are more or less fertile. The presence of a single fertile stamen characterises the great majority of the genera; in the genus *Cypripedium*, not found in Jamaica, the lateral stamens of the inner whorl are fertile, while the median stamen of the outer whorl is represented by a large staminode.

Cypripedium has three functional stigmas; but in Jamaican orchids only the lateral pair are functional, the third is sterile and forms the *rostellum*, which often plays an important part in transference of the pollen. The stigmas form a smooth, viscid, flat or cushion-like area on the inner face, or sometimes on the end of the column, or two distinct processes which in *Habenaria* are borne on projecting style-like structures.

The form of the anther, its position on the column and the means by which the pollen is transferred, vary widely in the different genera. The anther contains two, four, or eight cells, each of which contains a mass of pollen. The pollen masses or *pollinia* are granular or powdery, breaking up into grains on removal, as in *Pogonia*, *Spiranthes*, *Vanilla*, etc.; or the grains are united into packets by an elastic web as in *Habenaria*; or waxy pollinia are produced by the formation of a strong common covering round each mass as in *Epidendrum* and allied genera.

The anther may stand erect and free on the top of the column, as in *Habenaria* where it is strongly united to the *rostellum* by its base, and does not separate from the column when the pollinia are removed. Usually, however, the anther is attached to the rostellum by its apex. In this case it may be erect and stand behind the stigma in such a way that the rostellum reaches its apex, as in *Stenorrhyncos*, and the filament is generally thin but strong enough to hold the anther in place after removal of the pollen. In most cases, however, the anther is not erect, but is placed horizontally or hangs on the inner face of the column; it usually separates so easily from the thin filament that it falls when the pollen is removed. It lies along the rostellum or is more or less enclosed within a special cavity of it (*clinandrium*), or is in contact with it only at the apex.

The removal of the pollinia is generally effected by the help of insects, which visit the flower for the sake of a sweet juice contained in the sac or spur-like development or in the various excrescences of the lip. Insects are attracted by the colour, form and smell of the flower, the lip of which often forms an effective landing-stage. The pollinia and stigma are so placed in reference to the position assumed by the insect as to favour the carrying off of the pollinia from one flower and its deposition upon the stigma of another of the same species.

To ensure its transference the pollinia are often attached to a stalk, which is associated with a sticky mass, the so-called "gland." Thus in *Habenaria* each anther-cell is produced into a long process (*anther-canal*), the contents of which form stalks (*caudicles*), to which are attached above the numerous packets of grains forming the pollinia. Below, the caudicle is attached to a gland formed from the rostellum. The anthers split open longitudinally when mature, and the pollinia are drawn out by the adhesion of the gland to the head or proboscis of an insect visiting the flower. In other genera no definite gland is produced from the rostellum, but a quantity of viscid matter exudes from it when touched, and serves to glue the protruding tips of the caudicles to the retreating object. In many genera the pollinia are connected to the rostellar gland by a stalk (*stipes* or *pedicel*), which is not a development of the anther, but of the upper face of the rostellum (as in *Maxillaria*). In some cases short true caudicles are formed, by which the pollinia are attached to the pedicels.

The ovary is generally somewhat cylindrical-ellipsoidal or spindle-shaped, and often bears longitudinal lines, ridges, or wings, which become further developed in the fruit.

The fruit is a dry capsule, except in *Vanilla*, crowned by the withered remains of the flower, and opening usually by six longitudinal slits, forming three broad and three narrow valves, which remain united above and below; in *Harrisella* they are free above. In *Pleurothallis* there are only two slits. In *Vanilla* the fruit is fleshy and pod-like. The seeds are innumerable and minute, and contain a small rudimentary few-celled embryo enveloped by a thin loose membranous coat, which varies much in shape and colour. They are scattered by aid of hygroscopic hairs on the interior of the valves, and are carried by the wind.

The family falls into two groups, a small one, *Diandræ*, not represented in Jamaica, with two or rarely three fertile stamens, and a large one, *Monandræ*, with one fertile stamen.

The subdivision of the *Monandræ* was based by John Lindley (The Genera and Species of Orchidaceous Plants, 1840) solely on characters derived from the anther, the pollen, or the pollen-distributing apparatus. Lindley recognised six tribes. E. Pfitzer, who more recently elaborated the family in Engler and Prantl's *Pflanzenfamilien* (ii. part 6, 1889), criticises this arrangement, which was practically followed by Bentham and Hooker in the *Genera Plantarum* (iii. 1883), as depending too much on the relation of the floral structure to insect visits for the purpose of pollination, and too little on a general study of the whole plant. He retains the *Ophrydeæ* which are characterised by the persistent terminal erect anther (to which belongs *Habenaria*), but subdivides the remainder into twenty-eight tribes, based on the terminal or lateral character of the inflorescence, the development of the leaf and stem, and the form and relative size of the lip. We have followed the sequence of genera in Pfitzer's arrangement.

KEY TO GENERA.

I. Terrestrial Plants (growing in soil).

Anther-cells distant, diverging, with the stalks of the pollinia in special tubes and each ending in a naked gland .... 1. *Habenaria.*

Anther-cells closely approximate.

i. Leaves membranous, not plaited. Pollinia granular (*i.e.* breaking up into granules on removal), 2 or 4.

Anther deciduous, opercular, on the margin of the clinandrium. Pollinia 2. Leaves along the stem or wanting.................. 2. *Pogonia.*

Anther persistent, parallel to the back of the long rostellum. Pollinia 2 or 4. Leaves generally radical.

a. Lip on the lower side of the flower.

Lateral sepals combined into a spur.

Spur enclosing the elongated base of the lip................................ 4. *Pelexia.*

Spur not enclosing base of lip. Bracts coloured........................ 5. *Stenorrhyncos.*

Lateral sepals free.

Lip not spurred .... 6. *Spiranthes.*

Lip spurred .... 7. *Erythrodes.*

b. Lip on the upper side of the flower.

Lateral sepals forming a spur.

Spur short. Leaves wanting ...... 8. *Wullschlægelia.*

Spur very long. Leaves present... 9. *Pseudocentrum.*

Lateral sepals not forming a spur.

Lip inserted at base of column, free 10. *Cranichis.*

Lip inserted on the cup formed by the base of the sepals, which have grown together .... 12. *Prescottia.*

Lip and petals inserted on the long column .... 13. *Ponthieva.*

ii. Leaves membranous, not plaited. Pollinia waxy (not breaking up into granules), 4.

Column long, curved. Anther terminal, opercular, incumbent, deciduous .... 17. *Liparis.*

Column short, erect. Anther enclosed by the clinandrium, persistent .... 16. *Malaxis.*

iii. Leaves papery, plaited.

Flowers small, in terminal and axillary panicles. Pollinia 2, granular.

Sepals cohering into a tube. Lip linear from base. Column long .... 14. *Corymborchis.*

Sepals, lateral connate into a short chin. Lip saccate at base. Column short... 15. *Tropidia.*

Flowers large, racemose on a scape. Pollinia 4 or 8, waxy.

Pollinia 8.

[Lip continuous with base of column, spurred. Flowers white and brown... 39. *Phaius.*]

Lip jointed with base of column, not spurred. Flowers purple or rose ...... 41. *Bletia.*
Claw of lip connate with column, forming a tube, spurred. Flowers white .... 40. *Calanthe.*
Pollinia 4.
Lip produced into a wide spur .... 18. *Galeandra.*
Lip forming a chin between the lateral sepals. Flowers greenish, with purple lip.. 42. *Eulophia.*
Lateral sepals forming a chin. Flowers white .... 43. *Govenia.*

iv. Leaves fleshy, linear, semi-terete. Flowers small, pink; lip 3-lobed .... 36. *Tetramicra.*

II. Climbing Plants with fleshy Fruit .... 3. *Vanilla.*

III. Plants growing on Trees, Rocks or Stones, not in the Soil.

(Sometimes it may be difficult to decide whether a plant should be included under I. or III., as it may be growing amongst moss on rocky ground and apparently terrestrial, but unless it is actually growing in soil, it should be classed amongst the epiphytic orchids.)

i. Leaves plaited.
Pollinia 8.
Raceme from base of pseudobulb. Ovary 3-winged. Pseusobulbs present .... 27. *Cœlia.*
Flowers with large bracts in terminal heads or spikes. No pseudobulbs ...... 38. *Elleanthus.*
Pollinia 4.
Peduncles several, with one flower on each.
Flowers greenish. Pseudobulbs present 44. *Lycaste.*
Flowers whitish, with a blue variegated lip. No pseudobulbs .... 46. *Zygopetalum.*
Raceme with several flowers. Pseudobulbs 45. *Xylobium.*

ii. Leaves not plaited.
Lip on upper side of flower (see also some species of *Epidendrum* and *Oncidium*).
Plant very small. Flowers in head with large bracts .... 11. *Stenoptera.*
Plant not very small. Flowers in panicles, bracts small .... 19. *Polystachya.*
Lip on lower side of flower (but some species of *Epidendrum* and *Oncidium* have lip on upper side).

A. Pollinia 4.
a. Column with foot, with which the sepals more or less form a chin.
1. Lip free, jointed to foot of column.
Flowers small, several, on stalk from base of pseudobulb .... 47. *Bulbophyllum.*

Flowers large, solitary, axillary to large leaves. With or without pseudobulbs .... 48. *Maxillaria.*
Flowers small, solitary, axillary to small two-ranked leaves. No pseudobulb ...... 59. *Dichæa.*
Flowers small, in clusters, axillary in sheaths of large leaves. No pseudobulbs ... 49. *Neo-urbania.*

2. Lip not jointed to foot of column; erect.
Stem covered with imbricated sheaths in two ranks, with pseudobulbs here and there 50. *Ornithidium.*
Stem bearing short linear leaves in two ranks; no pseudobulbs .... 25. *Isochilus.*

b. Column without foot.
Claw of lip as long as column, with which it is connate into a cup. No pseudobulbs. Pollinia connected in pairs by a viscous substance which stretches elastically into two threads .... 28. *Seraphyta.*
Claw of lip shorter than column, with which it is connate into a cup. Pseudobulbs present. Pollinia connected by a viscous substance..................... 32. *Hormidium.*
Claw of lip erect, adpressed to column, and more or less connate with it into a tube, but sometimes quite free. Peduncle terminal, simple or branching. With or without pseudobulbs. Pollinia connected in pairs by a granular-viscous linear appendage ...... 30. *Epidendrum.*
Lip free from column, and embracing it at base, produced below into an adnate spur. Pseudobulbs present .... 31. *Broughtonia.*
Lip free from the column and continuous with its base, spreading, claw long and keeled-crested above. No pseudobulb .... 58. *Cryptarrhena.*
Lip free, subsessile, somewhat similar to sepals and petals ... 37. *Homalopetalum.*

B. Pollinia 8.
Without pseudobulbs.
Flowers numerous, mauve, in a long dense spike .... 29. *Arpophyllum.*
Flowers greenish or whitish-yellow, corymbose .... 35. *Brassavola.*
Flowers creamy-white, racemose ... 26. *Octadesmia.*

Flowers few, orange-coloured, racemose ... 33. *Lælia.*
With pseudobulbs.
Flowers purple-spotted, several, racemose ... 34. *Schomburgkia.*

C. Pollinia 6.
Sepals, petals and lip with long, tailed apices ... 24. *Brachionidium.*

D. Pollinia 2.
a. Without pseudobulbs.
1. Stem with one leaf at the apex.
Flowers opening only by a chink at the side... 20. *Cryptophoranthus.*
Flowers opening normally.
Petals and lip free, much alike, and spreading ... 21. *Stelis.*
Petals free, differing from the lip. Lip jointed with column ... 22. *Pleurothallis.*
Claw of petals and lip adnate to base of column... 23. *Lepanthes.*
2. Stem with two or more leaves or leafless.
Flowers with a short spur.
Lip longer than sepals; sepals forming the spur ... 52. *Ionopsis.*
Lip not longer than the sepals; lip forming the spur.
Flowers numerous, on spikes. Capsule dehiscing in middle. Sometimes leafless ... 60. *Campylocentrum.*
Flowers few, on racemes. Capsule dehiscing from apex. Leafless ... 61. *Harrisella.*
Flowers with a very long spur.
Stemless and leafless ... 62. *Dendrophylax.*
Flowers not spurred ... 56. *Oncidium.*
b. With pseudobulbs.
Flowers spurred ... 53. *Comparettia.*
Flowers not spurred.
Lip with large lateral lobes, embracing the column... 51. *Macradenia.*
Lip more or less entire.
Flowers large, sepals long, narrow ... 55. *Brassia.*
Flowers small.
Scape shorter than leaves ... 54. *Trichopilia.*
Scape longer than leaves... 57. *Leochilus.*

1. HABENARIA Willd.

Terrestrial herbs, with tubers or fleshy roots. Stem erect, simple, with a few leaves which are sheathing at their base. Flowers shortly stalked, in a raceme. Sepals unequal, free; median hooded; lateral spreading or deflexed. Petals smaller, 2-partite, or simple. Lip continuous with the column, spreading or pendulous, spurred, 3-partite, or simple. Column very short, without a foot; stigma produced into two papillose processes; anther-cells distant from one another; pollinia in each cell granular, with long stalks lying in the anther-canals and each ending in a naked gland. Capsule narrowly ellipsoidal.

There are about 500 known species, natives mainly of the warmer regions of both hemispheres.

§ 1. Petals 2-partite; lip 3-partite.
Stem leafy upwards.
Spur much longer than ovary .... 1. *H. macroceratitis.*
Spur not much longer, or shorter, than ovary.
Stem erect, with short roots .... 2. *H. monorrhiza.*
Stem creeping at base, with long roots 3. *H. repens.*
Stem leafy at base only .... 4. *H. distans.*
§ 2. Petals entire; lip simple.
Spur much longer than ovary .... 5. *H. Purdiei.*
Spur not much longer, or shorter, than ovary.
Sepals and petals acute .... 6. *H. alata.*
Sepals and petals obtuse.
Spur tapering towards apex .... 7. *H. socialis.*
Spur thickened towards apex............... 8. *H. troyana.*

§ 1. Petals 2-partite; lip 3-partite.

1. **H. macroceratitis** *Willd. Sp. Pl. iv.* 44 (1805); leaves reticulately many-nerved, oval to elliptical-oblong; spike few-flowered; spur much longer than ovary, curved.—*Griseb. Fl. Br. W. Ind.* 643; *Cogn. in Symb. Ant. vi.* 299; *Ames Orch. iv.* 222. H. macroceras *Spreng. Syst. iii.* 692 (1826); *Hook. in Bot. Mag. t.* 2947; *Lindl. Gen. & Sp. Orch.* 308. Orchis habenaria *L.** *Syst. ed.* 10, 1242 (1759) & *Herb. Linn.; Sw. Obs. Bot.* 319, *t.* 9. (Pl. 1, f. 1.)

On stony ground; in fl. Oct.–Jan.; *Shakespear*! *Swartz*! *Bancroft*! *Waters*! Alligator Pond, *Purdie*! Fairfield, *Wullschlaegel*, 1050! Moneague, *Prior*! *March*! *Lady Blake*! Red Hills, J.P. 467, *Morris*! Stony Hill, J.P. 2410, *Syme*! near Browns Town, *Miss T. M. Barrett*! Lacovia,

* The species name is generally spelt with a capital initial, *Habenaria* being the name of a genus. Linnæus, however, did not take the trivial name from a generic name, as *Habenaria* was founded much later by Willdenow (1805). As therefore the species name is not taken from a generic name (see International Rules of Botanical Nomenclature (1905) Rec. x) it should begin with a small letter as in Linnæus's original description.

Tomlinson! Red Hills, 800 ft.; near Hope Mines, 800 ft.; Long Mt., 900 ft. *Harris*! St. Ann, *Miss Stennett*! Fl. Jam. 6901, 7789, 9106, 10,404, 10,415, 10,416.—Cuba, Florida, Mexico, Guatemala, Costa Rica, B. Guiana, Surinam.

Plant 1–2½ ft. high. *Stem* leafy, 2–5 dm. l., 7 mm. br. *Leaves* oval below; elliptical-oblong in middle, 10–12·5 cm. l., 2·5–3·5 cm. br.; ovate-acuminate above. *Racemes* 10–15 cm. l., numerous-fld. *Bracts* ovate-lanceolate, acuminate, as long as the ovary, 3·5–4·5 cm. l., 1·3 cm. br. *Flowers* greenish-white. *Sepals*, median ovate-suborbicular, very obtuse, 1·5 cm. l., lateral ovate, obtuse, 1·6 cm. l. *Petals*, posterior segment oblong-falcate, acute to sub-acute, 1·3 cm. l., 3 mm. br., anterior linear-setaceous, curving upwards, 3·5 cm. l. *Lip* 3-partite nearly to base, middle lobe linear-lingulate, acute, barely 2 cm. l., about 2 mm. br., lateral linear-setaceous, curved-spreading, 4 cm. l. *Spur* pendulous, curved, filiform, slightly clavate, apex acute 10 cm. l. and more. *Stigmatic processes* short, obtuse; anther-canals longer, curved upwards.

2. **H. monorrhiza** *Reichb. f. in Ber. Deutsch. Bot. Ges. iii.* 274 (1885); leaves 3-nerved, lanceolate to ovate-lanceolate; sheaths marked with purple spots and lines; lip with lateral lobes erect, longer than sepals.—*Ames op. cit. iv.* 229. H. brachyceratitis *Willd. op. cit.* 44. H. brachyceras *Spreng. op. cit.* 692. H. maculosa *Lindl. op. cit.* 309; *Griseb. op. cit.* 643; *Cogn. op. cit.* 304. Orchis monorrhiza *Sw. Prodr.* 118 (1788) & *Fl. Ind. Occ.* 1391. (Pl. 1, f. 8–10.)

On damp clayey shady banks; in fl. Nov.–April; *Swartz; Macfadyen*! Hagley Gap, *Purdie*! *Wilson*! Mt. Diabolo, *Prior*! Fairfield, *Wullschlaegel*, 1049! *March*; Mt. Airy, J.P. 2052, *Syme*, and "common" *Morris*! Port Antonio & Blue Mt. Peak, *Hitchcock*; Browns Town, *Miss T. M. Barrett*! Cinchona, 5000 ft.; Silver Hill Gap; near Castleton; *Fawcett*! Westphalia, 4000 ft.; Cinchona; Wallenford, 3500 ft.; *Harris*! Fl. Jam. 7525, 7771, 7884, 10,407, 10,500 (partly).—Cuba, Porto Rico, St. Thomas, Guadeloupe, Dominica, Martinique, St. Lucia, St. Vincent, Grenada, Trinidad, Guatemala, Colombia, Ecuador, Peru, Venezuela, Brazil.

Plant to 2 ft. high. *Stem* leafy, 2·5–5 dm. l., to about 6 mm. br. *Leaves* 4–8 cm. l., 2–3 cm. br. *Raceme* 10–13 cm. l., densely-fld. *Bracts* ovate-lanceolate, acute or shortly acuminate, about as long as the ovary, 1·2–2·2 cm. l., 4–5 mm. br. *Pedicels* very short. *Flowers* white. *Sepals* reticulate-5-nerved, nerves prominent outside; median ovate or ovate-suborbicular, very obtuse, keeled, 7–8 mm. l., 4–5·5 mm. br.; lateral ovate, obtuse, apiculate, 7–8 mm. l., 3–4 mm. br. *Petals* posterior segments narrowly oblong, tapering to base and apex, apiculate, 2–3-nerved, connivent or spreading, 6·5–7·5 mm. l., 1·8–2·5 mm. br.; anterior segment linear, acuminate, generally about 4 mm. l., erect, sometimes reduced to a short, descending process, about 1 mm. l. *Lip*, lobes subequal; middle lobe linear-ligulate, 7–8 mm. l., 1–2 mm. br.; lateral filiform, 8–9 mm. l., barely ·5 mm. br. *Spur* pendulous or curved anteriorly, filiform, apex acute, 1·5–2 cm. l. *Stigmatic processes* short, fleshy; anther-canals much longer, ascending; rostellum small, triangular.

Cogniaux regards *O. monorrhiza* Sw. as distinct from *H. maculosa* Lindl. and identical with *H. alata* Hook., and describes the lip as undivided. Swartz, however, says "labio tripartito" Prodr. & "Labium ad basin fere tripartitum" Flor., in which *H. maculosa* agrees, but *H. alata* differs.

3. **H. repens** *Nutt. Gen. Amer. ii.* 190 (1818); stem creeping at base; leaves 3-nerved, oblong-lanceolate; lip with lateral lobes pendulous, shorter than sepals.—*Lindl. op. cit.* 310;

Rolfe in Hook. Ic. Pl. t. 2686; *Cogn. op. cit.* 305; *Ames Orch. i.* 51, *t.* 14 & *iv.* 216. H. tricuspis *A. Rich. Fl. Cub. iii.* 249 (1853).

In wet places; St. Ann, *McNab*! Worthy Park, *Prior*!—Cuba, Porto Rico, Florida, Georgia, Carolina, Alabama, Louisiana, Guatemala, Nicaragua, B. Guiana, Venezuela, Brazil.

Plant ½–2½ ft. high. *Stem* without tubers, 1·5–7·5 dm. *Leaves* 1–3 dm. l., ·7–2 cm. br. *Raceme* 5–20 cm. l., dense, many-fld. *Bracts* linear-lanceolate, acuminate, lowest longer than flower, upper shorter than ovary, to 1·5 cm. l. *Pedicels* 4–6 mm. l. *Flowers* erect, yellowish-green. *Sepals*, median broadly ovate, apiculate, 5 mm. l.; lateral ovate, apiculate, nearly 6 mm. l. *Petals* 4 mm. l., posterior segment falcate, lanceolate, acute; anterior segment linear-filiform, acute, about as long as the posterior. *Lip* 4–5 mm. l., lobes sub-equal, linear-filiform, median broader than lateral. *Spur* incurved, linear, 9 mm. l. *Stigmatic processes* oblong, thickish; anther-canals short, directed forwards. *Capsule* 10 mm. l.

4. **H. distans** *Griseb. Cat. Cub.* 270 (1866); steam leafy at base only; leaves reticulate-many-nerved, broadly oblong or lanceolate, acute.—*Cogn. op. cit.* 300; *Ames op. cit.* 202.

On damp shady banks; in fl. Dec.; Moody's Gap, 3000 ft., *Harris*! Fl. Jam. 7768.—Cuba, Porto Rico, Florida.

Plant 8–16 in. high. *Stem* 2–4 dm. *Leaves* 8–16 cm. l., 2·5–5 cm. br. *Raceme* laxly 6–10-fld., 6–12 cm. l. *Bracts* foliaceous, ovate-lanceolate, long-acuminate, 2–1·5 cm. l. *Flowers* greenish. *Sepals* 3-nerved, median ovate-oblong, somewhat obtuse, 5–6 mm. l.; lateral deflexed, oblong, acute, about 8 mm. l. *Petals*, segments somewhat equal; posterior linear-ligulate, abruptly acute, suberect, as long as the median sepal; anterior somewhat longer, filiform. *Lip*, middle lobe about equal to the lateral sepals, narrowly linear; lateral lobes generally longer, filiform, spreading. *Spur* incurved, filiform, gradually becoming spindle-shaped, somewhat acute, 1·3–1·6 cm. l. *Stigmatic processes* extended, thickened, altogether free from the lip; anther-canals shorter, ascending.

Var. **jamaicensis** *Cogn. loc. cit.* (1909); plant smaller than the typical species; leaves roundish-oval to oblong-lanceolate. Habenaria jamaicensis *Fawc. & Rendle in Journ. Bot. xlvii.* 126 (1909); *Ames op. cit. iv.* 204, *t.* 67. (Pl. 1, f. 2, 3.)

On rocky or clayey, shady banks; in fl. Dec.–March; below Guava Ridge, J.P. 483, *Morris*! Mt. Hybla, 4000 ft.; Clydesdale, 4000–4500 ft.; near Cinchona, 4500 ft.; *Harris*! Fl. Jam. 7851, 7854, 10,499.—Hispaniola.

Plant ½–1 ft. high. *Stem* 1–2 dm. *Leaves* 3–9 cm. l., 1·5–4·5 cm. br. *Raceme* laxly 4–12-fld., 4–15 cm. l. *Bracts* membranous, ovate-lanceolate or lanceolate, acute or acuminate; sterile 2–4, more or less distant, 5–2 cm. l.; floral about as long as the stalked ovary, 2·5–1·5 cm. l. *Pedicels* about 5 mm. l. *Sepals* 3-nerved, median ovate-suborbicular, very obtuse, 6–7 mm. l., 4·5–6 mm. br.; lateral half-ovate, subfalcate, obtuse, apiculate, 6·5–9 mm. l., 3·5–4 mm. br. *Petals*, segments equal or unequal; posterior oblong-lanceolate, falcate, acute, connivent, 5·5–7 mm. l., 1·5–2 mm. br.; anterior segment a little to many times shorter than the posterior, 1·5–4 mm. l. *Lip*, middle lobe linear-ligulate, 6·5–10 mm. l., about 1 mm. br.; lateral lobes filiform, erect, much longer than the petals, 9–12 mm. l. *Spur* 1·3–1·6 cm. l. *Stigmatic processes* short, truncate-roundish; anther-canals about as long; somewhat curved; staminodes shortly triangular.

§ 2. Petals entire ; lip simple.

5. **H. Purdiei** *Fawc. & Rendle in Journ. Bot. xlvii.* 263 (1909); leaves lanceolate; bracts lanceolate, shorter than the stalked ovary; sepals very obtuse; spur much longer than ovary.—*Ames op. cit. iv.* 270. (Pl. 2, f. 1–3.) Type in Herb. Kew.

Hollis's Savanna, Clarendon, *Purdie*!

Plant 1½ ft. high. *Stem* about 3 dm. l., 3·5 mm. br. *Leaves* thin, conspicuously reticulate-11–15-nerved, acute, to 8 cm. l., to 1·9 cm. br. *Raceme*, about 12 cm. l., somewhat laxly many-fld., oblong in outline. *Bracts* acuminate, lower 1·7 cm. l. *Pedicel* about 5 mm. l. *Ovary* 1–1·8 cm. l. *Flowers* somewhat spreading. *Sepals* reticulate-3-nerved, nerves prominent outside; median suborbicular, margin minutely denticulate, 8·5 mm. l., 7·7 mm. br.; lateral oval-oblong-falcate, concave, about 11 mm. l., about 5 mm. br. *Petals* oblong, very obtuse, 2-nerved, 6 mm. l., 2 mm. br. *Lip* linear-ligulate, margin revolute, with a small tooth on each side above the base, pendulous, half as long again as the sepals, 13 mm. l., nearly 2 mm. br. *Spur* slender, compressed, tapering to apex, nearly 3 cm. l. *Stigmatic processes* thick and fleshy at apex, about 2 mm. l.; anther-canals nearly twice as long, about 3·3 mm. l., ascending; anther acutely crested (in the specimen examined).

Near *H. obtusa* Lindl., from Brazil and Surinam, but differs in the smaller lanceolate bracts and in other details.

6. **H. alata** *Hook. in Exot. Fl. t.* 169 (1827); bracts ovate to lanceolate, acuminate, apex shortly awned, about as long as the stalked ovary; sepals acute; petals acute, linear-lanceolate.—*Griseb. op. cit.* 644; *Ames op. cit. iv.* 273. H. brachyceras *Lindl. op. cit.* 315 (1835), non *Spreng.* H. monorrhiza *Cogn. op. cit.* 307 excl. syn. (in part). (Pl. 1, f. 4–7.)

In damp clayey ground; in fl. Nov.–April; *Macfadyen*! *Purdie*! *Prior*! Golden Spring, J.P. 466, *Morris*! Farm Hill, *Miss B. Gosset*! Salt Hill Road; Westphalia, 4000 ft.; Halls Delight, 1000 ft.; below Cinchona, 4000 ft.; *Harris*! Fl. Jam. 7526, 7806, 10,470, 10,500.—Cuba, Porto Rico, St. Thomas, Antigua, St. Vincent, Grenada, Mexico, Guatemala, Costa Rica, Venezuela, Bolivia.

Plant reaching sometimes to 3½ feet. *Stem* leafy, covered by the leaf-sheaths, 2–5 dm. l. *Leaves* lanceolate, acute, passing from a sheath at the base into bracts above, 7–16 cm. l., 1–2·4 cm. br. *Raceme* somewhat laxly many-fld., 10–15 cm. l. *Bracts* 2·7–2 cm. l., 4–6 mm. br. *Flowers* green. *Sepals* reticulate-7-nerved, keeled, median broadly ovate, 6–8 mm. l., 4·5–6 mm. br.; lateral ovate, acute, 7–9 mm. l., 3–4·5 mm. br. *Petals* reticulate-3-nerved, with a small projecting lobe on the anterior side near the base, 6–7·5 mm. l., 1·75–2 mm. at greatest width. *Lip* linear-ligulate, margin revolute, with or without a small tooth on each side above the base, spreading or ascending, 6–8 mm. l., about 1 mm. br. *Spur* pendulous, clavate, slightly shorter than the ovary, 1·2-1·4 cm. l. *Stigmatic processes* short, compressed; anther-canals subequal, ascending. *Capsule* 6-winged.

7. **H. socialis** *Fawc. & Rendle loc. cit.*; bracts lanceolate, acuminate, about as long as the stalked ovary; lateral sepals 2-nerved, oblong, retuse, about 4·3 mm. l., about 2 mm. br.; petals oblong, apex emarginate, the nerves ending in two pro-

jecting points, 3·5 mm. l., about 1 mm. br.; spur tapering towards apex.—*Ames op. cit. iv.* 276. Type in Herb. Kew. (Pl. 2, f. 4–7.)

Growing with *H. alata* Hook. in marshy soil, near Mandeville, *Purdie*!

Plant 12 in. high, glabrous. *Stem* erect, leafy, covered by leaf-sheaths, about 2 dm. l., 3 mm. br. *Leaves* lanceolate, thin, 9-nerved, acute, passing into the bracts above, blade to 8·5 cm. l., to 2 cm. br. *Raceme* somewhat laxly many-fld, about 10 cm. l. *Bracts*, lower 1·3 cm. l. *Pedicel* about 3 mm. l. *Ovary* about 1 cm. l. *Flowers* green, erect. *Sepals* median, without nerves, suborbicular, very obtuse, apiculate, about 3·5 mm. l. and br. *Petals* reticulate-2-nerved, with a projecting lobe on the anterior side near the base, 3·5 mm. l., about 1 mm. br. *Lip* linear-ligulate, with a small projecting lobe just above the narrow base, margin revolute, simple, about 5·5 mm. l., about 1·2 mm. br. *Spur* slender, compressed, nearly as long as ovary, about 1 cm. l. *Stigmatic processes* very short, capitate, glandular; anther-canals ascending, about twice as long as processes.

Differs from *H. alata* Hook. in the texture and venation of the leaves; in the flowers being about half the size; in the form of the sepals and petals, and the relatively longer lip. Differs from the Brazilian species, *H. quadrata* Lindl., in form of petals, length of spur, which is 1·5 cm. l. in *H. quadrata*, and in other details.

8. **H. troyana** *Fawc. & Rendle op. cit.* 264; bracts ovate-lanceolate, subacuminate, like the leaves at base of raceme, lowest longer than flower, upper shorter than ovary; lateral sepals reticulate-nerved, falcately oblong-ovate, obtuse, the nerve on the lower face projecting as an apiculus just below the apex, 7 mm. l., 3·5 mm. br.; petals oblong, retuse, oblique at base and rather broader there than at the apex; spur increasing in size upwards from a slender base.—*Ames op. cit. iv.* 260. (Pl. 2, f. 8–12.)

In damp shady forest; in fl., Nov.; near Troy, 2500 ft., *Harris*! Fl. Jam. 10,432.

Plant somewhat robust, 2 ft. high. *Stem* leafy, 3 dm. l. *Leaves* lanceolate, acute, reticulate-11–15-nerved, passing into bracts above, about 10 cm. l., about 2·5 cm. br. *Raceme* densely-fld., 30 cm. l. *Bracts* 3·5–1·2 cm. l., 1·3–·6 cm. br. *Pedicels* short. *Sepals* reticulate-nerved, median ovate-suborbicular, very obtuse, 6 mm. l., 5 mm. br.; lateral spreading, 7 mm. l., 3·5 mm. br. *Petals* with an obsolete lobe near base anteriorly, 5 mm. l., 2·3 mm. br. *Lip* linear-ligulate, angled at the base, 7·5–8 mm. l., 1·5–1·75 mm. br. near apex, 1·5–2 mm. br. near base. *Spur* pendulous, slightly curved, shorter than the stalked ovary, 9 mm. l. *Stigmatic processes* short, compressed; anther-canals about as long, slightly projecting upwards.

Since the publication of *H. troyana*, we have had the opportunity, through the courtesy of Prof. I. Urban, of comparing flowers from the type specimen of *H. eustachya* Reichb. f. in Ber. Deutsch. Bot. Ges. iii. 274 (1885) from Porto Rico (Sintenis, no. 511b); this differs from *H. troyana* in being smaller, with thinner, non-clavate spur, and much smaller lip. When describing *H. troyana* we compared it with *H. Sanbornii* Ames, which we suggested might prove identical with *H. eustachya*. Mr. Oakes Ames now (*op. cit.* 257) refers *H. Sanbornii* to *H. eustachya*, and considers *H. troyana* as probably the same species.

2. POGONIA Juss.

Terrestrial herbs, with fleshy roots or a tuber. Leaves wanting, or appearing on the stem with the flowers; stalk sheathing, blade broad. Flowers medium-sized or small, corymbose, erect. Sepals equal, free, erect. Petals similar to, and nearly as long as, the sepals, erect. Lip erect from the base of the column, free, not spurred, contracted below, narrowly embracing the column, 3-lobed, the face with 1–3 raised lines. Column elongate, at the apex slightly thicker, with a short rostellum; stigma under the rostellum oblong or broad; clinandrium more or less elevated, entire or denticulate. Anther more or less incumbent, obtusely conical or semiglobose; pollinia two, thinly granular. Capsule ellipsoidal, erect.

Species about 90, widely dispersed over the world.

Stem leafy .... 1. *P. macrophylla.*
Stem leafless................................ 2. *P. gentianoides.*

1. **P. macrophylla** *Lindl. in Ann. & Mag. Nat. Hist. ser.* 3, *i.* 335 (1858); stem leafy; flowers greenish.—*Cogn. in Symb. Ant. vi.* 317. P. physurifolia *Reichb. f. in Nederl. Kruidk. Arch. iv.* 324 (1859); *Griseb. Fl. Br. W. Ind.* 637. (Pl. 1, f. 11.)

In damp shady places; probably in fl. all the year, our specimens were collected in Feb., April, June & Dec.; woods above Dunrobin Castle, St. Thomas in East, *Purdie*! J.P. 2090, *Morris*! Mabess River, 2500 ft., *G. Nichols*! Vinegar Hill, St. George, 4000 ft.; ridge below Vinegar Hill, 3800 ft.; near Mabess River, 3500 ft.; *Harris*! Fl. Jam. 6252, 7840, 9013, 10,097.—Cuba, Hispaniola, Porto Rico, Guadeloupe, Dominica, St. Lucia, St. Vincent, Grenada, Venezuela, Guiana, Colombia.

Plant 9–12 in. high, glabrous. *Stem* often prostrate at the base and rooting at the nodes, erect above, purple, 1–3 dm. l. *Leaves*, blade ovate, acute or obtuse, at the base rounded, sometimes somewhat cordate and clasping, green above, deep purple beneath, conspicuously 3-nerved, 3–7 cm. l., 2–3 cm. br. *Racemes* 1-several, with numerous broad bracts, short or becoming long, with 1–6 flowers open at one time at the apex, 2–13 cm. l. *Flowers*, perianth ½ in. or more l. *Bracts*, floral broadly ovate, obtuse, clasping, much shorter than the ovary, 5–1 cm. l. *Pedicels* 4 mm. l., elongating often to 5·5 mm. in fruit. *Ovary* slender with thin longitudinal ridges, 11 or 12 mm. l. *Sepals* oblanceolate, 3-nerved, margins somewhat involute, strongly keeled, keel ending in a pointed tip, about 1·5 cm. l., 3·5 mm. br., lateral falcate. *Petals* oblanceolate, acute or somewhat obtuse, 3-nerved, about 1·4 cm. l., 2·2 mm. br. *Lip* in outline obovate, a little shorter than the petals (by 1 mm.), nearly 5 mm. br. (including lobes), many-nerved, middle vein prominent, 3-lobed above; middle lobe much larger than the lateral, somewhat roundish-rhomboid, margin somewhat eroded; lateral lobes oblong, obtuse, with a very narrow sinus. *Column* narrowly winged; clinandrium with 2 pointed lobes. *Capsule* nearly 2·5 cm. l., 7 mm. br.

2. **P. gentianoides** *Spreng. Syst. iii.* 706 (1826); stem leafless; flowers dark purple.—*Lindl. Gen. & Sp. Orch.* 413; *Griseb. op. cit.* 637; *Cogn. op. cit.* 315. Limodorum gentianoides *Sw. Prodr.* 119 (1788). Arethusa gentianoides *Sw. Fl. Ind. Occ.* 1436 (1806) & *Ic. ined. t.* 18. (Pl. 2, f. 13–15.) Type in Herb. Mus. Brit.

In shady places; in fl. June; Morant Bay, *Swartz*! J.P. 482, *Morris*!

Plant growing from a tuber, glabrous 5–15 ins. high. *Scape* erect, slender, with a few sheaths, 1–3 dm. l., purplish. *Sheaths* about 5, clasping, rounded at the apex, dark- or reddish-green, 2–1 cm. l. *Flowers* at the end of the scape "dark purple, glaucescent, like the calyx of *Gentiana Amarella*" (Sw.). *Peduncles* 6–12, somewhat crowded, very long, 2·5–7·5 cm. l., filiform, 1-fld., lax, red. *Bracts* ovate, acuminate, entire, concave, 3-nerved, green, 5–9 mm. l. *Sepals* 3-nerved, narrowly oblong, hooded at tip, connivent, erect, dark purple, 9 mm. l., 1·75 mm. br. *Petals* nearly as long as the sepals, white, 3-nerved, lanceolate, obtuse, about 8 mm. l., 1·5 mm. br. *Lip* narrow at the base, erect, half-embracing the column, nearly as long as the petals, 3-lobed; middle lobe larger than lateral, oblong, blunt; lateral ovate, entire; 3 raised lines marked with green spots run from the base to the apex of the middle lobe. *Column* somewhat stout, semi-terete. *Anther* bluntly apiculate. *Capsule* with 6 shallow keels, 1·1 cm. l., ·5 cm. br.

3. VANILLA Sw.

Herbs; strong, high-climbing, branching; stems with or sometimes without leaves, emitting roots. Leaves leathery, papery, or membranous. Racemes or spikes axillary, generally short. Flowers large. Bracts ovate. Sepals about equal, free, spreading. Petals somewhat similar to the sepals. Lip, claw adnate to the column; limb broad, concave, embracing the column at the base. Column long, without a foot; stigma transverse under the short rostellum; clinandrium short or obliquely raised. Anther attached to the margin of the clinandrium, incumbent, convex, semiglobose or obtusely conical, with separate cells; pollinia powdery-granular, free, or at length sessile on the rostellum. Capsule generally long, fleshy, not dehiscing or only slightly later.

Species about 20, widely dispersed throughout the tropics of the whole world.

Leaves, wanting on main stem, lanceolate-subulate on terminal shoot .... 1. *V. claviculata.*
Stem leafy.
  Bracts large, foliaceous. Leaves much longer than internodes .... 2. *V. inodora.*
  Bracts small, not foliaceous. Leaves about as long as, or shorter than, internodes.
    Leaves ovate, small, shorter than internodes ... 3. *V. Wrightii.*
    Leaves oblong or elliptical, large, about as long as internodes.
      Flowers 2¼–3¼ in. l. Capsule thick.
        Capsule rounded and somewhat flattish in section .... 4. *V. phæantha.*
        [Capsule somewhat 3-cornered in section *V. pompona.*]
      [Flowers 2 in. l. Capsule slender .... *V. planifolia.*]

1. **V. claviculata** *Sw. in Nov. Act. Upsal. vi.* 66, *t.* 5, *fig.* 1 (1799) (figure * bad); leaves wanting on main stem; on the

* This figure does not agree with description of Fl. Ind. Occ., but it may refer to *V. Wrightii.*

terminal shoots lanceolate, acuminate, rigid, recurved at apex, concave, sessile, half-clasping at base.—*Sw. Fl. Ind. Occ.* 1515; *Lindl. Gen. & Sp. Orch.* 434; *Reichb. f. in Ber. Deutsch. Bot. Ges. iii.* 275; *Griseb. Fl. Br. W. Ind.* 638; *Rolfe in Journ. Linn. Soc. xxxii.* 471 (1896); *Cogn. in Symb. Ant. vi.* 324. V. barbellata *Reichb. f. in Flora xlviii.* 274 (1865). Cereo affinis scandens planta &c. *Sloane Hist. ii.* 160, *t.* 224, *fig.* 3, 4. Cactus parasiticus *L. Syst. ed.* 10, 1054 (1759) (in part, *i.e.* so far as concerns reference to Sloane Jam.). Epidendrum claviculatum *Sw. Prodr.* 120 (1788). (Pl. 2, f. 18, 19.) Type in Herb. Mus. Brit.

Greenwithe.

On shrubs and trees, growing on limestone rocks; in fl. and fr. June; Angels, near Spanish Town, and on hills near, *Sloane* Herb. vii. 86! in very dry, calcareous mountainous places, *Swartz*! Retirement, St. Elizabeth, J.P. 460, *Morris*! Stony Hill, J.P. 460 (2613) *Syme*! Constant Spring 650 ft., Stony Hill, 900 ft., *Harris*! Fl. Jam. 5827, 10,411 (*March*! in Hb. Kew.; owing to the unsatisfactory condition of the flower, which has apparently a glabrous lip, we have had some hesitation in referring it to this species).—Cuba, Hispaniola, Porto Rico, St. Thomas, Bahamas.

Stem leafless on main stem, thick, 1 cm. or more br., shrivelling very much, to 5 or 6 mm., when dry; internodes about 10 cm. l.; adventitious roots often twisting spirally like tendrils. *Leaves* 3–8 cm. l. *Raceme* with 8–12 sessile flowers, to 14 cm. l. *Bracts* below like the leaves, but much smaller, above triangular, acuminate, 2–·5 cm. l. *Ovary* ¾ as long as the sepals, cylindrical. *Flowers* fragrant, glaucous green sepals with a white and purplish lip, perianth nearly 2 ins. l., crowded towards apex of raceme. *Sepals* narrowly elliptical, obtuse, 4–4·7 cm. l., 1–1·1 cm. br. *Petals* as long as sepals, keeled, 1·3 cm. br. *Lip* roundish, crenate and curled on upper margin, attached to column for two-thirds its length, with ramentaceous hairs inside along the central line, and at the sides below, 5 cm. l., about 4·5 cm. br. when flattened out. *Column* 3 cm. l. *Capsule* ellipsoidal-cylindrical, thickest in the upper part, narrowing gradually to the base, 10 cm. l., 1·6 cm. br.

Swartz says, "flores albi . . . odor fortis Orchidearum." Grisebach quotes Wright as saying "perigonium rufescenti-virens, labello rubromarginato et punctato, mediano intus barbato." We have not taken the species name from Linnæus's *Cactus parasiticus* owing to the great confusion in his references. In the *Systema* (1759) he describes *C. parasiticus* as "repens, teres, striatus, muticus," referring to (1) Sloan. Jam. 224, f. 3, 4, the specimen of which is in Herb. Sloane, and is *Vanilla claviculata* Sw.; (2) Plum. Ic. 197, f. 2, probably a species of *Rhipsalis*. In *Species Plantarum*, ed. 2, i. 668 (1762) he repeats the diagnosis from the *Systema*, but drops the reference to Sloane, citing instead *Cactus parasiticus inermis &c.* Browne Jam. 238. In the Linnean Herbarium there is a specimen from Browne named *Cactus parasiticus* by Linnæus; this has no flower, and is either *Dendrophylax funalis*, or a leafless species of *Campylocentrum*.

2. **V. inodora** *Schiede in Linnæa iv.* 574 (1829); leaves much longer than the internodes, elliptical, shortly acuminate, membranous, very shortly stalked; bracts in two ranks, foliaceous, membranous, like the leaves but much reduced and varying in size.—*Rolfe op. cit.* 449; *Cogn. op. cit.* 320. Vanilla flore viride et albo, fructu nigrescente, *Plum. Nov. Gen. Pl. Amer.* 25, *t.* 28

(1703) & *Pl. Amer.* (*Burm.*) 183, *t.* 188 (1758). V. volubilis siliquosa Mexicana plantaginis folio *Catesby Nat. Hist. Carol. App. t.* 7 (1743). V. anaromatica *Griseb. op. cit.* 638. Epidendrum Vanilla, *L. in Herb.* (Pl. 2, f. 16.)

On trees, in rocky woodland; in fl. Dec., in fr. March–Aug.; *P. Browne* (in Herb. Linn.)! Manchester, *Purdie*! J.P. 461 (2614) *Syme*! Claverty Cottage, J.P. 461, *Hart*! in Swift River course, near Hope Bay; near Troy, 2600 ft.; *Harris*! Middle Quarters, *Farquharson*! Pedro, St. Ann, *Sir H. Blake*! Fl. Jam. 8555, 8829.—Cuba, Hispaniola, Porto Rico, Guadeloupe, Dominica, Martinique, Grenada, Trinidad, Mexico, Nicaragua, B. Guiana, Surinam.

Stem leafy, somewhat thick, 8 mm. br., shrivelling to 5 or 6 mm. in drying; internodes 3·5–10 cm. l. *Leaves* 10–20 cm. l., 5–10 cm. br., nerves numerous, prominent in drying. *Racemes* axillary, with 6 or 7 sessile flowers, to 15 cm. l. *Bracts* usually 1·3–2·5 cm. l., sometimes to 8 cm. l. *Ovary* ¾ as long as sepals, or nearly as long, cylindrical. *Flowers* greenish, with white lip with orange-yellow crest, perianth 1¾–2 in. l. *Sepals* and *petals* lanceolate to lanceolate-oblong, somewhat obtuse, 5 cm. l., to 1·5 cm. br. *Lip* 3-lobed, nearly as long as sepals and petals, united to column only a short distance at base (about 5 mm.); lateral lobes oblong, obtuse, median lobe spreading and then erect, ovate-oblong, subacute; disk 3-keeled, without a tuft of hairs, 4·5 cm. l., about 3 cm. br. (when flattened out). *Column* 2·5–3 cm. l. *Capsule* slender, elongate-linear, slightly tapering at apex, not fragrant, 12–18 cm. l.

3. **V. Wrightii** *Reichb. f. in Flora xlviii.* 273 (1865); leaves shorter than the internodes, ovate or elliptical, shortly acuminate, shortly stalked, papery.—*Griseb. Cat. Cub.* 267; *Rolfe op. cit.* 454; *Cogn. op. cit.* 321. V. claviculata *Lindl. in Ann. & Mag. Nat. Hist., ser.* 3, *i.* 334 (1858), non *Sw.* Vanillas piperis arbori jamaicensis innascens *Pluk. Alm.* 381, *t.* 320, *f.* 4. (Pl. 2, f. 17.)

Climbing over trees; in fl. March, in fr. August; Mt. Airey, near Mt. Moses, 2300 ft., J.P. 2088, *Syme*! Bull Head, *Hart*! Tweedside; Mt. Moses, 2200 ft.; *Harris*! Fl. Jam. 7885.—Cuba, Trinidad, B. Guiana, Surinam.

Stem leafy, slender, 3–4 mm. br., shrivelling to 2 mm. in drying; internodes 5–10 cm. l.; roots 2–4 from the same node. *Leaves*, nerves numerous, prominent in drying; blade 4–7 cm. l., 2–3·3 cm. br., stalk 5–12 mm. l. *Racemes* axillary, with 5, or more, sessile flowers, to 7 cm. l. *Bracts* triangular, obtuse, half-clasping, 5–7 mm. l. *Ovary* about ¾ as long as sepals, cylindrical. *Flowers*, perianth 2–2¼ in. l., with dark reddish sepals (in bud reddish-brown). *Sepals* elliptical-lanceolate, obtuse, 4·7 to about 5 cm. l., 8–9 mm. br. *Petals* keeled, 4·7 cm. l., 7 mm. br. *Lip* entire, united to column about 2 cm. above base, roundish-obovate, shortly acute, margin crenulate, with ramentaceous hairs on the disk especially along the central line, about as long as the sepals. *Column* 3·7 cm. l., hairy below the stigma, winged on each side of anther. *Anther* somewhat rectangular in outline. *Capsule* elongate, linear, somewhat flattened, slender, slightly tapering at apex, 12–14 cm. l., 4 mm. br.

Reichenbach describes the lip as "apice bilobulo (disco lævi?)"; Grisebach as "barba destituto," and Rolfe as "disco lævi." Wright, quoted by Lindley, says of his Cuban specimen (672): "Flowers light dull red. Labellum light-yellow with a white margin. Column adhering to the labellum more than half its length, reddish towards the base."

4. **V. phæantha** *Reichb. f. op. cit.* 274; leaves about as long as the internodes, oblong, tapering to an acute apex, with very short stalk, clasping at the base, fleshy; flowers 2¼–3¼ in. l.; capsule linear-cylindrical, somewhat flattened, tapering to base.—*Rolfe op. cit.* 464. *Cogn. op. cit.* 323. V. planifolia, var. macrantha *Griseb. loc. cit.* (Pl. 2, f. 20.)

In fl. May; Askenish, 800 ft., near Dolphin Head, *Harris*! Fl. Jam. 10,489.—Cuba, Bahamas, St. Vincent, Trinidad.

Stem leafy, thick, 5–8 mm. br. when dry; internodes 13 to 14 cm. l. *Leaves* several-nerved (when dry), 2·5–3 cm. br. *Raceme* with several flowers, 3–5 cm. l. *Bracts* ovate, acute, 6–14 mm. l., 4–8 mm. br. *Flowers* fleshy, perianth 3¼ in. l. *Ovary* cylindrical, slightly 2-angled, 5 cm. l., 6–7 mm. br. *Sepals* oblanceolate, about 11-veined; lateral somewhat acute, slightly falcate, 8·2–8·8 cm. l., 1·6–1·8 cm. br.; median somewhat blunt, 8·5–9·1 cm. l., 1·7 cm. br. *Petals* about 10-nerved, distinctly broadly keeled, 8·1–8·7 cm. l., 1·2–1·4 cm. br. *Lip* attached to column almost to its apex, limb very broadly ovate, apex retuse, apiculate just below the apex on the back, distinctly keeled along the back, the keel ending in a long mucro at 5 mm. below the apex, nerves prominent in the tube and on disk, with a tuft of transverse, flat, toothed crests pointing downwards opposite the column near the apex, margin somewhat curled and crenulate, 8·1–8·3 cm. l., about 4·5 cm. br. *Column* 6·3 cm. l., bearded below the stigma; in the specimen examined 3 anthers are present, in the lateral one cell incomplete. *Capsule* (unripe), 8 cm. l., 1·2 cm. br.

[V. planifolia *Andrews Bot. Rep.* viii. 538 (1808); leaves like those of *V. phæantha*; flowers 2 in. l.; capsule slender.—*Bauer Ill. Orch., Gen. tt.* 10, 11; *Lindl. Gen. & Sp. Orch.* 435; *Griseb. Fl. Br. W. Ind.* 638; *Rolfe op. cit.* 463; *Cogn. op. cit.* 322; *Salisb. Ic. ined. vi. fol.* 582, 591.

S.E. Mexico, Yucatan, B. Honduras, Guatemala, Costa Rica.

Stem leafy, somewhat thick, shrivelling to 5 or 6 mm. in drying; internodes 8 or 9 cm. l. *Leaves* longer than the internodes, with short thick stalks, oblong to subovate-oblong, acute or shortly acuminate, thick and fleshy, nerves obscure in fresh specimens; 10–18 cm. l., 4–6 cm. br.; stalks half-clasping, about 1 cm. l. *Racemes* axillary, many-flowered (to 14), 5–7 cm. l. *Bracts* oblong, subacute or obtuse, 5–10 mm. l. *Flowers* greenish-yellow, perianth 2 in. l. *Sepals* and *petals* oblong-oblanceolate, somewhat obtuse, about 5 cm. l. *Lip* convolute, attached by its margins to the column almost to its apex, with longitudinal crenulate keels, and a hairy tuft in the tube pointing downwards; somewhat 3-lobed above, apex retuse; with revolute margin, somewhat longer than the sepals. *Column* hairy anteriorly, about 3 cm. l. *Capsule* long, linear, 15–25 cm. l.

This species furnishes the best "vanilla beans" of commerce, and is cultivated to some extent in Jamaica. It may consequently occur here and there, in a naturalised state, producing perfect fruit, as it has done in Hope Gardens without being artificially pollinated, and on that account we have included it.

Miller's *V. mexicana* (Gard. Dict. ed. 8, 1768), as well as Swartz's *V. aromatica* (in Nov. Act. Upsal. vi. 66, 1799 & in Schrad. Journ. ii. 208, t. 1, fig. 1b, 1799) was founded on Linnæus's

Epidendrum Vanilla, and included the vanilla beans of commerce and the leaves and flowers of another species.

The pods of both *V. pompona* Schiede and *V. planifolia* are used in commerce, and as *V. pompona* is much more widely spread than *V. planifolia*, it is probable that its pods were more common in early times before cultivation was resorted to. *V. pompona* is not so much cultivated at present as *V. planifolia*, as its pods are not nearly so valuable commercially.

Judging from the habitat given by Miller for the plant cultivated by him, from the thickness of the pod ("scarce an inch broad"), and from the shape of the leaves "oblongo-ovatis," it seems more probable that his plant was *V. pompona* rather than *V. planifolia*.

V. inodora Schiede has much the same distribution on the mainland as *V. pompona*, and extends also to the West Indies. The stem and leaves of this species may have been sent to England as specimens of Vanilla, and certainly some of the old drawings suggest *V. inodora* rather than *V. pompona* or *V. planifolia, e.g.* Catesby's plate (Nat. Hist. Carol., App. t. 7), which is quoted by Miller as his *V. mexicana*. Miller also states in his generic description that the lip "runs out into a long point," and this cannot apply to either *V. pompona* or *V. planifolia*, but may well refer to *V. inodora*.]

[V. pompona *Schiede in Linnæa iv.* 573 (1829); leaves and flowers like those of *V. phæantha* but larger; capsule thick, somewhat 3-cornered in section.—*Rolfe op. cit.* 465; *Cogn. op. cit.* 323. V. grandiflora *Lindl. Gen. & Sp. Orch.* 435 (1840). Banille *Merian Pl. Ind. t.* 25 (1768).

Trinidad, Mexico, Nicaragua, Panama, Colombia, Venezuela, B. Guiana, Surinam, Cayenne.

Like *V. planifolia* but larger. *Leaves* ovate-oblong, 14-25 cm. l., 5–11 cm. br. *Racemes* 6–8 fld., 2–5 cm. l. *Bracts* broadly elliptico-oblong, 10–18 mm. l. *Flowers* greenish-yellow, perianth 3½ in. l. *Sepals* and *petals* 8–8·5 cm. l., oblong-lanceolate, subspathulate. *Lip* 9–9·5 cm. l., adnate to the column almost to the middle, constricted for some distance below, limb obscurely 3-lobed, veins somewhat thick, with tuft pointing downwards, median lobe broadly rounded, apex emarginate and apiculate. *Column* hairy anteriorly, 6–7 cm. l. *Capsule* linear-oblong, obtusely 3-cornered, 10–15 cm. l.

The description of this species, which is cultivated in the Hope Gardens, is inserted for comparison.]

4. PELEXIA Lindl.

Terrestrial herbs, with roots clustered. Leaves radical, long-stalked, at base of a simple stem, with several sheaths. Flowers of medium size, with very short stalks, in a somewhat lax spike. Bracts narrow or lanceolate. Sepals, median erect, connate with the petals; lateral linear, connate at the base into a long linear spur-like appendage, partly open on the upper side, adnate to ovary or half-free. Lip erect, concave, embracing the column above, involute below the column and attached at its base to the inside of the spur of the sepals. Column short, without a foot; stigma broad, at the base of the erect, long, subulate-acuminate rostellum; clinandrium short. Anther erect, behind the rostellum, oblong-linear, cells contiguous; pollinia narrow, powdery-granular, on the dehiscence of the anther pendulous from the apex of the rostellum. Capsule erect, ellipsoidal.

Species about 20, natives of tropical America, from the West Indies and Central America to Brazil.

Sepals acute. Spur adnate .... 1. *P. adnata.*
Sepals produced into a long, awn-like apex.
Spur half-free .... 2. *P. setacea.*

1. **P. adnata** *Spreng. Syst. iii.* 704 (1826); flowers barely ½ in. l. from apex of sepals to point of spur; sepals acute; spur adnate to ovary almost to apex; lip tubular below, terminal lobe large, kidney-shaped.—*Cogn. in Symb. Ant. vi.* 329. P. spiranthoides *Lindl. Bot. Reg. t.* 985 (1826) & *Gen. & Sp. Orch.* 481; *Griseb. Fl. Br W. Ind.* 642. Satyrium adnatum *Sw. Prodr.* 118 (1788). Neottia adnata *Sw. Fl. Ind. Occ.* 1409 (1805); *Salisb. Ic. ined. fol.* 587.

In shady woods; in fl. May; *Masson*! Bahamas, Cuba, Hispaniola, Porto Rico, Guadeloupe, St. Vincent, Grenada, Margarita.

Plant 1½ ft. high. *Stem* downy above, glabrous below, about 3 dm. l. *Leaves* 2 or 3, oval, acute, blade about 10 cm. l., 5 cm. br.; stalks red, 14 cm. l., sheathing at the base. *Sheaths* coloured, shortly tubular below, free part lanceolate, 4·5–2 cm. l. *Spike* downy, many-flowered, somewhat lax. *Bracts* linear-lanceolate, as long as the ovary, 10–17 mm. l. *Flowers* more or less erect, green or whitish. *Sepals*, lateral linear, somewhat spreading, acute, 1-nerved, green outside, white inside, spotted, from attachment at base of column about 6 mm. l., about 1 mm. br.; median elliptical, obtuse, very concave, 3-nerved, pale green, about 6 mm. l., (with the attached petals) about 3·5 mm. br. *Petals* linear, white, closely attached to the median sepal, and nearly as long as it. *Lip* white, about 5·3 mm. l. from base of column, terminal lobe 2 mm. l., 3·6 mm. br. *Capsule* 6-keeled, 1·5 cm. l., ·5 cm. br.

2. **P. setacea** *Lindl. Gen. & Sp. Orch.* 482 (1840); flowers 1½–2 in. l. from apex of sepals to point of spur; sepals produced into a long awn-like apex; spur half-free from ovary; lip oblong, broadening upwards, terminal lobe lanceolate, fimbriate below, with awn-like apex.—*Griseb. op. cit.* 642; *Cogn. op. cit.* 327. P. setacea var. glabra *Cogn. in Fedde Repert. vii.* 123 (1909). Neottia calcarata *Hook. in Bot. Mag. t.* 3403 (1835). (Pl. 2, f. 21, 22.)

In stony land in shade or woodland; in fl. and fr. Dec., Jan.; Mts. above Negril, *Purdie*! Moneague, *Prior*! *Jenman*! Stony Hill, J.P. 470, *Morris*! & *Syme*! near Halls Delight, 1000 ft.; Knapdale, near Browns Town; Stony Hill, 1000 ft.; *Harris*! Fl. Jam. 7077 (partly) 7863.—

Cayman Brac, Bahamas, Cuba, Hispaniola, Porto Rico, Trinidad, Colombia, Brazil.

Plant 15–18 in. high. *Stem* somewhat stout, downy above, glabrous below, 3 dm. and over. *Leaves* 1 or 2, oval, acute, blade 11–15 cm. l., 3-5 cm. br.; stalks 12-18 cm. l. *Sheaths* 1 or 2, radical, below the leaves, scarious, 2–5 cm. l.; cauline tubular below, free part triangular, acuminate, 2–3 cm. l. *Spike* downy, lax, with several flowers, 3–15 cm. l. *Bracts* ovate, apex acuminate, awn-like, 5-nerved, longer than the ovary, 2–2·5 cm. l., about ·5 cm. br. *Ovary* spindle-shaped (with stalk) 2 cm. l. *Flowers* greenish-white. *Sepals* linear-lanceolate, 7-nerved below (from point of attachment at base of column) 2·8 cm. l., barely 2 mm. br., median 2·5 mm. br.; spur tapering, obtuse. *Petals* similar to sepals, attached throughout their length to the median sepal, nearly 2 cm. l., 1·4 mm. br. *Lip* white, many-nerved. *Column* (below stigma) 3 mm. l., from base of column to apex of anther 9 mm. l. *Capsule* broadly ellipsoidal, with six shallow keels, dehiscing at some little distance below the apex, 2 cm. l., 8 mm. br.

Var. *glabra* Cogn. is based on Harris Fl. Jam. 7863, the author says "scapus usque ad apicem glaberrimus," but all the specimens which we have seen bear the usual pubescence.

5. STENORRHYNCOS Spreng.

Terrestrial herbs, root fibres generally fleshy, clustered, with radical leaves, or flowering stems leafless. Leaves elliptical to ligulate, radical, tapering at base into long stalks which are flattish, concave, sheathing at base. Flowers medium size to small, in a spike. Sepals free, subequal; median erect, cohering with the petals; lateral obliquely affixed at the apex of the ovary, and decurrent with the column, produced at the base into a chin or distinct spur, erect. Lip sessile or clawed, erect, concave, embracing the column and sometimes adherent to it, spreading at the apex, entire or indistinctly 3-lobed, without calli. Column terete, at the base with a long foot; stigma broad under the erect, long rostellum; clinandrium posteriorly short, margins sometimes raised and continuous with the rostellum. Anther erect, oblong-lanceolate, 2-celled; pollinia powdery-granular, pendulous from a gland of the rostellum. Capsule obovoid or ellipsoidal, 3-keeled.

Species about 25, natives of tropical America.

Leaves, several, elliptical .... 1. *S. speciosus.*
Leaves, 1 or 2, or wanting during flowering season.
  Plant with glandular-pubescent hairs. Perianth $\frac{1}{2}$–$\frac{3}{4}$ in. l. Spur about $\frac{1}{2}$ length of ovary ...... 2. *S. lanceolatus.*
  Plant scurfy with white scales. Perianth 1–1$\frac{1}{4}$ in. l. Spur about $\frac{2}{3}$ length of ovary .... 3. *S. squamulosus.*

1. **S. speciosus** *Lindl. Gen. & Sp. Orch.* 477 (1840); plant glabrous; leaves several, elliptical; bracts crimson or rosy-pink, elliptical, acuminate, nerves 3, lateral branching, nearly as long as or longer than the flower; perianth crimson or rosy-pink, $\frac{2}{3}$–$\frac{3}{4}$ in. l.; sepals forming a short chin.—*Griseb. Fl. Br. W.*

Ind. 642. Cardamomum minus pseudo-asphodeli foliis *Sloane Cat.* 61 & *Hist. i.* 166, *t.* 103, *f.* 3. Neottia speciosa *Jacq. Collect. iii.* 174 (1789) & *Ic. Pl. Rar. iii. t.* 600 (1786–1793); *Sw. Fl. Ind. Occ.* 1400; *Hook. Exot. Fl. tt.* 3 & 4; *Salisb. Ic. ined. fol.* 586, 587.

On banks and in rocky places in damp woods; in fl. Oct.–Feb.; Guanaboa, *Sloane*, Herb. iii. 38, iv. 115*! *Shakespear*! *Bancroft*! Browns Town, *Purdie*! Fairfield, *Wullschlaegel*, 1075! Mt. Moses, 2500 ft., J.P. 2059, *Syme*! Clydesdale, *Moore*! Cinchona, 5000 ft.; Holly Mount, Mt. Diabolo, 2600 ft.; *Harris*! Fl. Jam. 7632, 7640, 7757, 8872, 9898.—Cuba, Porto Rico, Mexico, Guatemala, Costa Rica, Panama, Colombia, Venezuela.

Plant 9 in. to nearly 2 ft. high, with thick, clustered roots. *Scape* erect, somewhat stout, 2·5–6 dm. l. *Leaves* acute, with several nerves, 6–17 cm. l., 3·5–6 cm. br., stalks 3·5–10 cm. l. (incl. sheath). *Sheaths* several at base, 5–7 above, membranous, lax; free portion lanceolate, acuminate; upper red; 4–8 cm. l. *Spike* erect, dense, somewhat one-sided, generally short, 4–11 cm. l. *Bracts* 3·5–2 cm. l. *Ovary* oblong, tapering to base, oblique at apex, 3-ribbed, purplish. *Flowers* numerous. *Sepals* slightly spreading at apex, nerves 3, sometimes branching; lateral oblong-lanceolate to narrowly lanceolate, acute to acuminate, 1·6–2 cm. l., about 4·5 mm. br.; median oblong, subacuminate, to triangular-lanceolate, concave, 1·25–1·5 cm. l., about 4·5 mm. br. *Petals* 3-nerved, oblong tapering to an acute apex, or linear-lanceolate, slightly falcate, oblique at base and there attached both to median and lateral sepals, 1·25–1·5 cm. l., 2·5–3 mm. br. *Lip* 7-nerved, pubescent on the disk, shortly pubescent on the lower face, somewhat rhomboidal, with a recurved oblong acute or lanceolate apex, concave, slightly saccate at each side at base, 1·25–1·85 cm. l., about 5 cm. br. *Column* 3·5 mm. l. from junction with lip to attachment of anther, 1·1 cm. to apex of rostellum; rostellum obovate, with a long beaked apex (beak 3·5 mm. l.), pubescent below the stigma, 2-winged below, wings united with stalk of anther. *Anther* 5–6 mm. l., ovate-lanceolate, shortly stalked.

2. **S. lanceolatus** *Griseb. loc. cit.*; plant with glandular-pubescent hairs; leaves glabrous, 1 or 2, ligulate or elliptical-lanceolate, often disappearing before flowering; bracts lanceolate to linear-lanceolate, long acuminate, nerves 3, lateral branching, about as long as the ovary; perianth dirty yellow, between ½ and ¾ in. l.; spur not more than half as long as ovary, bluntly oblong, apex obtuse, almost wholly adnate to ovary, not more than about 1 mm. free.—S. orchioides *Lindl. op. cit.* 477 (in part). Helleborine purpurea &c. *Plum. Pl. Amer.* (*Burm.*) 176, *t.* 181, *f.* 2. Limodorum lanceolatum *Aubl. Pl. Guian. ii.* 821 (1775). Satyrium Orchioides *Sw. Prodr.* 118 (1788). Neottia Orchioides *Sw. Fl. Ind. Occ.* 1411, *t.* 28, *fig. a. b.* (1806). (Pl. 3, f. 22.)

On banks amongst grass; in fl. and fr., March and April; *Macfadyen*! *Distin*! Port Royal Mts., *Purdie*! Fairfield, *Wullschlaegel*, 1382 (in part)! Mt. Moses, 2800 ft., J.P. 2400, *Syme*! below Vinegar Hill, Portland, 3000 ft.; Mt. Airy, 2000 ft.; *Harris*! Fl. Jam. 7876.—Cuba, Bahamas, Hispaniola, Antigua, Guadeloupe, Dominica, Martinique, St. Lucia, St. Vincent, Barbados, Grenada, Trinidad, Mexico, Guatemala, Colombia, Brazil, Argentine, Paraguay.

Plant 1 ft.-2 ft. 4 in. high, with thick, clustered roots. *Scape* erect, somewhat stout, glandular-pubescent above, glabrescent below, 2·5-6·5 dm. l. *Leaves* acute, entire, with several nerves and reticulate veins, 7-20 cm. l., 1·8-2·3 cm. br.; stalk 1-5 cm. l. (including sheath). *Sheaths* several at base, 5 or 6 above; free portion ovate, acute, below, to acuminate, above; glabrous, 2-3·5 cm. l. *Spike* erect, short or elongate, glandular-pubescent, 4-8 cm l. *Bracts* 9-13 mm. l., to 2·75 mm. br., sparsely glandular-pubescent, and sometimes margin ciliolate. *Ovary* obovate-oblong, very oblique at the apex, slightly glandular-pubescent, or glabrescent. *Flowers* close together. *Sepals* glandular-pubescent outside; lateral 5-nerved, lanceolate, acute, greatest length about 1·5 cm. l., length anteriorly about 1 cm. l., 2·5-3 mm. br.; median 7-nerved, triangular-lanceolate, acute, about 1 cm. l., 3·2-4·2 mm. br. *Petals* 7-nerved, oblong, tapering above middle to apex, subacute, somewhat falcate, about as long as the median sepal, 2·2-2·75 mm. br. *Lip* several-nerved, dilated and saccate about the middle; oblong, acute above, below convolute, slightly tapering to base; pubescent below the middle on inside, and on outside on margin, lighter in colour than sepals and petals, 12-14 mm. l., 3·2-4·5 mm. br. in middle, 2·5 mm. br. in lower portion. *Column* 6-7 mm. to apex of rostellum; beak of rostellum about 4 mm. l. *Anther* sessile at base of column, about 4 mm. l. *Capsule* with a few scattered hairs, 1-1·2 cm. l., 5 mm. br.

3. **S. squamulosus** *comb. nov.*; plant scurfy with white scales; leaves wanting during flowering, according to Macfadyen (fide Grisebach) "serotinous, long, broadly lanceolate, pointed"; bracts lanceolate to linear-lanceolate, acuminate, somewhat longer than the ovary, 3-5-nerved with reticulate veins; perianth brick-red colour, 1-1¼ in. l.; spur about ⅔ as long as ovary, ovate, apex obtuse, free from ovary near apex, 2-3 mm. free.—S. orchioides *Lindl. loc. cit.* (in part); *Griseb. loc. cit.* Neottia squamulosa *H. B. & K. Nov. Gen. & Sp. i.* 332, *t.* 71 (1815). N. orchioides *Sims Bot. Mag. t.* 1036 (1807), non *Sw.*; *Bot. Reg. t.* 701; *Salisb. Ic. ined. fol.* 585. (Pl. 3, f. 21.)

On dry banks; in fl. and fr. April, May; St. Mary; Pedro, St. Ann.; Port Royal Mts.; *Purdie*! *Shepherd*! *Wilson*! Fairfield, *Wullschlaegel* 1382 (in part)! Liguanea Plain, J.P. 8, *Morris*! *Campbell*! Chapelton, *Fawcett*! Cinchona, *Harris*! Fl. Jam. 10,395, 10,396, 10,410, 10,496.—Cuba, Colombia.

Plant 1-2 ft. high, with clustered thick fleshy roots. *Scape* erect, somewhat stout, scurfy with white scales, 3-6 dm. l. *Sheaths* several at the base, 5-8 above, free portion ovate, acute below, acuminate above, sometimes scurfy, 2·5-4·5 cm. l. *Spike* erect, long, scurfy, 6-14 cm. l. *Bracts* scurfy, sometimes with ciliolate margin, 16-23 mm. l., 3·5-6 mm. br. *Ovary* oblong, tapering to base and apex, oblique at apex, scurfy. *Flowers* shortly stalked, crowded towards the apex of the spike. *Sepals* scurfy outside; lateral 5-nerved, lanceolate, long acuminate, very oblique and decurrent at the base, greatest length 2-2·35 cm. l., 3-3·5 mm. br.; median 5-nerved, ovate-lanceolate, acuminate, 1·5-1·75 cm. l., 3·5-4·5 mm. br. *Petals* 5-7 nerved, oblong-lanceolate, subacute, somewhat falcate, about as long as the median sepal, 2·5-3 mm. br. *Lip* several-nerved, dilated and saccate about the middle, lanceolate, acuminate, above; below convolute, linear; pubescent below the middle on inside and on outside on margin, 2-2·35 cm. l., 5-6 mm. br. in middle. *Column* about 11 mm. to apex of rostellum. *Capsule* scurfy, 1·1-1·3 cm. l.

6. SPIRANTHES L. C. Rich.

Terrestrial herbs, roots fibrous or tuberous, stems leafy or the flowering leafless. Leaves linear, ovate or elliptical. Flowers small, secund, in often twisted spikes. Sepals free, subequal, median erect, cohering with the petals; lateral gibbous at the base. Lip sessile or clawed, erect, concave, embracing the column and often adherent to it, spreading at the apex, entire or 3-lobed, base concave, sometimes with calli inside. Column short, terete, base generally decurrent on the ovary; stigma broad under the rostellum, which is erect, obtuse, or elongate; clinandrium short, margins sometimes raised and continuous with the rostellum. Anther erect, ovate or oblong, 2 celled; pollinia powdery-granular, pendulous from a gland of the rostellum. Capsule ellipsoidal, erect.

Species about 85, widely dispersed through the north temperate zone and tropical Asia and America to Chile.

Leaves linear or linear-lanceolate, generally decaying before flowering .... 1. *S. tortilis.*
Leaves ovate or elliptical, present during flowering.
  Leaves all radical .... 2. *S. elata.*
  Leaves radical and along the stem ... 3. *S. Fawcettii.*

1. **S. tortilis** *L. C. Rich. in Mém. Mus. Par. iv.* 59 (1817) (in part, excl. *S. quadridentata*); leaves radical, generally decaying before flowering, linear or linear-lanceolate, erect, clasping at base, elongate; lip shortly clawed, ovate-elliptical, scarcely 3-lobed, slightly cordate at base, emarginate at apex, margin of upper half curled, crenulate, with numerous nerves, median nerve prominent below, with two conical calli at base, green shaded off into white at margin.—*Lindl. Gen. & Sp. Orch.* 468; *Griseb. Fl. Br. W. Ind.* 640; *Cogn. in Symb. Ant. vi.* 342. S. apiculata *Lindl. op. cit.* 474. Satyrium spirale *Sw. Prodr.* 118 (1788). Neottia tortilis *Sw. in Vet. Handl. Stockh.* 1800, 226, & *Fl. Ind. Occ. iii.* 1406, *t.* 28, *f. g, h.* (Pl. 3, f. 1–5.)

On damp shady banks and in pastures; in fl. and fr. March–June; Hollis Savanna, Clarendon; Guy's Hill, St. Mary; *Purdie*! Fairfield, *Wullschlaegel*, 1067! Hanover, J.P. 2312, *Morris*! Cinchona, 4900 ft., *Moore*! Castleton, 490 ft., *Fawcett*! *Harris*! Farm Hill, *Miss B. Gosset*! below Vinegar Hill, Portland, 3000 ft.; between Cinchona and Morse's Gap, 4900 ft.; *Harris*! Fl. Jam. 7314, 7853, 9553, 10,495.—Louisiana, Florida, Bermuda, Bahamas, Cuba, Hispaniola, Porto Rico, St. Thomas, Antigua, Guadeloupe, Dominica, Martinique, St. Lucia, Grenada, Trinidad.

Plant 1–1½ ft. high. *Scape* erect, slender, sparsely pubescent near apex, glabrous below, 2–4 dm. *Leaves* striate, smooth, to 3 dm. *Sheaths* 6 or 7, tubular below, free portion lanceolate, acute, 3–1 cm. l. *Spike* erect, spirally twisted, somewhat dense, puberulous, 8–13 cm. l. *Flowers* on one side of spike, nodding, white, perianth $\frac{1}{8}$–$\frac{1}{4}$ in. l. *Bracts* puberulous, ovate, subacuminate, as long as, or longer than, the ovary, 6–7 mm. l. *Ovary* pubescent, broader above. *Sepals* slightly puberulous, white, 3-nerved, lateral oblique at base, oblong tapering to an obtuse apex, 5 mm. l., nearly 2 mm. br.; median elliptical, obtuse, 4·5–6 mm. l., 2 mm. br. *Petals* 3-nerved, oblong, apex rounded, slightly tapering to base, as long as the median sepal, white. *Lip* 5·5 mm. l., 3·75 mm. br. *Column* very short, green; rostellum and clinandrium connected by broad wings, rostellum on removal of linear gland with pollinia, 2-dentate. *Anther* ovate, cordate, somewhat acute, sessile, brown. *Capsule* 4–6 mm. l., 3–4 mm. br.

2. **S. elata** *L. C. Rich. loc. cit.*; leaves glabrous, radical, long stalked, oval, elliptical or oblong, erect, somewhat abruptly acute, tapering somewhat abruptly into the stalks, which are rather flattish-concave, dilated and clasping at base; lip, lower half somewhat fiddle-shaped with rounded corners, 5-nerved, abruptly contracted above into a neck and then expanding into a roundish or transversely elliptical apical lobe, minutely curled and crenulate, rather broader than the lower half, base slightly saccate with a pair of short calli, white, faintly tinged with pink or with faint tinge of brown at apex and striped with green about middle.—*Lindl. op. cit.* 470; *Griseb. op. cit.* 641; *Cogn. op. cit.* 336 (excl. syn. Sloane and Browne) & under S. bicolor *Lindl.* (as regards Jamaican spec.). Satyrium elatum *Sw. Prodr.* 119 (1788). Neottia elata *Sw. in Vet. Handl. Stockh.* 1800, 226 & *Fl. Ind. Occ.* 1403; *Bot. Mag. t.* 2026. P. spiranthoides *Griseb. Cat. Cub.* 269 (1866) (non *Lindl.* nec *Griseb. Fl.*). (Pl. 3, f. 10–14.) Type in Herb. Mus. Brit.

On stone walls, on rocky banks and in peaty soil in woodland; in fl. and fr. Feb.–April; *Macfayden*! *Distin*! Manchester and St. Ann, *Purdie*! Bethabara, *Wullschlaegel*, 1073, 1074! *Prior*! near Cinchona J.P. 36; Portland Gap, 5550 ft. J.P. 235 (somewhat abnormal); Mt. Moses, 2500 ft. J.P. 2191; lowlands, J.P. 472; *Morris*! near Browns Town, *Miss T. M. Barrett*! Green River, *Sullivan*! Knapdale, near Browns Town, 1000 ft.; near Mabess River, 4000 ft.; near Tweedside, south St. Andrew, 2500 ft.; near John Crow Peak, 5500 ft.; Wallenford, 2500 ft.; Holly Mount, Mt. Diabolo, 2400 ft.; Newhaven Gap, 5600 ft.; between Cinchona and Morse's Gap, 4900 ft.; near Morse's Gap, 4500 ft.; Iron Face, Chester Vale, 3500 ft.; *Harris*! Fl. Jam. 7077 (partly), 7631, 7845, 7856, 7857, 7882, 8910, 10,458, 10,492, 10,497.—Bahamas, Cuba (Wright 624, 3293 in Herb. Mus. Brit.), Hispaniola, Porto Rico, St. Thomas, St. Croix, Tortola, St. Martin, Antigua, Guadeloupe, Dominica, Grenada, Trinidad, Venezuela, B. Guiana, Brazil, Argentine.

Plant 6 in. to 2 ft. high, with clustered, thick, fleshy root-fibres. *Scape* erect, somewhat slender, pubescent above, glabrate below, 1·5–4 dm. l. *Leaves* 2–6; blade 3–13 cm. l., 1·5–5·5 cm. br.; stalks 2–7 cm. l. *Sheaths* 5–7, free portion oblong-lanceolate, acute or acuminate, 7–1 cm. l. *Spike* erect, loosely spirally twisted, sometimes somewhat secund, puberulous, 2–18 cm. l. *Flowers* very shortly stalked, nodding, perianth $\frac{1}{4}$–$\frac{1}{5}$ in. l., green or brownish-green. *Bracts* glabrate, dotted, oblong-lanceolate, acuminate, longer than the ovary, 8–12 mm. l., about 2 mm. br. *Ovary* sparsely puberulous, cylindrical. *Sepals* green or brownish-green, puberulous, dotted, 1-nerved, slightly gibbous at base; lateral linear-oblong, dilated and oblique at base, slightly falcate, tapering to a somewhat obtuse apex, 5–7 mm. l., ·75–1·3 mm. br., expanding to 1·5–1·8 at base; median oblong, tapering to a somewhat obtuse apex, 4·75–6 mm. l., 1·3–1·5 mm. br. *Petals* brownish-green or green with pinkish tips, dotted, 1-nerved, oblanceolate with rounded apex, shorter than median sepal, 4–5 5 mm. l., ·6–·75 mm. br. *Lip* 5–6·3 mm. l., 1·6–2·5 mm. br. below; neck about 1 mm. l. and 1 mm. br. *Column* pubescent on the upper half anteriorly, to base of anther 1–1·4 mm. l., to apex of rostellum 3·2–3·6 mm. l. *Anther* ovate-elliptical, subsessile. *Capsule* 8–9 mm. l., 4 mm. br.

3. **S. Fawcettii** *Cogn. in Fedde Repert. vii.* 123 (1909); leaves conspicuously 3-nerved with ascending connecting branches, ovate, sometimes subcordate, obtuse to subacute, long-stalked, glabrous, stalks long, flattish-concave, sheathing at the base; lip 3-nerved, lateral nerves with 1 or 2 branches, oblong, suddenly contracting near apex and with somewhat square corners there, pilose inside about the middle of oblong part, apical lobe rounded-kidney-shaped, with small crenulations on margin, striped with green.—Sauroglossum tenue ***Lindl.*** *in Ann. & Mag. Nat. Hist.* (1858) 334; *Griseb. op. cit.* 642. (Pl. 3, f. 6–9.)

In shade, in leaf-mould in woods and amongst mosses on rocky banks; in fl. and fr. April; Pedro, St. Ann, *Purdie*! near Mabess River, 3500 ft.; Cascade, 3000 ft., 2500 ft.; between Cinchona and Morse's Gap, 4900 ft.; *Harris*! Fl. Jam. 7877, 7879, 10,493.—Cuba, Porto Rico.

Plant 2½–8½ in. high, leafy in the lower portion, with few sheaths, and tuberous roots. *Scape* erect, somewhat weak, pubescent above, glabrous below, 4–16 cm. l. *Leaves*, blade 1·5–3·5 cm. l., 1–2 cm. br.; stalks 1·5–4 cm. l. Sterile sheathing *bracts* generally 3 (2–4), loose, free part lanceolate, acuminate, 2–2·5 cm. l. *Spike* erect, short, pubescent, 1–3 cm. l. *Flowers* sessile, erect or spreading, perianth about ⅓ in. l., greenish-white. *Bracts* oblong-subulate, 1-nerved, glabrous, dotted, as long as, or a little longer than the ovary, the lower 1·2 cm. l., nearly 3 mm. br. *Ovary* linear-oblong, slightly pubescent towards the apex, 7 mm. l. *Sepals* pubescent at the base, greenish-white; lateral linear, spathulate, with one nerve which is slightly prominent on the outside, oblique and somewhat gibbous at the base, free, dotted, about 8 mm. l., about 1 mm. br., 1·4 mm. near apex; median oblong, with obtuse apex, 1-nerved, dotted, about 8 mm. l., 1·6 mm. br. *Petals* linear-spathulate, dilated at base, acute, with margin slightly denticulate near apex, somewhat shorter than median sepal, striped with brown, nearly 8 mm. l., ·5 mm. br., ·75 mm. br. near apex. *Lip* about 9 mm. l., 2 mm. br., apical lobe nearly 4 mm. br. *Column* long, about 4 mm. l. to junction of anther, about 6 mm. to apex of rostellum; rostellum with 2 acuminate membranous lobes. *Anther* subsessile. *Capsule* about 1 cm. l.

7. ERYTHRODES Blume.

Terrestrial leafy herbs. Roots thickish, clustered at base of stem and from the nodes in the lower part of the stem. Stem ascending or erect, simple, sometimes prostrate and rooting in the lower portion, often pubescent above. Leaves stalked, ovate or lanceolate, membranous, with many converging nerves with netted veins. Flowers small or minute, subsessile, in a lax or dense terminal spike which is often elongate. Sepals subequal, free, erect or somewhat spreading. Petals cohering with the median concave sepal. Lip erect from the base of the column which it embraces, produced below into a descending obtuse spur. Column short; stigma at the base of the erect rostellum. Anther erect, shortly acuminate, cells contiguous, distinct; pollinia granular, pendulous from an oblong gland of the rostellum, which is deeply 2-fid on the fall of the pollinia. Capsule oblong or ellipsoidal.

Species about 60, natives of the warmer regions of America and Asia.

The name *Erythrodes* (Blume Bijdr. Flor. Ned. Ind. 410, t. 72, 1825) must be adopted for this genus,—*Physurus* (L. C. Rich. in Mém. Mus. Par. iv. 55, 1818) being a *nomen nudum*. We cannot follow Schlechter (Schum. & Lauterb. Nachtr. Fl. Deutsch. Südsee, 87) in separating as *Erythrodes* the Old World species as a distinct genus from the species of the New World. His distinction rests on the bilobed character of the spur in the Old World species, which in our opinion is insufficient. The habit of the species from both hemispheres is similar, and the general structure of the flowers also closely corresponds. In the West Indian species, *E. plantaginea*, the spur sometimes shows an indication of lobing in a slight median depression.

Perianth, including spur, about $\frac{1}{3}$ in. l. .... 1. *E. plantaginea.*
Perianth, including spur, about $\frac{1}{5}$ in. l.
  Plant more or less hairy (except leaves).
    Lip with a lunate apical lobe........... 2. *E. hirtella.*
  Plant glabrous. Lip with a roundish 3-lobed apex, the middle lobe long and narrow.. 3. *E. jamaicensis.*

1. **E. plantaginea** *comb. nov.*; stem trailing along the ground, attaining a length of 6 ft. and emitting roots at the nodes like Vanilla, stout, leafy and glabrous below, slender, hairy, and with a few scales above; leaves glabrous, oval-elliptical, shortly acuminate, tapering into the stalk, sometimes the uppermost leaf much smaller in transition to the sheaths; lip oblong, concave, enclosing the column, with an apical, ovate, often crenulate lobe (but the lip appears to be sometimes undivided), 5–7-nerved.—Physurus plantagineus *Lindl. Gen. & Sp. Orch.* 503 (1840); *Griseb. Fl. Br. W. Ind.* 643; *Cogn. in Symb. Ant. vi.* 346. Orchis elatior &c. *Sloane Cat.* 119 & *Hist. i.* 250. *t.* 147, *f.* 2. Satyrium plantagineum *L. Syst. ed.* 10, 1244 (1759); *Sw. Obs. Bot.* 320; *Salisb. Ic. ined. fol.* 585. Orchis plantaginea *Sw. in Vet. Handl. Stockh. xxi.* 207 (1800). (Pl. 3, f. 15.)

In damp soil in woods; in fl. and fr. Nov.-April; Mt. Diabolo, *Sloane Herb.* iv. 116! Hector's River, Manchester, *Purdie*! Fairfield, *Wullschlaegel*, 1072! Mt. Moses, 2500 ft. J.P. 2034, *Syme*! John Crow Peak, *Moore*! Mt. Diabolo, *Fawcett & Harris*! near Mabess River, 3000 ft.; Moody's Gap, 3000 ft.; near Woodcutter's Gap, 4000 ft.; Morse's Gap, 5000 ft.; near Hardware Gap, 4200 ft.; *Harris*! Fl. Jam. 7175, 7573, 7837, 7875, 7886, 9794, 10,101, 10,129.—Cuba, Porto Rico, Saba, St. Eustatius, St. Kitts, Guadaloupe, Dominica, Martinique, St. Vincent, Grenada, Trinidad.

Plant, erect portion 1 to 2 ft. high. *Stem*, erect portion 3–6 dm. l. *Leaves*, blade 6–12 cm. l., 3–5·5 cm. br. *Sheaths* tubular below, free part triangular, shortly acuminate, 3-nerved, glabrous, 2·5–1 cm. l. *Spike* hairy, with numerous flowers, cylindrical, 5–17 cm. l. *Bracts* triangular, acuminate, hairy, 1-nerved, concave, lower 1 cm. l.; becoming smaller upwards. *Flowers* white (including spur), about ⅜ in. l. *Ovary* tapering downwards, shortly stalked, very hairy, (with stalk) 6·5–8 mm. l. *Sepals* oblong to oblong-lanceolate, obtuse, 1-nerved, hairy; lateral slightly oblique, 6–7·5 mm. l., 1·6–1·9 mm. br.; median 6–7 mm. l., 2·5 mm. br. *Petals* oblanceolate-linear, obtuse or acute, slightly oblique, rounded at base, 1-nerved, as long as the median sepal, maximum breadth 1·5 mm. *Lip* about 5·5 mm. l., 2 mm. br. *Spur* ellipsoidal, not as long as ovary, 3·5–5 mm. l., 1·6–2 mm. br. *Column* (including rostellum) about 4·5 mm. l. *Capsule* ellipsoidal, with 6 shallow keels, 8 mm. l., 3 mm. br.

2. **E. hirtella** *comb. nov.*; stem somewhat sparsely pilose, with several leaves crowded at the base; leaves glabrous, ovate to lanceolate, tapering into the stalk, the leaves near the middle the largest, those above and below smaller; lip oblong, with an apical lunate lobe, apex mucronate.—Physurus hirtellus *Lindl. op. cit.* 501 (1840); *Griseb. op. cit.* 642; *Cogn. in Symb. Ant. vi.* 347. Satyrium hirtellum *Sw. Prodr.* 118 (1788). Orchis hirtella *Sw. in Vet. Handl. Stockh. xxi.* 207 (1800) & *Fl. Ind. Occ. iii.* 1394 (1805). (Pl. 3, f. 16–19.)

In damp places in woodland; in fl. and fr. April; *Macfadyen*! *March*! Mt. Moses, 2500 ft., J.P. 2107, *Syme*! near Mabess River, 3000 ft., *Harris*! Fl. Jam. 8110 (fide Cogn.) 9010.—Cuba, Porto Rico, St. Eustatius, Guadeloupe, Dominica, Martinique, St. Lucia, St. Vincent, Grenada, Trinidad.

Plant 9–12 in. high. *Stem* 2–2·5 dm. l. *Leaves* 5–6 cm. l., 1·5–2·5 mm. br. *Sheaths* 2–3, tubular below, free part oblong, acute, 1–2 cm. l., glabrous. *Spike* hairy, many-flowered, less than 4 cm. l., elongating in fruit to 15 cm. *Flowers* white, shortly stalked, erect-spreading, (including spur) about ⅓ in. l. *Bracts* lanceolate, acuminate, 4–6·5 mm. l. *Ovary* somewhat clavate, hairy, including stalk 6 mm. l. *Sepals* 1-nerved, ovate-oblong, obtuse, sparsely hairy, green tipped with white, 3·5 mm. l.; lateral 1 mm. br., median broader. *Petals* linear-oblanceolate, obtuse, 1-nerved, 3 mm. l., about ·75 mm. br. *Lip* about 3 mm. l. *Spur* oblong, 1·5 mm. l. *Capsule* oblong, tapering to base, 6-keeled, sparsely hairy, 8 mm. l., 2·5–3 mm. br.

3. **E. jamaicensis** *comb. nov.*; stem glabrous, leafy nearly to apex; leaves lanceolate, acute, stalked, highest leaf quite small, succeeded by one or two scarious bract-like sheaths; lip 3-nerved, constricted below the middle, upper portion broadly roundish in outline, 3-lobed, lateral lobes somewhat semicircular, middle lobe

subulate, lower portion somewhat ovate, very concave, somewhat cordate at base.—Physurus jamaicensis *Fawc. & Rendle in Journ. Bot. xlvii.* 264 (1909). (Pl. 3, f. 20.)

In shady woods; in fl. Oct.; Olive River, near Christiana, 3000 ft., *Harris*! Fl. Jam. 10,472.

Plant glabrous, about 8 in. high. *Stem* about 1½ dm. l. *Leaves*, blade 3·5–5 cm. l., 11–16 mm. br.; stalk with sheath 7–17 mm. l.; highest leaf, blade 1·2–1·7 cm. l. *Spike* many-flowered, cylindrical, about 4 cm. l. *Bracts* ovate-triangular, acuminate, 1-nerved, as long as the ovary, 6–4 mm. l. *Ovary* slender. *Flowers* spreading, (with the spur) about ⅙ in. l. *Sepals* spotted, 1-nerved; median lanceolate, obtuse, margin undulate above the middle, concave, 3·2 mm. l., 1 mm. br.; lateral oblong-elliptical, obtuse, slightly falcate, 3·4 mm. l., 1–1·1 mm. br. *Petals* spotted, 1-nerved, linear-lanceolate, margin undulate, 3·1 mm. l., ·75 mm. br. *Lip*, middle portion about 1·4 mm. l., 1·6 mm. br., middle lobe ·75 mm. l., lower portion about 1 mm. l. *Spur* roundish-elliptical, about half as long as ovary, 1·6–1·8 mm. l., 1–1·3 mm. br.

8. WULLSCHLÆGELIA Reichb. f.

Terrestrial leafless herb, with clustered fibrous roots slightly thickened. Stem simple, very slender, with a few small scales. Flowers very small, subsessile in a lax spike. Bracts very small. Sepals thin, erect, connivent; median free; lateral a little broader, united at the base into a saccate chin. Petals similar to the median sepal, but smaller. Lip on the upper side of the flower, sessile at the base of the column, erect, broad, concave, undivided, produced at the base into a sac within the chin of the sepals. Column very short; stigma entire, beneath the short broad erect rostellum; clinandrium short. Anther erect, sessile, ovate, scarcely acuminate, with distinct cells; pollinia powdery-granular, affixed to a small gland of the rostellum. Capsule small, ovoid, erect.

Species 3, natives of the West Indies and Brazil.

W. aphylla *Reichb. f. in Bot. Zeit.* (1863) 131; *Griseb. Fl. Br. W. Ind.* 639; *Cogn. in Fl. Bras. iii. pt.* 4, 242, *t.* 57, *f.* 1; & *in Symb. Ant. vi.* 351.—Cranichis aphylla *Sw. Prodr.* 120 (1788), & *Fl. Ind. Occ. iii.* 1421, *t.* 29, *f.* 1, & *Ic. ined. t.* 21; *Lindl. Gen. & Sp. Orch.* 450. (Pl. 3, f. 23, 24.) Type in Herb. Mus. Brit.

In damp deep-shady woods; in fl. Sept., Oct.; mountains of Clarendon, *Swartz*! Bethabara, *Wullschlaegel*, 1071! St. Ann, *Purdie*!—Cuba, Dominica, Brazil, Paraguay.

Plant 8–14 in. high, with numerous, very long roots, thickened near base. *Stem* erect, slender, leafless, puberulous, 16–24 cm. l., with numerous minute scales. *Scales* triangular-lanceolate, acuminate, half-clasping, 4–5 mm. l. *Raceme* somewhat long, many-flowered, 4·5–13 cm. l. (in fruit). *Pedicels* 2 mm. to 3 mm. l. (in fruit). *Bracts* similar to the scales,

3 mm. l. *Flowers* very small, pale white, erect, somewhat spreading. *Ovary* 3–4 mm. l., thicker upwards, puberulous, hairs bifurcate, composed of 3 cells, one of which is the stalk, the other two, forming the fork, are clavate. *Sepals* sparsely puberulous, median oblong, concave, 1·75 mm. l.; lateral transversely elongated, oblate, with subacute apex, asymmetrical, greatest length about 3·5 mm., 1·5 mm. br. *Petals* oblong, obtuse, 1·75 mm. l. *Lip* incumbent, oblong, conduplicate-concave, 3 mm. l. *Column* with a long foot; clinandrium cup-shaped. *Capsule* 5–7 mm. l., 3–3·5 mm. br.

9. PSEUDOCENTRUM Lindl.

Terrestrial, tall herbs; rhizome short with clustered fleshy roots. Stems leafy, simple, erect. Leaves several, radical, long-stalked, membranous; a few on the stem, reduced above to sheaths. Flowers of medium size, numerous, almost sessile on a long spike. Bracts membranous. Sepals, median free, narrow, incurved; lateral shorter, broader, connate at the base and produced into a very long cylindrical ascending spur. Petals narrow, free. Lip turned upwards, adnate to the base of the column, produced anteriorly into a long narrow channelled process, inclosed in the sepaline spur with apex inflexed and saccate, auriculate with two erect lobes near the column. Column very short; rostellum erect, broadly membranous, acuminate in the middle, with the deeply-hollowed stigma at the base; clinandrium short, membranous-dilated, the margins continuous with the rostellum. Anther erect on the margin of the clinandrium, ovate, with contiguous cells; pollinia oblong, powdery-granular, pendulous from a terminal gland of the rostellum. Capsule ellipsoidal.

Species 5, natives of Jamaica, Central America, and the Andes of South America.

P. minus *Benth. in Hook. Ic. Pl. xiv.* 63, *t.* 1382 (1882); *Cogn. in Symb. Ant. vi.* 352. (Pl. 4.) Type in Herb. Kew.

In shady places; in fl. Sept.–Oct., in fr. Feb.; Portland Gap, 5400 ft., J.P. 228, *Morris*! Morse's Gap; near John Crow Peak, 5000 ft.; between Newhaven Gap and Vinegar Hill, 5500 ft.; *Harris*! Fl. Jam. 6058, 7756, 10,083.

Plant 1–2 ft. high. *Stem* stout, glabrous, 3–4 dm. l. *Leaves* radical, elliptical to lanceolate-oblong, tapering to an acute or shortly acuminate apex, glabrous, blade 10–14 cm. l., 3–6 cm. br., one or two higher up much smaller. *Spike* dense, 8–21 cm. l. *Bracts* 1–3-nerved, ovate-lanceolate, acuminate, glabrous, 1–1·5 cm. l. *Ovary* narrowly cylindrical, somewhat densely covered with several jointed pellucid hairs, 10–12 mm. l. *Sepals* sparsely pilose on the outside, median oblong-lanceolate, 5·5–6 mm. l., 1·5 mm. br.; lateral, spur somewhat longer than the ovary, 15 mm. l., the limb connate half way, each sepal semicircular, with an obtuse apex. *Petals* oblanceolate-oblong, apex acute, half as long as the median sepal. *Column* 1 mm. l. *Capsule* 1·5 cm. l.

10. CRANICHIS Sw.

Terrestrial herbs, with roots somewhat clustered. Leaves radical, broad, spreading, membranous, many-nerved, reticulately veined, with long stalks, flattish, sheathing at base. Scape generally slender, simple, with a few sheaths. Flowers small, in a spike. Bracts short or narrow. Sepals free, subequal, more or less connivent. Petals a little shorter than the sepals, and less than half as broad, spreading, free from the column. Lip on the upper side of the flower, sessile or shortly clawed at the base of the column, erect, concave, embracing the column, undivided. Column short; rostellum ample, erect, often longer than the column, posteriorly (towards the anther) concave and 2-winged; stigma anterior, near the apex; clinandrium short. Anther on the margin of the clinandrium, sessile or stalked, erect, parallel to the rostellum and somewhat shorter than it, with distinct cells; pollinia powdery-granular, acuminate, pendulous from a gland of the rostellum. Capsule erect, ovoid or oblong.

Species about 30, natives of the West Indies, Mexico, and the Andes of tropical America.

Sheaths foliaceous on scape, leaves 4-6 .... 1. *C. muscosa.*
Sheaths not foliaceous, leaves 1 or 2.
Leaves ovate, cordate; petal glabrous ... 2. *C. diphylla.*
Leaves roundish-ovate, sub-cordate; petals hairy on margin .... 3. *C. pilosa.*

1. **C. muscosa** *Sw. Prodr.* 120 (1788); glabrous; sheaths foliaceous, passing into bracts above; leaves few (4–6), and often one just above base, oval, oval-roundish, elliptical or ovate, acute or obtuse; petals narrowly linear-lanceolate, obtuse to subacute, glabrous, white.—*Sw. Fl. Ind. Occ.* 1429, *t.* 29, *f.* 3, & *Ic. ined. t.* 25; *Lindl. Gen. & Sp. Orch.* 451; *Griseb. Fl. Brit. W. Ind.* 639; *Cogn. in Symb. Ant. vi.* 354; *Ames Orch. S. Fla.* 14, *t.* 3. Type in Herb. Mus. Brit.

On damp shady banks, in fl. Nov.–Feb.; *Shakespear*! *Swartz*! *Wiles*! *Bancroft*! *Macfayden*! *McNab*! Ponds, St. Elizabeth; Woodside, St. Mary; Hectors River, Manchester, *Purdie*! *Wilson*! Fairfield, *Wullschlaegel* 1070! *Prior*! J.P. 2068, *Morris*! Mt. Moses, J.P. 2068, *Syme*! below Whitfield Hall Works, *Sullivan*! near Morse's Gap, 4500 ft.; near Vinegar Hill, 4500 ft., 3950 ft.; Tweedside, 3000 ft.; Silver Hill Gap, 3800 ft.; Moody's Gap, 3000 ft.; Troy, 2000 ft.; *Harris*! Fl. Jam. 7469, 7551, 7563, 7800, 10,093, 10,660.—Florida, Cuba, Porto Rico, St. Kitts, Dominica, Martinique, St. Vincent, Grenada, Trinidad.

Plant ½–1 ft. high. *Scape* 15-30 cm. l. *Leaves*, blade 4-7 cm. l., 2·5-3·5 cm. br.; stalks 2-6 cm. l. *Sheaths*, blade roundish-oval, obtuse, near base of stem, to lanceolate, acuminate, above, 3·5–·75 cm. l. *Spike* dense-flowered, 3-5 cm. l. *Flowers* white, perianth about $\frac{1}{10}$ in. l. *Bracts* triangular-ovate or lanceolate, acute or acuminate, 1-nerved, shorter than the ovary, 4 mm. l. *Ovary* 5 mm. l. *Sepáls*, median narrowly ovate-oblong, obtuse, 1-nerved, 2·6 mm. l., 1·3 mm. br.; lateral broadly semi-ovate, somewhat acute or obtuse, 2-nerved, 2·7-

3 mm. l., 1·6 mm. b. *Lip* sessile, roundish in outline, obtuse, with two lateral folds near the base, 2·6–2·8 mm. l., white, spotted with green. *Column* (including rostellum) 1·5 mm. l., dilated above, towards the anther concave and broadly 4-winged, wings broader above, apex of rostellum obtusely pointed; stigma somewhat roundish. *Anther* stalked, nearly as high as apex of rostellum. *Capsule* ellipsoidal, with six shallow keels, 7–8 mm. l., 3–4 mm. br.

2. **C. diphylla** *Sw. Prodr.* 120 (1788); glandular-pubescent above; leaves generally 2, sometimes 1 or 3, one smaller, ovate, cordate, shortly acuminate or acute; petals oblanceolate, obtuse, glabrous, white.—*Sw. Fl. Ind. Occ.* 1423; *Lindl. Gen. & Sp. Orch.* 450; *Griseb. op. cit.* 639; *Cogn. in Symb. Ant. vi.* 357. Type in Herb. Mus. Brit.

In peaty soil on damp shady banks; in fl. and fr. Dec.–Feb.; *Swartz*! Mt. Moses, 3000 ft., J.P. 2440, *Syme*! Morse's Gap; Stony Hill, J.P. 477; *Morris*! Morse's Gap, 5000 ft., *C. Nicholls*! *Fawcett*! Morse's Gap; near Mabess River, 4000 ft.; Clydesdale, 4000 ft.; *Harris*! Fl. Jam. 7555, 7557, 7855, 9797, 10,080.

Plant 4–14 in. high. *Scape* slender, 10–40 cm. l., about 1 mm. br. *Leaves*, blade 2·5–6·5 cm. l., 1–3 cm. br.; stalk, 1–3 cm. l. *Sheaths* several (4 or 5), linear-lanceolate, shortly acuminate, 2·5–1 cm. l. *Spike* lax-flowered, glandular-pubescent, 1–6·5 cm. l. *Flowers* greenish-white or green, perianth about $\frac{1}{10}$ in. l. *Bracts* ovate, acuminate, 1-nerved, with a few glandular hairs, much shorter than the ovary, 3–4 mm. l. *Ovary* spindle-shaped, glabrous or with occasional glandular hairs, 5·5 mm. l. *Sepals* oval, obtuse, 1-nerved, 2·5 mm. l., 1·3 mm. br.; lateral slightly shorter and broader than the median. *Petals* a little shorter than the median sepal, ·6 mm. br. *Lip* very slightly keeled posteriorly, sessile, roundish in outline, obtuse, 3-nerved, with 2 lateral folds near the base, about 2 mm. l., greenish-white spotted with green. *Column* (including rostellum) 1·5 mm. l., on the posterior side towards the anther concave and winged, with a long acuminate apex to the rostellum; stigma roundish. *Anther* stalked, slightly shorter than the rostellum. *Capsule* ellipsoidal, slightly 3-keeled, 6–7·5 mm. l., 3 mm. br.

3. **C. pilosa** *Fawc. & Rendle in Journ. Bot. xlvii.* 265 (1909); glandular-pubescent above; leaves 1 or 2, roundish-ovate or roundish, subcordate, acute; petals linear, acute, margins pilose, light pink with white hairs or pale reddish-brown. (Pl. 5, f. 1–3.) Type in Herb. Mus. Brit.

In damp shady places; in fl. Nov., April; John Crow Peak, 5500 ft., J.P. 240, *Syme*! Morse's Gap, 5000 ft.; Vinegar Hill, Portland, 3000 ft.; Cascade, 3000 ft.; near John Crow Peak, 5800 ft.; between Cinchona and Morse's Gap, 4900 ft.; *Harris*! Fl. Jam. 7749, 10,503.

Plant 4–10 in. high. *Scape* 10–25 cm. l. *Leaves*, blade 4–10 cm. l., 2·5–7·5 cm. br.; stalks 1·5–3 cm. l. *Sheaths* (1 or 2) scarious, below the leaves, and several (5 to 9) loose, elliptical, shortly acuminate, 3–1 cm. l. *Spike* dense-flowered, glandular-pubescent, 1–2 cm. l. *Flowers* greenish tinted with pink, perianth about $\frac{1}{8}$ in. l. *Bracts* as long as, or shorter than, the ovary, 4–7 mm. l., ovate, acuminate, 3-nerved, with short lateral nerves, glabrous, green or white. *Ovary* spindle-shaped, broadest below the middle, glabrous, greenish or white, 6–7 mm. l. *Sepals* 3-nerved, oval, very shortly acuminate, greenish below, flesh-coloured in the upper half, or greenish tinted with umber; median 3·5–4 mm. l., 1·5–2 mm. br.; lateral 3·2–3·5 mm. l., 1·5–2 mm. br. *Lip* shortly and broadly clawed, broadly ovate, strongly 3-nerved, the lateral nerves with descending branches on the outer side, with a narrow triangular apex, 2·6–2·8 mm. l., 2 mm. br., white tinted with green outside and pink inside, or white with reddish markings. *Column* white, tinted, 1·5 mm. l., on the posterior side towards the anther winged at the base, wings broadest below, with small wings on the sides at the base, with a rather long capitate rostellum; stigma roundish. *Anther* very shortly stalked. *Capsule* (not ripe) spindle-shaped, about 1 cm. l.

11. STENOPTERA Presl.

Epiphytic herb; rhizome short with clustered fleshy roots. Stem leafy, simple, with a sheath above the leaves. Leaves crowded at the base of the stem, tapering into a short clasping stalk. Flowers very small in a dense ovoid bracteate spike. Perianth united below to form a tube, which is attached to back of the column. Sepals free above, incurved, somewhat erect, median shorter. Petals shorter than the sepals. Lip on the upper side of the flower, erect from the tube. Column straight, grooved anteriorly, not winged; rostellum erect, ample, conical, with a broad stigma at the base; clinandrium with membranous sides which are connate with the margins of the rostellum into a cup. Anther erect on the margin of the clinandrium, roundish-ovate, cells contiguous; pollinia powdery-granular, on the dehiscence of the anther remaining attached to the apex of the rostellum. Capsule oblong, erect.

Species 8 or 9, natives of the mountains in tropical America from the West Indies to Brazil and Peru.

S. ananassocomos *Reichb. f. in Bot. Zeit. xxi.* 131 (1863); *Griseb. Fl. Br. W. Ind.* 640; *Cogn. in Fl. Bras. iii. pt.* 4, 254, & *in Symb. Ant. vi.* 358. (Pl. 5, f. 5, 6.)

On trees, pendent; in fl. March; Beaufort, *Wullschlaegel*, 1077! *Prior*! Iron River, Golden Spring, 900–2500 ft., J.P. 2283, *Syme*!—Cuba, Brazil, Venezuela.

Plant very small, about 1½ in. l. *Leaves* elliptical, acute, ciliate, with 3 branching nerves, fleshy, semi-transparent, waxy, 1·5–2·5 c.m. l., 7–11 mm. br. *Scape* erect, densely clothed with short hairs, as long as the leaves. *Spike* very short, capitate, many-flowered (5–11 or more), 6–10 mm. l. *Bracts* ovate, acuminate, glabrous, margin ciliate or fimbriate, longer than the flowers, 8–10 mm. l., about 5 mm. br. *Sepals* lanceolate, 1-nerved, with short hairs at apex and at back, lateral 5 mm. l., median shorter. *Petals* oblanceolate, adhering to, and shorter than, the median sepal. *Lip* similar to petals, shorter than lateral sepals. *Column* 3·5 mm. l.

12. PRESCOTTIA Lindl.

Terrestrial herbs, roots clustered on a rhizome, fibrous, sometimes fleshy; stem leafy, slender or tall, simple, with several sheaths above the leaves. Leaves crowded at the base of the stalk or radical, sessile or with long stalks, small or ample, membranous. Flowers small, sometimes very small, numerous in a thick or slender spike, subsessile. Bracts shorter than the flower. Sepals membranous, connate at the base in a short cup or a somewhat long tube, at the apex spreading or generally revolute. Petals narrow, thin, adnate to the sepaline cup just above the median sepal. Lip on the upper side of the flower; claw adnate to the sepaline cup, blade erect, broad, somewhat fleshy, very concave, arched, hooded or almost closed, at the base 2-auriculate, often enclosing the column. Column very short, adnate to the sepaline cup; stigmatic surface beneath the broad, membranous, retuse rostellum; clinandrium erect, acuminate or produced, as it were, into a filament, margins connate with rostellum. Anther erect on the margin or point of the clinandrium, generally short, with cells diverging; pollinia powdery-granular. Capsule small, erect, ovoid or oblong.

Species about 32, natives of tropical America from the West Indies and Mexico to Brazil and Ecuador.

Flowers green; sepals revolute; lip 4 mm. l. ...... 1. *P. stachyodes.*
Flowers white or rosy, minute; sepals connivent;
lip 1 mm. l. .... 2. *P. oligantha.*

1. **P. stachyodes** *Lindl. in Bot. Reg. sub. t.* 1915 (*err. typ.* 1916) (1836); leaves large, long-stalked; flowers green, changing to a tawny yellow; sepals revolute, narrowly oblong; lip long, 4 mm. l.—*Lindl. Gen. & Sp. Orch.* 454. P. myosurus *Reichb. f. in Griseb. Fl. Br. W. Ind.* 639 (1864); *Cogn. in Mart. Fl. Bras. iii. pt.* 4, 258, *t.* 61, & *in Symb. Ant. vi.* 359. Cranichis stachyodes *Sw. Prodr.* 120 (1788), *Fl. Ind. Occ.* 1427, *t.* 29, *f.* 4, & *Ic. ined. t.* 24. Type in Herb. Mus. Brit.

On stony ground or peaty soil in damp shady places; in fl. and fr. Nov.–Feb.; Blue Mts., *Swartz*! *Macfadyen*! Christiana Woods, Manchester, *Purdie*! Tiddenham, St. Ann, *Prior*! Blue Mts., J.P. 234, *Morris*! Morse's Gap, 4900 ft., *Fawcett*! *Moore*! *Harris*! near Mabess River, 3000–4000 ft.; near John Crow Peak, 5800 ft.; ridge below Vinegar Hill, 3800 ft.; road to Vinegar Hill, 4500 ft.; *Harris*! Fl. Jam. 7553, 7556, 7572, 7750, 7769, 7846, 10,094 (partly), 10,096, 10,408, 10,480, 10,481.—Cuba, Hispaniola, Porto Rico, St. Eustatius, Guadeloupe, Dominica, Martinique, St. Vincent, Trinidad, Costa Rica, Guatemala, Colombia, Brazil.

Plant 1–2 ft. high. *Stem* slender or stout, with several radical leaves, invested by numerous, membranous sheaths, 2–6 dm. l. *Leaves* broadly oval, ovate or elliptical, shortly acuminate, base sometimes subcordate, tapering into the stalk, entire or minutely crenulate, veins slightly prominent on the under surface, blade 7–14 cm. l., 3·5–8 cm. br.; stalks slender, narrowly winged, 2–17 cm. l. *Sheaths* at base of stem, few, scarious, loose, and soon decaying. *Cauline sheaths* loose, tubular below, free part triangular, acuminate, with several nerves and branching veins, 4·5–1·5 cm. l. *Spike* long, crowded with flowers, to 10 or 12 cm. l. *Bracts* elliptical, acuminate, 1-nerved, glabrous, as long as, or longer than, the ovary, 6·5–8 mm. l., 1·7–2·5 mm. br. *Ovary* obovoid-oblong, glabrous. *Sepals* obtuse, 1-nerved; lateral slightly falcate, free part 3·1–3·4 mm. l., 1–1·3 mm. br.; median, free part 2·5–2·9 mm. l., 1 mm. br. *Petals* revolute, narrowly linear, obtuse, 1-nerved, as long as the median sepal, ·4 mm. br. *Lip* projecting much beyond the revolute sepals, upper margin rolled in; nerves 7, branching; 3·2 mm. br.; auricles linear, about ·5 mm. br. *Capsule* ellipsoidal, with six shallow keels, about 1 cm. l., 3·5 mm. br.

2. **P. oligantha** *Lindl. Gen. & Sp. Orch.* 454 (1840); leaves small, short-stalked; flowers white or rosy; sepals erect, roundish; lip short, 1 mm. l.—*Cogn. in. Symb. Ant. vi.* 360. Cranichis oligantha *Sw. Prodr.* 120 (1788), *Fl. Ind. Occ.* 1425, *t.* 29, *f.* 2 & *Ic. ined. t.* 23. Prescottia myosurus *Reichb. f. in Griseb. Fl. Br. W. Ind.* 640 (1864). (Pl. 5, f. 4.) Type in Herb. Mus. Brit.

On damp shady banks; in fl. and fr. March; Blue Mts., *Swartz*! Orange Bay, Hanover, *Purdie*! St. Ann, *Prior*! Stony Hill, J.P. 471 (2606) *Syme*! Mt. Moses, J.P. 2156, *Morris*! near Cinchona, 4500 ft.; Clydesdale 4000 ft.; Tweedside, south St. Andrew; Morse's Gap 4900 ft.; *Harris*! Fl. Jam. 7612, 7844, 7856A, 10,094 (partly).—Cuba, Porto Rico, St. Thomas, Guadeloupe, Dominica.

Plant 8–12 in. high. *Stem* slender, with a few radical leaves and many sheaths, 15–23 cm. l. *Leaves* elliptical or oval, obtuse or acute, tapering into the flattish membranous stalks, sheathing below; blade 2·5–7·5 cm. l., 1·5–3 cm. br.; stalk 1–3·5 cm. l. *Sheaths* tubular below, free part oblong, tapering into a shortly acuminate apex, 3–1 cm. l. *Spike* densely crowded with numerous small flowers, 3–7 cm. l. *Bracts* ovate, long-acuminate, 1-nerved, as long as, or a little longer than, the ovary, 3–2 mm. l. *Ovary* obovoid, barely 2 mm. l. *Sepals* concave, 1-nerved, tinged with brownish-pink at the tips; lateral connivent; median spreading; free part about 1·3 mm. l., nearly 1 mm. br. *Petals* narrowly obovoid, very obtuse, 1-nerved, recurving, barely 1 mm. l., about ·5 mm. br. *Lip* 3-nerved, 1–1·2 mm. l.; auricles short, ·2 mm. l. *Column* winged laterally at the apex. *Capsule* ellipsoidal, with six shallow keels, 4 mm. l., 2·5 mm. br.

13. PONTHIEVA R. Br.

Terrestrial herbs with fibrous roots. Leaves radical, ovate, oval or elliptical, membranous, many-nerved, reticulately veined; stalks long or short, flattish, membranous, sheathing below. Scape simple, slender with a few sheaths. Flowers small, shortly stalked, in a lax raceme generally glandular-pubescent. Bracts narrow, generally small. Sepals free, spreading, either nearly equal or the lateral broader than the median. Petals raised on the column, spreading, very asymmetrical, triangular. Lip on the upper side of the flower;

raised on the column, ascending; blade abruptly dilated, concave and more or less folded on itself. Column short, somewhat terete, dilated at the apex; rostellum erect, dilated, posteriorly concave; stigma under the rostellum, broad, hollowed out; clinandrium short. Anther erect behind the rostellum and shorter than it, with cells contiguous; pollinia 4, joined in pairs, powdery-granular, pendulous from a gland of the rostellum. Capsule erect, obovoid-ellipsoidal or ellipsoidal.

Species about 22, natives of the warmer regions of America from the southern United States to Chile, including the West Indies.

Petals, margin ciliate.
  Lip roundish with linear apical lobe .... 1. *P. glandulosa.*
Petals glabrous.
  Lip obscurely 3 lobed at apex, with 2 minute erect lobes at base .... 2. *P. Harrisii.*
  Lip kidney-shaped with linear lobe at apex and 2 minute auricles at base .... 3. *P. pauciflora.*
  Lip 3-lobed at apex, saccate at base .... 4. *P. ventricosa.*

1. **P. glandulosa** *R. Br. in Ait. Hort. Kew. ed.* 2. *v.* 200 (1813); sepals glandular-pubescent, median 3-nerved, lateral incompletely 4-nerved; petals, margin ciliate; lip clawed, very broadly roundish, almost transversely oval, with linear apical lobe, disc with linear median crest.—*Lindl. Gen. & Sp. Orch.* 444; *Griseb. Fl. Br. W. Ind.* 638; *Cogn. in Symb. Ant. vi.* 363; *Ames Orch. S. Fla.* 14, *t.* 4. Neottia glandulosa *Sims in Bot. Mag. t.* 842 (1805).

On damp shady banks; in fl. and fr. Dec.-Feb.; *Macfadyen*! *McNab*! St. Ann, *Purdie*! Fairfield, *Wullschlaegel*, 1069! *Prior*! near Cinchona, J.P. 13, *Morris*! near Castleton, J.P. 2377, *Syme*! Browns Town, *Miss T. M. Barrett*! Cinchona, 5000 ft., *Fawcett*! Mt. Diabolo, 2300 ft., *Fawcett & Harris*! Cinchona; Hall's Delight, 1000 ft.; Clydesdale, 4000 ft.; *Harris*! Bloxburgh, Blue Mts., *Miss B. Gosset*! Fl. Jam. 7549, 7861, 7862 (7266, 7532, 7562, 9787, *fide* Cogniaux).—Cuba, Hispaniola, Porto Rico, St. Thomas, St. Kitts, Guadeloupe, Dominica, St. Vincent, Grenada, Carolina, Georgia, Virginia, Florida, Mexico, Costa Rica, Ecuador.

Plant 1–2 ft. high, glandular-pubescent above. *Scape* glandular-pubescent above, 2 to nearly 4 dm. l. *Leaves* broadly oval to oblanceolate-elliptical and oblanceolate, obtuse or acute; blade 4–15 cm. l., 2–4 cm. br. *Sheaths* (3–6), those near the base (1 or 2) generally foliaceous and passing into the oblong-lanceolate to lanceolate-acuminate sheaths above, glabrous at the lower part of the stem and glandular-pubescent above. *Raceme* 6–20 cm. l., glandular-pubescent, with few to many wide-spreading flowers. *Bracts* oval or elliptical, acuminate, 3-nerved, half as long as the stalked ovary, 10 mm. l. or less, 2·7 mm. br. *Ovary* densely glandular-pubescent, about 1–1·5 cm. l. *Pedicel* 3–4 mm. l. in flower, 7–8 mm. l. in fruit. *Flowers* white, about ½ in. l. *Sepals* glandular-pubescent; median elliptical, obtuse, 5–7 mm. l., 2–2·5 mm. br.; lateral obliquely ovate, obtuse, 5·5–7 mm. l., 3·3–4 mm. br. *Petals* on the outer side dilated and semi-cordate, curved inwards, clawed, several nerves springing from claw and more or less branching; extreme length of blade 4–6 mm. l., 2·5–3·5 mm. br.; claw attached to column for about 1 mm. above base. *Lip* folded on itself down the middle, each half somewhat round, 5-nerved; blade about 3–4 mm. l. to base of lobe (1 mm. l.), 4–5 mm. br.; claw 1·3 mm. l. above its attachment to the column. *Column* (including rostellum) 4–4·5 mm. l., curved, rostellum long-pointed, winged posteriorly on the side of the anther, wing broadest at the middle; stigma oblong. *Anther* stalked, nearly reaching apex of rostellum. *Capsule* obovoid-ellipsoidal, tapering to base, narrowly 6-keeled, 1·3–1·6 cm. l., 4–5 mm. br.

2. **P. Harrisii** *Cogn. in Fedde Repert. vi.* 304 (1909); sepals, median glabrous, 5–7 nerved, lateral sparsely glandular-pubescent, 7-nerved; lip boat-shaped, obscurely 3-lobed at apex, obtuse, with 2 minute, erect lobes near base, with a transverse callus near apex.—*Cogn. in Symb. Ant. vi.* 362. (Pl. 5, f. 7–11.)

In peaty soil in damp places in woodland; in fl. April, May; Greenhill Wood, 4000 ft.; near Mabess River, 3000 ft.; below Vinegar Hill; *Harris*! Fl. Jam. 7618, 7883.

Plant 14–24 in. high. *Scape* 3–5 dm. l., somewhat sparsely glandular-pubescent. *Leaves* several (3–8), ovate-elliptical, acute, glabrous, margins slightly wavy; blade 6–11 cm. l., 3 to about 5 cm. br.; stalks with sheath to 6 cm. l. *Cauline leaf* much smaller than the radical. *Cauline sheaths*, 2 or 3, subfoliaceous, tubular below, passing into the bracts above, acute, glandular-pubescent. *Raceme* 9–10 cm. l., glandular-pubescent, many flowered. *Bracts* elliptical, shortly acuminate, acute, sparsely glandular-pubescent, 3-nerved, about half as long as the stalked ovary, lower about 8·5 mm. l., 3·5 mm. br. *Pedicels* 4–5 mm. l. *Ovary* densely glandular-pubescent, 7–8 mm. l. *Flowers* erect-spreading, cream-colour or pale brick colour and lip marked with green. *Sepals*, median elliptical, shortly acuminate, 4–5 mm. l., 1·5–2 mm. br.; lateral subfalcately obovate-elliptical, obtuse, 5·5–7 mm. l., 3–4 mm. br. *Petals* attached by a short claw to the column at 2·7 mm. above its base, narrowly triangular, on outer side dilated and semi-auriculate, basal lobe obtuse, from apex to end of basal lobe 3 mm. l. *Lip* attached to column about 2 mm. above base, 2·5–3 mm. l., 1 mm. br. *Column* (to apex of rostellum) 3–4 mm. l., the long pointed rostellum much thickened upwards, hollow and broadly winged towards anther; stigma round.

3. **P. pauciflora** *comb. nov.*; sepals glandular-pubescent, median 3-nerved, lateral incompletely 5-nerved; lip clawed, kidney-shaped, with a sinus above from which projects a short blunt linear lobe, and with very small auricles at base.—Cranichis pauciflora *Sw. Fl. Ind. Occ.* 1431; *Griseb. op. cit.* 639; *Cogn. in Symb. Ant. vi.* 356. (Pl. 5, f. 12, 13.)

On damp shady banks, amongst mosses; in fl. and fr. Nov.–Jan.; *Swartz*; *Distin*! St. Mary, *McNab*! J.P. 13, 476, *Morris*! Cinchona, 5000 ft., *C. Nicholls*! Morse's Gap, 5000 ft., *Harris*! Fl. Jam. 7532, 7804, 9787.—Cuba.

Plant ½–1½ ft. high, glandular-pubescent above. *Scape* 1–3 dm. l., glandular-pubescent above, glabrous below. *Leaves* few (about 3), oval or elliptical, obtuse, blade 1·5–6·5 cm. l., 1–2·5 cm. br., stalk 5–13 mm. l. *Sheaths* few (about 3), tubular below, free portion ovate, obtuse near base of stem to ovate-lanceolate, acuminate, above, 1–2·5 cm. l. *Raceme* somewhat one-sided, few-flowered, glandular-pubescent, 2–10 cm. l. *Bracts* broadly oval to obovate, shortly acuminate, 3-nerved, with scattered glandular hairs, about half as long as the stalked ovary, 4–5·5 mm. l., 2–2·5 mm. br. *Pedicel* 4 mm. l., glabrescent. *Ovary* thicker above, glandular-pubescent, 7·5–8 mm. l. *Flowers* white, erect-spreading, about ⅓ in. l. *Sepals* white, slightly tinged with green outside; median elliptical or lanceolate-elliptical, very shortly acuminate, 5·2–5·5 mm. l., 1·8–2 mm. br.; lateral ovate, slightly oblique, very shortly acuminate, 4·5–5·5 mm. l., 2·6–2·7 mm. br. *Petals* dilated on the outer side, slightly falcate, 2–4-nerved, 4 to nearly 5 mm. l., 1·6–2·5 mm. br., clawed, attached to column ·5–1 mm. above its base. *Lip* folded on itself, white with 2 green spots on either side, greenish where it joins the column, 3-nerved, very concave at base and thickened on outside, 2·5–3·5 mm. l. (without claw), 3·2–4 mm. br., terminal lobe 1 mm. l.; claw thickened and broad above, rather more than ·5 mm. l. *Column* (including rostellum), 2–2·5 mm. l., the short pointed rostellum dilated above, concave and winged on the posterior side next the anther, wings forming a cup from the edge of which the anther springs; stigma large. *Anther* shortly stalked, about as high as apex of rostellum. *Capsule* (unripe) cylindrical-ellipsoidal, somewhat wider near apex.

We have transferred this and the following species to *Ponthieva*, as the petals and lip are raised on the column.

4. **P. ventricosa** *comb. nov.*; sepals glandular-pubescent, median 3-nerved, lateral 5-nerved; lip sessile, forming a very short saccate spur, 3-lobed above, lateral lobes rounded, median lobe narrow, blunt.—Cranichis ventricosa *Griseb. loc. cit.*; *Cogn. loc. cit.*

On rocky shady banks; in fl. and fr. Dec.; Fairfield, *Wullschlaegel*, 1068! *Prior*! Browns Town, *Miss T. M. Barrett*! Farm Hill; Tweedside, 2500 ft.; Guava Ridge Road, 2500 ft.; *Harris*! Anchovy, *Cradwick*! Fl. Jam. 5538, 6924, 7814, 10,403, 10,469.—Cuba, Hispaniola, Porto Rico.

Plant 7–15 in. high. *Scape* 1·5–3 dm. l., glabrate. *Leaves* 3 or 4, oblong, elliptical or obovate, often tapering into the stalks, 2–7 cm. l., 1–2·5 cm. br.; sometimes the stalk of the upper leaf completely sheathing. *Sheaths* 3–6, lower sometimes a leaf-like blade, varying greatly in size; the others tubular below, free portion oblong, acute, 7–16 mm. l. *Raceme* glandular-pubescent, 3–7 cm. l., lengthening in the fruit. *Bracts* ovate, acute, 3-nerved, glandular-pubescent, 5 mm. l. *Ovary* thicker above, tapering to base, glandular-pubescent, 1 cm. l. *Flowers* white, few or several, small, erect-spreading, shortly stalked, nearly ¼ in. l. *Sepals* median oval, obtuse, tapering to apex and base, 5·2 mm. l., 2·2 mm. br., lateral forming a chin round short spur of lip, broadly ovate, oblique on the side forming the chin, 6 mm. l., 3·5 mm. br. *Petals* with 3-nerves branching, somewhat triangular, unequally 2-lobed, about 5 mm. l., nearly 2·5 mm. br. *Lip* attached to column shortly above the base, 5-nerved, folded on itself, 4·5 mm. l. *Column* thicker above, with a sharp pointed rostellum, 2·5 mm. l., concave and slightly winged at base posteriorly towards the anther; stigma roundish. *Anther* stalked, exceeding the apex of the rostellum. *Capsule* ellipsoidal, tapering to base, 8 mm. l., 3·5–4 mm. br.

14. CORYMBORCHIS Thou.*

Terrestrial, high-growing herbs, leafy, sometimes branching, with thick, woody roots clustered on the short rhizome. Sheaths decaying below, spathaceous or passing into leaves above. Leaves ample, plaited, chartaceous, petioles sheathing. Flowers medium-sized, subsessile, in short terminal or axillary sub-corymbose panicles. Bracts small, ovate. Sepals and petals oblong. Lip erect from the base of the column, ovate, channelled, with a recurved apical lobe. Column long, erect, at the apex thicker, terminated by two erect lobes or auricles; stigma at the base of an erect acuminate rostellum; clinandrium short. Anther erect, narrow, nearly as long as the rostellum, acuminate, cells contiguous; pollinia granular, attached to a subulate stalk which depends from a peltate gland of the rostellum; on the fall of the pollinia the rostellum remains 2-fid. Capsule cylindrical, subterete, crowned by the remains of the perianth.

Species about 13, widely dispersed through the tropical regions of the world, one in West Indies and Mexico.

C. flava *Kuntze Rev. Gen. Pl.* 658 (1891).—Serapias flava *Sw. Prodr.* 119 (1788); Neottia flava *Sw. Fl. Ind. Occ.* 1417 (1806). Chloidia flava *Reichb. f. in Walp. Ann. Bot. vi.* 644 (1863) & *Beitr. Orch. Centr. Amer.* 5; *Griseb. Fl. Br. W. Ind.* 643. Corymbis flava *Hemsl. in Biol. Centr. Amer. Bot. iii.* 297 (1884). (Pl. 5, f. 14.)

In damp woods; in fl. Oct.–April; mountains of Clarendon, *Swartz;* Hector's River, Manchester, *Purdie*! *Prior*! Olive River, Christiana district, 3000 ft., *Harris*! Fl. Jam. 10,398.—Cuba, Hispaniola, Porto Rico, Guadeloupe, Mexico.

Plant to 6 ft. high. *Stem* about 6 dm. high, simple. *Leaves* oblong-lanceolate, acuminate, about 7-nerved, blade 18–26 cm. l., 4–6 cm. br. *Panicle* 4–5 cm. l., with several lax flowers. *Bracts*, at base of branches oblong, acuminate, about 1 cm. l.; floral ovate, shortly acuminate, 4 mm. l. *Flowers* yellow, about ⅗ in. l., subsessile. *Ovary* cylindrical, about 8 mm. l. *Sepals* 7-nerved, oblong-linear, acute, nearly 1·5 cm. l., 2–2·5 mm. br. *Petals* 5-nerved, oblong-elliptical, somewhat blunt, 1·2–1·3 cm. l., about 3 mm. br. *Lip* ovate, conduplicate below, ovate-oblong above, as long as the petals. *Column* (including anther) as long as petals.

15. TROPIDIA Lindl.

Like *Corymborchis*, except in the following particulars: Flowers small, not so lax, in few short dense spikes at the apex of the branches. Bracts linear-subulate. Sepals connivent, median oblong-lanceolate, lateral oblong, connate at the base

---

* *Corymborchis* is the generic name given by Thouars (Orch. Iles Afr. t. 37). *Corymbis* is a uninomial composed from the generic and specific names; it is, therefore, inadmissible to substitute *Corymbis* for the true generic appellation.

under the lip into a short chin. Lip saccate at base, oblong. Column short.

Species about 15, natives of the East Indies, Malaya, China and Japan, and one in W. Indies and Florida.

T. polystachya *Ames Orch. ii.* 262 (1908).—T. Eatoni *Ames Orch. S. Fla.* 14, *t.* 5 (1904). Serapias polystachya *Sw. Prodr.* 119 (1788). Neottia polystachya *Sw. Fl. Ind. Occ.* 1415 (1806). Chloidia vernalis *Lindl. Gen. & Sp. Orch.* 484 (1840). C. polystachya *Reichb. f. in Walp. Ann. vi.* 644 (1861), & *in Griseb. Fl. Br. W. Ind.* 643. Polystachya membranacea *A. Rich. Fl. Cub. ii.* 248 (1853). Corymborchis polystachya *Kuntze Rev. Gen. Pl.* 658 (1891). (Pl. 5, f. 15.) Type in Herb. Mus. Brit.

In damp woods; in fl. in spring; mountains of Clarendon, *Swartz*! Manchester; Westmoreland; *Purdie*! *Prior*! Stony Hill, J.P. 469, *Morris*! —Cuba, Florida.

Plant nearly 2 ft. high. *Stem* 4 dm. high, branching. *Leaves* oblong-lanceolate, acuminate, about 7-nerved; blade 12-22 cm. l., 3–5 cm. br. *Panicle* 10–15 cm. l., with numerous lax flowers. *Bracts* linear-subulate; at base of branches 2·5–7 cm. l.; floral 2–7 mm. l. *Pedicels* very short, 1–2 mm. l. *Flowers* pale red, ¼ in. l. *Ovary* cylindrical, about 8 mm. l. *Sepals* 5-nerved; lateral, acute, somewhat gibbous at base, erect, falcate 6 mm. l., 2 mm. br.; median, with acute apex, concave, 7 mm. l., 2 mm. br *Petals* 3-nerved, oblong, somewhat falcate, acute, 6 mm. l., 1·5 mm. br. *Lip* very concave and embracing the column below, retuse and obscurely 3-lobed at apex, 2-lamellate on disk, as long as petals. *Column* 4·5 mm. l. *Capsule* 1 cm. l., 3 mm. br.

16. MALAXIS Solander ex Swartz.*

Terrestrial herbs, base of stem generally pseudobulbous. Leaves 1 or 2, broad, not jointed with the sheathing petiole. Scape slender, elongated. Flowers small or minute in a terminal raceme, sometimes contracted and corymbose. Bracts small, narrow. Sepals nearly equal, free or lateral slightly connate at base, spreading. Petals equal to sepals or a little shorter, narrow. Lip sessile, entire or lobed, with auricles embracing the column, not extending as far as the sepals, but, including the auricles, longer than the sepals. Column very short, terete, hollowed at apex. Anther terminal, incumbent, two-celled. Pollinia 4, waxy, ovoid. Capsule small, ovoid or turbinate.

Species about 140, widely distributed, but mainly in Central America and Indo-Malaya.

---

* Swartz adopted the name *Malaxis* from Solander and defined the genus in Prodr. 119 (1788), including therein the two Jamaican species *M. spicata* and *M. umbelliflora*. This name must therefore be retained for these and for the other species now included under *Microstylis*. The latter name is cited by Bentham & Hooker (Gen. Pl. iii. 494) as of Nuttall (Gen. N. Amer. Pl. ii. 196), but Nuttall uses it only for a section of *Malaxis*. *Microstylis* was first used as a generic name by Lindley (Orch. Scelet. & Gen. & Sp. Orch.).

Leaves two
  Flowers corymbose.. 1. *M. umbelliflora.*
  Flowers racemose.
    Lip with a distinct apical lobe.............. 2. *M. spicata.*
    Lip not lobed .... 3. *M. integra.*
Leaf one .... 4. *M. Grisebachiana.*

1. **M. umbelliflora** *Sw. Prodr.* 119 (1788); leaves two; raceme umbellately corymbose, very short, not exceeding 1 cm. but elongating in fruit; lip 3-lobed above, middle lobe 2 mm. l., with revolute margins, large, half the length of the lip, triangular, obtuse; side lobes small, roundish, obtuse; auricles short, rounded, obtuse.—*Sw. Ic. ined. t.* 20. M. umbellulata *Sw. Fl. Ind. Occ.* 1444, *t.* 28, *fig. a, b, c* (1806). Microstylis umbellulata *Lindl. Gen. & Sp. Orch.* 19 (1830); *Griseb. Fl. Br. W. Ind.* 612; *Ridl.* in *Journ. Linn. Soc. xxiv.* 324. M. umbelliflora *Hitchc.* in *Rep. Miss. Bot. Gard. iv.* 132 (1893); *Cogn. in Symb. Ant. vi.* 371. Type in Herb. Mus. Brit.

Peaty soil in damp shady places; in fl. and fr. Sept.-Jan.; *Macfadyen*! "Gap of St. Andrew's," *Purdie*! J.P. 2080, *Morris*! Moody's Gap and Second Breakfast Spring, J.P. 2080, *Syme*! Moy Hall, *Sullivan*! Cinchona, 4850 ft., *C. Nicholls*! *Fawcett*! Near Vinegar Hill; Green Hill Wood, 3500 ft.; Morse's Gap, 4800 ft.; Mabess River, 3500 ft.; Woodcutters' Gap, 4350 ft.; *Harris*! Fl. Jam. 7549, 7760, 7779, 7780, 9785, 9798, 10,482. —Cuba, St. Kitts, Guadeloupe, Dominica, St. Vincent, Trinidad, Mexico, Andes (Jameson), La Banca (10,000 ft., Pearce).

Plant 4–10 in. high. *Stem* pseudobulbous, with two sheaths; upper sheath 2·5–4 cm. *Leaves*, the lower larger, blade 5–10 cm. l., 3–6 cm. br., ovate, shortly acuminate, petiole long-sheathing. *Flowers* pale green, numerous, small; pedicels long, to 12 mm. *Bracts* 2–3 mm. l., lanceolate, acuminate, setaceous, much shorter than the lower pedicels. *Sepals* about 3 mm. l., half as broad, median ovate-elliptical, lateral ovate-oblong with revolute margins. *Petals* 3 mm. l., ·3 mm. br., linear. *Lip* 4 mm. l., 2·25 mm. br. *Column* ·75 mm. l. *Capsule* 8–9 mm. l., ellipsoidal, tapering to base, with three angles.

2. **M. spicata** *Sw. Prodr.* 119 (1788); leaves two; raceme somewhat lax; lip* broadly elliptical, with a prominent linear-oblong obtuse apical lobe; auricles elliptical, obtuse.—*Sw. Fl. Ind. Occ.* 1441 & *Ic. ined. t.* 19. Microstylis spicata *Lindl. op. cit.* 19; *Griseb. op. cit.* 612; *Ridl. op. cit.* 323; *Cogn. in Symb. Ant. vi.* 368. Type in Herb. Mus. Brit.

On damp shady banks and in woods; in fl. and fr. Sept.-Jan.; Westmoreland Mts., *Purdie*! J.P. 233, *Morris*! Stony Hill, J.P. 464,

---

* The lip is described and figured by Swartz as trilobed, but in all the Jamaican specimens examined by us, including Swartz's type, the lateral lobes are not evident. Swartz describes his figure as the flower of *M. spicata* or *M. umbelliflora*, and in his description of the latter omits the parts of the flower as being very similar to *M. spicata*. The drawing agrees best with *M. umbelliflora*, under which we have cited it. In his Icon. ined. tt. 19 & 20, the flowers are represented as precisely similar, and we can only suggest that a flower of *M. umbelliflora* has been drawn for each plate.

Syme! Moody's Gap, 3000 ft.; near Vinegar Hill, 4500 ft.; near Christiana, 3000 ft.; *Harris*! Trelawney, *Miss T. M. Barrett*! Fl. Jam. 7880, 10,471, 10,473.—Cuba, Hispaniola, Porto Rico, St. Eustatius, Martinique.

Plant 8–12 in. high. *Stem* thickened at the base, pseudobulbous in fruiting stage, with two sheaths at the base, upper sheath 2–3 cm. l., the outer smaller one often withering. *Leaves* distant, the lower larger, 6–10 cm. l., 3–5 cm. br., ovate, ovate-elliptical or ovate-lanceolate, petiole long-sheathing. *Raceme* 3–8 cm. l. *Flowers* numerous, small; pedicels 6–9 mm. *Bracts* 2–4 mm., lanceolate, setaceous, much shorter than the pedicels. *Sepals* about 3 mm. l., and half as broad; median ovate-lanceolate, acute; lateral shortly connate at base, lanceolate, obtuse, with revolute margins. *Petals* 2 mm. l., ·5 mm. br., linear, shorter. *Lip* 3·5–4 mm. l., 2·5 mm. br. *Capsule* (unripe) 8 mm. l., ellipsoidal, slightly tapering to base.

3. **M. integra** *comb. nov.*; leaves two; raceme lax; lip ovate, entire, obtuse, auricles roundish-elliptical.—Microstylis integra *Fawc. & Rendle in Journ. Bot. xlvii.* 7 (1909); *Cogn. in Symb. Ant. vi.* 370. (Pl. 6, f. 11–17.) Type in Herb. Mus. Brit.

On rocky shady banks; in fl. and fr. Nov.; Guava Ridge Road, 2500 ft.; Mount Hybla, 4000 ft.; *Harris*! Fl. Jam. 7735, 7740, 7741.

Plant 6 to 12 in. high. *Stem* thickened at the base, 2-leaved below, with a sheath at the base; sheath 1·5 cm. l. *Leaves* elliptical, obtuse or abruptly acute, petiole long-sheathing; lower leaf generally 3–5 cm. l., 2–2·5 cm. br. *Raceme* 3–8 cm. *Flowers* very small, stalked; pedicels 4-5 mm. *Bracts* 2 mm., linear-lanceolate, much shorter than the pedicels. *Sepals* 2·2 mm. l., 1·5 mm. br., obtuse with revolute margins, median ovate, lateral broadly elliptical. *Petals* linear, scarcely as long as the sepals. *Lip* 2·5 mm. l., 2 mm. br. *Capsule* 6 mm. l., ellipsoidal, slightly narrowed at the base.

4. **M. Grisebachiana** *comb. nov.*; leaf solitary; raceme oblong, with flowers crowded towards the apex; lip somewhat quadrate, 3-lobed above, median lobe very short, tooth-like; auricles roundish-elliptical.—Microstylis Grisebachiana *Fawc. & Rendle in Journ. Bot. xlvii.* 7 (1909); *Cogn. in Symb. Ant. vi.* 372. (Pl. 6, f. 18–23.) Types in Herb. Mus. Brit. & Jam. Herb.

On damp clayey shady banks; in fl. and fr. Oct.–Dec.; Cinchona, 4920–5000 ft.; Morse's Gap, 4800 ft.; *Harris*! J.P. 230 (in part), *Morris*! Fl. Jam. 7530, 7743, 7744, 9790, 9792.

Plant 4–10 in. high. *Stem* 1·5–3 cm., swollen underground into a small spherical corm, 1–2 cm. br., with one leaf affixed below the middle or near the base, with one or two sheaths at the base. *Leaf* 2·5–5 cm. l., 1·5–3 cm. br., elliptical to ovate-elliptical, obtuse, subcordate, with sheathing petiole. *Raceme* 3–4 cm., oblong, with flowers crowded towards the apex. *Flowers* yellowish-green, very small, stalked. *Bracts* triangular, acuminate, much shorter than the pedicels. *Sepals* ovate-oblong, obtuse, with revolute margins, median about 2 mm. l., ·6 mm. br., lateral slightly shorter. *Petals* 1·5 mm. l., linear shorter than the sepals. *Lip* 2·25 mm. l., 1·75 mm. br., lateral lobes ·6 mm. l., median ·25 mm. l., auricles ·5 mm. l.

Near *M. unifolia* Michx., but distinguished by the auriculate, not cordate, lip, and the position of the leaf nearer the base of the stem.

17. LIPARIS L. C. Rich.

Terrestrial herbs; stems leafy, short, thickened into a corm or narrow pseudobulb. Leaves one, two or few, springing either from the base of the corm or from the nodes of the pseudobulb, or from below it, with a sheathing leaf-stalk, membranous, conspicuously many-nerved. Flowers small or medium-sized, greenish or purple, in a terminal stalked raceme. Bracts small, narrow. Sepals free, spreading, about equal. Petals narrower. Lip attached to the base of the column, oblong or expanded into a broad spreading or reflexed blade, entire or emarginate. Column long, incurved, half-terete, without a foot, at the apex on both sides margined or appendiculate with a wing; clinandrium short. Anther terminal, opercular, incumbent; pollinia 4, waxy, ovoid, sometimes acuminate, segregated in pairs in the cells, each pair superposed. Capsule generally small, ellipsoidal or obovoid.

Species about 200, widely dispersed through the temperate and warmer regions of the globe.

Leaf one.
- Leaf oblong or lanceolate. Corm present... 1. *L. vexillifera.*
- Leaf roundish, cordate. Pseudobulb present 2. *L. neuroglossa.*

Leaves two.
- Lip greenish, ⅓ in. l. Pseudobulb present... 3. *L. Harrisii.*
- Lip purple, ½ in. l. Corm present .... 4. *L. Saundersiana.*

Leaves three or more .... 5. *L. elata.*

1. **L. vexillifera** *Cogn. in Fl. Bras. iii. pt.* 4, 289 (1895); with corm; leaf one, oblong-elliptical to oblong-lanceolate, acute to subobtuse; lip pale yellow with reddish-brown veins, or altogether reddish-brown, 3-nerved, middle nerve prominent, lower half broadly roundish, conduplicate; upper half broadly oblong-elliptical or roundish, recurved; at base sub-cordate, apex obtuse, somewhat apiculate.—*Cogn. in Symb. Ant. vi.* 377. L. elliptica *Reichb. f. in Walp. Ann. vi.* 218 (1861); *Griseb. Fl. Br. W. Ind.* 612; *Ridl. in Journ. Linn. Soc. xxii.* 265. L. jamaicensis *Lindl. ex Griseb. Cat. Cub.* 261 (1866). Cymbidium vexilliferum *Llave & Lex. Nov. Veg. Descr. ii.* 7 (1825).

On damp clayey banks; in fl. and fr. Nov.–Feb; *Wiles; Macfadyen*! Hollis Savanna, Clarendon, *Purdie*! below Cinchona, J.P. 237, *Morris*! Mt. Moses, 3000 ft., J.P. 2443, *Syme*! Clydesdale, 4500 ft., *Fawcett*! Cinchona, *C. Nicholls*! Westphalia, 4000 ft.; Clydesdale; Mt. Hybla, 4000 ft.; Cinchona, 4800–5000 ft.; *Harris*! Fl. Jam. 7528, 7751, 7842, 9512, 9791, 9788.—Cuba, Porto Rico, Trinidad, Mexico, Guatemala, Colombia, Venezuela, B. Guiana, Ecuador, Argentine (var.), Bolivia.

Plant 4–11 in. high. *Stem* swollen below the ground into a spherical corm, covered by the sheathing leaf-stalks and one or two sheaths, upper sheath sometimes foliate above. *Leaf* blade 7–12 cm. l., 1·5–3·5 cm. br.; stalk sheathing, sometimes very short. *Scape* 1–3 dm. l., angled or winged.

Raceme 5–18 cm. l., somewhat lax. *Flowers* pale yellow, 12–30, perianth nearly ⅓ in. l. *Bracts* lanceolate, acuminate, 16 mm. l. below, to about 5 mm. l. at apex. *Pedicels* 5–8 mm. l. (lengthening to about 1 cm. in fruit). *Ovary* about 5 mm. l., 3-winged. *Sepals* 3-nerved, oblong-lanceolate, apex somewhat obtuse; median 7·5 mm. l., 2·3 mm. br.; lateral attached at base to lip, 6·5–7 mm. l., 2·3–2·6 mm. br. *Petals* 1-nerved, linear, obtuse, about 7 mm. l., ·8 mm. br. *Lip* 7–8 mm. l., 5–6 mm. br. *Column* erect, curved near apex, channelled anteriorly and angled, broadly winged at apex, 3·8 mm. l. *Capsule* narrowly obovoid, 3-keeled, about 1·2 cm. l., 4 mm. br.

2. L. **neuroglossa** *Reichb. f. Xen. Orch. iii.* 26 (1900); with pseudobulb; leaf one, roundish, acute, cordate, many-nerved, with long, sheathing leaf-stalk; lip greenish with five purple nerves, triangular-cuneate with a broad præmorse-apiculate apex, the two upper corners præmorse and irregularly erose; shortly 2-lobed, crimson callus at base.—*Ridl. op. cit.* 280. L. cardiophylla *Ames Orch. iii.* 92, *t.* 59 (1908). L. rotundifolia *Cogn. in Fedde Repert. vi.* 305 (1909), & *in Symb. Ant. vi.* 378.

In peat and decaying leaves or mosses, in damp shady places; in fl. Nov., Dec., in fr. Feb.; near Cinchona, 5000 ft., *Cradwick*! *Harris*!; below John Crow Peak, 5500–6000 ft., J.P. 239, *Sullivan*! *Harris*! Fl. Jam. 7529, 7574, 7733, 9784, 9799.—Bolivia.

Plant 3–4½ in. high. *Stem* a creeping rhizome from which are given off at intervals secondary stems which become thickened and form pseudobulbs, 8–10 mm. l., covered by the sheathing leaf-stalk and sheaths. *Sheaths*, one at each node, papery, ·8–2 cm. l. *Leaf*, blade 3–4 cm. l., 2–4 cm. br.; stalk 2–3·5 cm. l. *Scape* 5–10 cm. l., angled. *Raceme* 1–4 cm. l., somewhat lax below, becoming crowded above. *Flowers* greenish, perianth ⅛ in. l. *Bracts* linear-lanceolate, 1-nerved, 2·5–4 mm. l. *Pedicels* 3 mm. l. *Ovary* 6-angled, about 2 mm. l. *Sepals* 1-nerved; lateral lanceolate, obtuse, about 4 mm. l., 1·4 mm. br.; median oblong-lanceolate, about 4 mm. l., 1·2 mm. br. *Petals* 1-nerved, linear, 4·3–5 mm. l., ·5 mm. br. *Lip* 3·5 mm. l., 3–3·5 mm. br. *Column* erect, curved towards apex, 2·5 mm. l. *Capsule* ellipsoidal, 5–6 mm. l.

The flowers of the type from Bolivia are rather larger than those from Jamaica, but we fail to see any specific difference. The distribution of *L. vexillifera* and *L. elata* shows that we may look for the occurrence of *L. neuroglossa* at intermediate stations.

3. L. **Harrisii** *Fawc. & Rendle in Journ. Bot. xlvii.* 7 (1909); with pseudobulb; leaves two, oval-elliptical, abruptly acute, with a somewhat long, sheathing stalk; lip greenish, veined and tinged with purple or brown, elliptical, obtuse, cordate, veins prominent in the middle and at the base converging into two calli.—*Cogn. in Symb. Ant. vi.* 375. (Pl. 6, f. 1–5.) Type in Herb. Mus. Brit.

In spongy peat and vegetable mould or amongst mosses in damp shady places; in fl. and fr. Sept.–Dec. Cinchona 5000 ft., J.P. 229, *Morris*! Green Hill Wood, 4000 ft.; near John Crow Peak, 5300–6000 ft.; *Harris*! Fl. Jam. 7732, 7737, 9786, 9795.

Plant 4 to 9 in. high. *Stem* at base slightly thickened and at length forming a pseudobulb 2·5 cm. l., 8 mm. br., covered by the sheathing petioles and two loose papery sheaths, upper sheaths 3·5–5 cm. l. *Leaves*, blade of upper leaf 7–12 cm. l., 4–6·5 cm. br. *Scape* 7–13 cm. l., 3-winged. *Raceme* 2·5·5 cm., lax. *Flowers* large, greenish, usually from 15 to 10, perianth nearly ½ in. l. *Bracts* 4–5 mm. l., lanceolate, 1-nerved. *Pedicels*, without ovary, 8–10 mm. *Sepals* 9 mm. l., 2·5 mm. br., greenish margined with purple, linear, acuminate. *Petals* equal in length to sepals, purplish, filiform. *Lip* 13 mm. l., 8·5 mm. br. *Column* 4·5 mm. l., greenish, slender, curved, enlarged at base and below the anther. *Capsule* 1·3 cm. l., ellipsoidal, narrowed at base.

Allied to *L. Saundersiana*, but a larger plant with evident stem, and leaves and flowers twice as long.

4. **L. Saundersiana** *Reichb. f. in Gard. Chron.* 1872, 1003; with corm; leaves two, the lower larger, roundish-ovate, cordate to oval, obtuse; leaf-stalk sheathing, one-third to about one-half the length of the blade; lip purple, oval, cordate, apex rounded, sometimes minutely apiculate.—*Ridl. op. cit.* 274. (Pl. 6, f. 6–10.)

Damp shady banks amongst mosses; in fl. and fr. Nov.–Dec.; Cinchona, 5000 ft., *Watt*! J.P. 5, 236, *Morris*! *C. Nicholls*! *Harris*! Fl. Jam. 7531, 7734, 7739, 9789.

Plant very variable in size, 1½–4 in. high. *Stem* swollen below the ground into a small spherical corm covered by the sheathing leaf-stalks and one or two sheaths, upper sheath 5 to 15 mm. *Leaves*, lower blade very variable, 12 × 11 mm. to 30 × 25 mm., sometimes to 69 × 40 mm. *Scape* angled, edges scabrid. *Raceme* with many flowers, 18–45 mm. l. *Flowers* green in bud, dark purple or sometimes crimson when expanded, dark purple when dried, perianth about ⅛ in. l. *Pedicels* 4–7 mm. *Bracts* 4–5 mm., thin, lanceolate, acuminate, 1-nerved. *Sepals* 4–5 mm. l., shorter than the lip, greenish in bud, colourless during flowering, lanceolate. *Petals* longer than the sepals, purple, narrower, linear, 6 mm. l., ·5 mm. br. *Lip* 5–6 mm. l., 4·5–5 mm. br. *Column* curved, enlarged at base and apex. *Capsule* (unripe) 8 mm. l., ellipsoidal, narrowed at base.

5. **L. elata** *Lindl. in Bot. Reg. t.* 1175 (1828), with pseudobulb; leaves few (3–5), erect, broadly elliptical, to lanceolate-acute, tapering at the base into sheathing stalks, many-nerved; lip purple, very broadly spathulate, emarginate, with a lunate callus at base.—*Lindl. Gen. & Sp. Orch.* 27 (excl. syn.); *Griseb. Fl. Br. W. Ind.* 612; *Ridl. op. cit.* 259; *Cogn. in Fl. Bras. iii. pt.* 4. 286.

In shady places; in fl. June–Dec., in fr. July–April; *Macfadyen*; *Distin*! *Wilson*! *March*! Near Castleton, J.P. 2395, *Syme*! *Morris*! Morse's Gap; Silver Hill; Cedar Hurst, 2000 ft.; *Harris*! Fl. Jam. 7608, 7659, 7878.—Cuba, Hispaniola, Porto Rico, St. Thomas, Guadeloupe, Dominica, St. Vincent, Trinidad, Costa Rica, Colombia, Venezuela, Brazil, Paraguay, Ecuador, Bolivia, Peru.

Plant 12–15 in. high (sometimes only 6 in.). *Stem* forming a pseudobulb, 2–5 cm. l., 6–8 mm. br., covered by the sheathing leaf-stalks and sheaths. *Sheaths*, 2 or more, passing into leaves. *Leaves* 1–3 dm. l., 4–10 cm. br. *Scape* 2–3 dm. l., angled. *Raceme* 8–11 cm. l., somewhat lax. *Flowers* purple and green, 10–20, perianth nearly ¼ in. l. *Bracts* lanceolate, 1-nerved, 5–8 mm. l. *Pedicels* 5 or 6 mm. l. (somewhat longer in fruit). *Ovary* 8 mm. l., narrowly 3-winged. *Sepals* greenish streaked with purple, 5-nerved, oval-elliptical, obtuse; median 5·5 mm. l., 2·5 mm.

br.; lateral shorter, nearly 5 mm. l., 2·5 mm. br., slightly falcate. *Petals* greenish streaked with purple, 3-nerved, linear-oblanceolate, obtuse, as long as the median sepal, 5·5 mm. l., about 1·5 mm. br. *Lip* 4 mm. l., and nearly as broad. *Column* suberect, curved, channelled anteriorly and winged, 3·5 mm. l. *Capsule* obovoid, 1·5 cm. l., 5–6 mm. br.

18. GALEANDRA Lindl.

Terrestrial herb. Raceme terminal. Sepals equal, free, spreading. Petals similar to sepals. Lip attached at base of column, orbicular, produced into a very short conical spur, erect, disk with longitudinal crests. Column short, without a foot. Anther terminal, opercular, incumbent, imperfectly 2-celled; pollinia 4, ovoid, waxy, on the dehiscence of the anther affixed to a gland of the rostellum.

Species about 25, natives of tropical America from Mexico and the West Indies to Brazil.

Galeandra Beyrichii *Reichb. f. in Linnæa xxii.* 854 (1849); *Cogn. in Fl. Bras. iii. pt.* 4, 308, *t.* 74 & *in Symb. Ant. vi.* 380. (Pl. 7, f. 1–3.)

In dry woodland; in fl. and fr. Sept.; St. Ann, J.P. 531, *Steer*! Hopeton, Westmoreland, 1300 ft., *Harris*! Fl. Jam. 9780.—Cuba, Venezuela, Colombia, Brazil, Bolivia, Paraguay, Peru.

Plant terrestrial, to 4 ft. high. *Stem* robust, with thickened pseudo-bulb-like joints, including raceme 8–12 dm. l. with no leaves during flowering, but covered with sheaths and terminated by a raceme of greenish flowers. *Rhizome* short, stout, covered with scales, and bearing numerous roots. *Sheaths*, 12–15, large, loose, scarious, somewhat overlapping, free portion triangular-oblong, acute or shortly acuminate, 3–10 cm. l., at the base of the stem small and embracing, generally tubular below, but becoming bract-like above. *Leaves* (not seen, but from figure and description in *Fl. Bras. loc. cit.*) appearing after flowering, erect, large, elongate-lanceolate, very acute, with 3 thick nerves, 3 dm. l., 3–3·5 cm. br., tapering insensibly below into a robust stalk about 1 dm. l. *Raceme* long, lax, several-flowered (10–12), a fruiting specimen measured 2·5 dm. l. *Bracts* oblong, tapering to an acute or shortly acuminate apex, about as long as the flower-stalks, 1–2 cm. l., 3–6 mm. br. *Flowers* greenish, erect-spreading, large, stalked, perianth nearly an inch l. *Ovary* terete, tapering gradually to base, about as long as sepals. *Sepals* linear-oblong, tapering to base and apex, keeled, 5-nerved, green, 2·3 cm. l., lateral rather broader than median (6 mm. br.), with a blunt apiculate apex; median 5 mm. br., obtuse. *Petals* oblanceolate, abruptly acute, 5–7-nerved, light green, a little shorter than the sepals but as broad as the lateral. *Lip* with numerous nerves, almost orbicular, shorter than the sepals, broader than long, about 2 cm. l., 2·5 cm. br. when flattened, green with lines of crimson, disk with 4 crests running from the base to the middle, pubescent between the crests, in the middle and along the nerves above the crests. *Column* barely 1 cm. l., stout, thicker upwards, concave in front with short hairs on apex; anther with short hairs, shortly conical. *Capsule* reflexed, ellipsoidal, about 2·5 cm. l.

19. POLYSTACHYA Hook.

Epiphytic herbs. Stems leafy, short, sheathed at the base, sometimes thickened into slender pseudobulbs. Leaves few, in two ranks, oblong-ligulate, many-nerved but not plaited, contracted at the base into sheaths and jointed with them. Peduncle terminal, with a few sheaths above the leaf. Raceme terminal on the leafy stem, simple, or several short racemes along a common rachis forming a narrow lax panicle. Flowers small. Bracts small. Sepals connivent, median free, lateral broader, at the base attached to the foot of the column. Petals narrow. Lip on the upper side of the flower, jointed with the foot of the column, at the base contracted, incumbent, then erect; 3-lobed; lateral lobes but little prominent, erect, median spreading or recurved, undivided. Column very short, not winged, produced at the base into a foot; clinandrium short, truncate. Anther terminal, opercular, incumbent, very convex, 1-celled or imperfectly 2-celled; pollinia 4, waxy, broadly ovate, lightly adpressed in pairs, or connate into 2, not appendiculate, affixed to a short stalk or gland. Capsule oblong.

Species about 170, natives chiefly of tropical and south Africa, some of India and Malaya, a few of tropical America.

Flowers with very prominent chin .... 1. *P. minuta.*
Flowers smaller, with chin not prominent 2. *P. minor.*

1. **P. minuta** *Britton in Small Fl. S.E. Un. St.* 328 (1903); lateral lobes of lip oblong, sometimes subfalcate, obtuse; sinus above middle of lip; callus extending along the median nerve from the base to the middle.—P. luteola *Hook. Exot. Fl. t.* 103 (1825); *Lindl. Gen. & Sp. Orch.* 73 (in part, excl. syn. P. mauritiana); *Griseb. Fl. Br. W. Ind.* 628. P. extinctoria & P. luteola *Reichb. f. in Walp. Ann. Bot. vi.* 638, 639 (1861). Helleborine ramosa &c. *Plum. Pl. Amer.* (*Burm.*) *t.* 185, *f.* 1. Epidendrum minutum *Aubl. Pl. Guian. ii.* 824 (1775). Dendrobium polystachyum *Sw. in Vet. Handl. Stockh. xxi.* 247 (1800); *Lindl. Coll. Bot. t.* 20 (non *Thouars & A. Rich.*). Cranichis luteola *Sw. Fl. Ind. Occ.* 1433 (1806). (Pl. 7, f. 4, 5.)

On trees and rocks; in fl. and fr. Nov. to April; *Wilson*! Fairfield, *Wullschlaegel* 1055 (partly)! J.P. 21, 22, *Morris*! Moneague, *Lady Blake*! Content Road, *Campbell*! Mavis Bank, 2000 ft., 7765, *Harris*!—Cuba, Hispaniola, Porto Rico, St. Thomas, St. Eustatius, Dominica, Martinique, St. Lucia, St. Vincent, Grenada, Trinidad, Florida, Mexico, Guatemala, Colombia, B. Guiana, Surinam, F. Guiana, Brazil, Peru.

Plant 8-14 in. high. *Pseudobulb* about 3 cm. l. *Leaves* acute or somewhat obtuse, sometimes slightly oblique, 10-20 cm. l., 12-28 mm. br. *Panicle* 1·5-3 dm. l., generally longer than the leaves. *Bracts* triangular, acuminate, 2 mm. l. *Flowers* yellowish-green, perianth ¼ in. l., and somewhat broader; buds very obliquely 3-cornered, apex somewhat acute.

Ovary cylindrical, 4-5 mm. l. *Sepals* 3-nerved, apiculate; median triangular-ovate, 3 mm. l., 2 mm. br.; lateral deltoid, very oblique, 3·5 mm. l., 4 mm. br.; chin obtuse. *Petals* linear-subspathulate, subobtuse, apiculate, 3 mm. l. *Lip*, apex of lateral lobes extending to or beyond middle of median lobe; median larger, roundish-elliptical, margin curled, apex rounded, emarginate, apiculate; disk with very short glandular hairs, and a callus extending along the median nerve from the base to the middle; 4 mm. l., 3·3 mm. br. *Column*, incl. anther, 1 mm. l. *Capsule* 1 cm. l.

2. **P. minor** *Fawc. & Rendle in Journ. Bot. xlviii.* 106 (1910); lateral lobes of lip small, rounded; sinus not above middle of lip; callus basal, projecting, conical.—P. foliosa *Griseb. loc. cit.* (as regards spec. from Jamaica) non *Reichb. f.* (Pl. 7, f. 6, 7.)

On trees and walls; in fl. and fr. Sept.-Jan.; road to Guanaboa and Red Hills, *Sloane* Herb. vii. 86! Fairfield, *Wullschlaegel*, 1055 (partly)! J.P. 523, *Morris*! near Browns Town, *Miss T. M. Barrett*! Claremont, 1700 ft., *Fawcett & Harris*! Lancaster, 2200 ft., 2500 ft., 3000 ft.; Belvedere, Hanover, 500 ft.; *Harris*! Ramble, Hanover; below Guava Ridge; *Fawcett*! Fl. Jam. 7568, 7777, 7805, 10,434, 10,457.—B. Guiana.

Plant 5-12 in. high. *Pseudobulb* ·5-2 cm. l. *Leaves* obliquely emarginate, 7-20 cm. l., 11-25 mm. br. *Panicle* 1 to nearly 3 dm. l., generally shorter than the leaves. *Bracts* triangular, shortly acuminate, 1·5 mm. l. *Flowers* yellowish-white, perianth ⅛ in. l. *Sepals* 3-nerved, apiculate; median oblong-elliptical, 2·5-2·7 mm. l., 1·5 mm. br.; lateral triangular-ovate, slightly oblique, 3 mm. l., 2 mm. br.; chin rounded. *Petals* linear-subspathulate, acute, 2·5 mm. l. *Lip*, apex of lateral lobes scarcely extending beyond sinus; median much larger, roundish-elliptical, obscurely apiculate, margin curled; disk with very short, mealy hairs, about 3 mm. l.; median lobe 2 mm. br. *Column*, incl. anther, about 1 cm. l. *Capsule* 6-8 mm. l.

Differs from *P. foliosa* Reichb. f. in being a larger plant, with larger flowers, also in other details, especially in petals and lip; the conical projecting callus at the base of the lip is characteristic.

20. CRYPTOPHORANTHUS Barb. Rodr.

Epiphytic herbs, with creeping rhizome without pseudobulbs. Stems with one leaf, short, with scarious sheaths. Leaf leathery or fleshy, more or less erect. Flowers rather small, shortly stalked, solitary, or sometimes a few clustered, at the base of the leaf. Sepals connate altogether, except for an open chink on each side between the median and the lateral sepals. Petals much smaller than the sepals. Lip free, a little longer than the petals, parallel with the column, very shortly clawed, jointed to the foot of the column, 3-lobed. Column erect, produced at the base into a rather long foot. Anther terminal, opercular, incumbent, hooded, 1-celled; pollinia 2, waxy, narrowly obovoid, compressed. Capsule broadly ellipsoidal.

Species 9, natives of W. Indies and tropical America.

C. atropurpureus *Rolfe in Gard. Chron. ii.* 693 (1887). Specklinia atropurpurea *Lindl. in Bot. Reg. under t.* 1797 (1835). Pleurothallis atropurpurea *Lindl. in Bot. Reg. xxviii. Misc.* 81

(1842). Masdevallia fenestrata *Lindl. ex Hook. in Bot. Mag. t.* 4164 (1845). (Pl. 7, f. 8–10.)

On trees, banks and decaying logs; in fl. Aug.–Dec., in fr. Apr., Aug.; *Purdie*! J.P. 484, *Morris*! Mansfield, near Bath; Mabess River, 3000–4000 ft.; Holly Mount, Mt. Diabolo, 2600–3000 ft.; *Harris*! Fl. Jam. 7580, 7626, 7817, 8911, 9014, 9891, 10,466.—Cuba.

Plant 3–6 in. high. *Stem* 3–5 cm. l., 5–7-jointed, covered with sheaths; sheaths increasing in size from below upwards, ·5–3 cm. l., dilated at mouth, glabrous. *Leaf* obovate-elliptical, tapering into a short stalk (1 cm. l.) which is enclosed in the top sheath, 3–9 cm. l., 1·5–3 cm. br. *Pedicel* 2-jointed, 13 mm. l. *Ovary* 7–8 mm. l., 12-winged. *Bracts* at each joint of pedicel, sheathing, one at upper joint 6–7 mm. l., with 3 or 4 smaller at base. *Flowers* about ½ in. l., deep crimson. *Sepals* 14–15 mm. l., 5 mm. br., conduplicate; chink 5 mm. from apex, 2 mm. l., ·6–·7 mm. br. *Petals* nearly 4 mm. l., 2 mm. br., oblong, truncate, 3-toothed at apex. *Lip* 4·5 mm. l., ·75 mm. br. at base, hastate, lateral lobes folded in front, a minute auricle at each side at the base. *Column* 2–2·5 mm. l.; clinandrium 2-toothed in front and 1-toothed behind.

21. STELIS Sw.

Stems erect, clustered, simple, with one leaf at the apex, 2-jointed, with a sheath to each joint and small basal sheaths. Leaf leathery, contracted at the base into a short stalk, jointed near the base. Flowers very small, numerous on a long raceme springing from the base of the leaf, enclosed by a spathaceous bract and smaller inner bracts. Floral bracts small, sheathing below, about equalling the pedicel. Sepals generally subequal, lateral slightly oblique, spreading, shortly or deeply connate. Petals much shorter, broad, with thickened margins. Lip sessile at the base of the column, equal and somewhat similar to the petals. Column short, broad, generally thicker above; clinandrium with a lobe at each side, and a posterior lobe which is 3-lobulate; rostellum tongue-like. Anther terminal, opercular, incumbent, semiglobose, incompletely 2-celled; pollinia 2, waxy, pyriform, generally connected at the apex by slight viscum. Capsule small, ovoid or oblong, 3-cornered.

Species about 210, natives of tropical America, from the West Indies and Mexico to Brazil and Peru.

Closed flowers rounded; sepals elliptical ... 1. *S. micrantha.*
Closed flowers 3-cornered; sepals deltoid ... 2. *S. ophioglossoides.*

1. **S. micrantha** *Sw. in Schrad. Journ. Bot. ii.* 240 (1799); racemes somewhat longer than leaf; closed flowers rounded; sepals elliptical-ovate, longer than broad, shortly connate at base, nerves slightly prominent below, the central nerve generally forming a keel in the lateral sepals.—*Sw. Fl. Ind. Occ.* 1553; *Smith Exot. Bot. t.* 75; *Hook. Exot. Fl. t.* 158; *Griseb. Fl. Br. W. Ind.* 611; *Cogn. in Symb. Ant. vi.* 386. Epidendrum micranthum *Sw. Prodr.* 125 (1788) & *Icon. ined. t.* 50. Dendrobium

micranthum *Sw. in Nov. Act. Upsal. vi.* 84 (1799). (Pl. 7, f. 15–28.) Type in Herb. Mus. Brit.

On trees and rocks; in fl. all the year; *Swartz*! *Macfadyen*! Blue Mt. Peak, *Prior*! Mt. Moses, J.P. 23, *Morris*! Blue Mt. Peak, *Fawcett*! New Haven Gap, *Moore*! Morse's Gap, 5000 ft.; Woodcutters Gap near Newcastle, 4000 ft.; ridge from Newhaven Gap to Vinegar Hill, 4000 ft.; Greenwich woodland, 4500 ft., *Harris*! Fl. Jam. 7537, 7815, 7835, 10,087, 10,141.

Stem 7–12 cm. l., 1·5–2 mm. br., sometimes as short as 2 or 3 cm., or as long as 15 cm. *Sheaths* striate, apex truncate, apiculate, generally covering stem, scarious, soon decaying. *Leaf* lanceolate-oblong or ligulate, apex obtuse, sometimes minutely toothed, generally somewhat longer than stem, 11 or 12 cm. l., about 1·5 cm. br., sometimes as short as 7 cm., or as long as 16 cm. *Raceme* solitary (rarely 2), many-flowered, 12–14 cm. l., rarely longer; spathaceous bract at base about 12 mm. l. *Bracts* sheathing at base, apex rounded, acute or apiculate to shortly acuminate, 1·5–2 m. l. *Flowers*: the size of the flower varies considerably: in Swartz's original specimen the median sepal is 2 3 mm. l., 1·4–1·5 mm. br., and the petal ·6 mm. l. and br.; in Fl. Jam. 7537 the median sepal is 2·5 mm. l. and br., and the petal ·75 mm. br.; in Fl. Jam. 7835 the median sepal is 3 mm. l., 2·5 mm. br., and the petal 1 mm. br. Measurements of parts of the flower from dried specimens are somewhat misleading. In specimens put into spirit when collected (Fl. Jam. 10,087) the median sepal is 3·75 mm. l., 3 mm. br., and the sepals are not distinctly keeled on back (petal 1·2 mm. br.); sepals from the dried specimen (moistened in hot water) measure 3·5 mm. l., 2·4 mm. br., and are obviously keeled. The flowers from the type specimen, and also from the Marquis of Blandford's specimen figured by Sowerby (*J. E. Smith, Exot. Bot. t.* 75), are much smaller, but these are evidently not fully developed. *Sepals* pale green, 3-nerved, the central nerve in the median sepal sometimes forming a dorsal mucro just below apex, lateral nerves generally meeting the central nerve above in lateral sepals, but not always in the median sepal, lateral shorter but as broad or broader. *Petals* deep purple, sub-quadrangular, upper margin subdeltoid-lunate. *Lip* deep purple, somewhat like petals but broader above, upper surface semicircular with a narrowly elliptical median depression, with the upper margin lunate. *Column* deep purple.

2. **S. ophioglossoides** *Sw. in Schrad. Journ. Bot. ii.* 239 (1799); raceme much longer than leaf; closed flowers 3-cornered; sepals deltoid-ovate, broader than long, connate nearly to outer basal angle, nerves only slightly prominent below, angles at base and apex somewhat rectangular.—*Sw. Fl. Ind. Occ.* 1551; *Lindl. Fol. Orch. Stelis*, 7 (in part) not *in Orch. Cub. Wright*; *Griseb. loc cit.*; *Cogn. op. cit.* 387 (in part). S. polystachya *Cogn. loc. cit.* Ruscus? foliis ovatis petiolatis &c. *Plum. Pl. Amer.* (*Burm.*) 172, *t.* 176 *f.* 3 (1758). Epidendrum ophioglossoides *Jacq. Enum. Pl. Carib.* 29 (1760) & *Sel. Stirp.* 225, *t.* 133, *f.* 2, *Ed. pict. t.* 211; *Sw. Obs. Bot.* 331. E. trigoniflorum *Sw. Prodr.* 125 (1788) & *Icon. ined. t.* 49. Dendrobium ophioglossoides *Sw. in Nov. Act. Upsal. vi.* 83 (1799). (Pl. 7, f. 11–14.) Type in Herb. Mus. Brit.

On trees and rocks; in fl. all the year; *Swartz*! J.P. 12, *Morris*! J.P. 2024, *Syme*! near Morse's Gap, 5000 ft., *Fawcett*! near Moy Hall, *Sullivan*! Greenhill, *Moore*! Mabess River, 3000 ft.; Cinchona, *Harris*! Fl. Jam. 7764a, 7764b.

Similar to *S. micrantha* with the following differences: *Stems* 3–9 cm. l., 1–2 mm. br. *Leaf* longer than the stems, 5–11 cm. l., 1–2 cm. br. *Racemes*, 1–3, much longer than leaf, 12–22 cm. l. *Bracts* shortly acuminate, 2–3 mm. l. *Sepals* greenish-yellow with purplish tinge at base, apex obtuse, mucronate, 3-nerved, lateral nerves generally meeting the central nerve above in lateral sepals, but not always in median, median 2–2·3 mm. l., 2·5–2·6 mm. br., lateral 1·6–2 mm. l., 2–2·1 mm. br. *Petals* deep purple, broadly roundish or obovate, upper margin lunate, with very thin points, ·7 mm. l., ·9–1·1 mm. br. *Lip* deep purple, angular-semicircular, ·75 mm. l., 1 mm. br.

22. PLEUROTHALLIS R. Br.

Stems clustered or branches of a creeping stem or rhizome, simple, somewhat long, short or very short, with one leaf at the apex, sometimes with one to numerous sheaths below the leaf, not pseudobulbous. Racemes, one to numerous, at the base of the leaf, within the sheath or spathe, simple, with several flowers or laxly few-flowered, or in a few species shortened, 1–3-flowered, and densely crowded, or sometimes solitary. Flowers small, sometimes very small, or in a few species of medium size or rather large, generally on one side of the raceme. Bracts small, shortly tubular. Sepals about the same length, erect or spreading; the median free or very shortly connate with the lateral at the base; lateral connate at the base or higher into one, entire, 2-fid or generally 2-partite, generally concave or slightly gibbous under the foot of the column. Petals shorter or narrower than the sepals, sometimes very small, more rarely about as long as the sepals. Lip shorter than the petals or more rarely a little longer, at the base generally contracted and jointed with the base of the column, 3-lobed; lateral lobes erect, embracing the column, or small, dentiform, or obsolete; the median somewhat spreading. Column as long as the lip or a little shorter, semiterete, the angles sometimes membranous-margined, not truly winged, at the base generally produced into a very short foot; clinandrium small or with a membranous extension, oblique, entire or shortly 3-lobed. Anther terminal or within the margin of the clinandrium, opercular, incumbent, 1-2-celled, pollinia 2, waxy, globose, ovoid or pyriform, free or cohering at the apex by a sparse viscid substance. Capsule sub-globose or ellipsoidal, 3-keeled or 3-cornered.

Species nearly 600, natives of the mountains of tropical America from the West Indies and Mexico to Brazil and Bolivia.

I. Spathe at base of racemes large and leathery.
  Racemes solitary, much longer than leaf.
    Sepals yellow, acute .... 1. *P. longissima.*
    Sepals purple, acuminate .... 2. *P. racemiflora.*
  Racemes two or more, clustered, not longer than leaf.
    Lip undivided; sepals white, gibbous......... 3. *P. gelida.*
    Lip 3-lobed; sepals pale yellow.................. 4. *P. velaticaulis.*

II. Spathe inconspicuous, scarious.

A. Sheaths of the stem straight, adpressed, not ciliate.

1. Stem evident, leaf distant from base.

a. Racemes many-flowered, longer than the leaf or about as long.

Sepals yellowish-green, shortly acute, connivent .... 5. *P. alpestris.*

Sepals brownish-purple, obtuse, connivent .... 6. *P. multirostris.*

Sepals deep purple, acuminate, divaricate .... 7. *P. laxa.*

b. Flower-stalks with few flowers or only one, shorter than leaf.

Leaf 4 or 5 inches l. Flower-stalks very numerous, short, clustered .... 10. *P. ruscifolia.*

Leaf less than 4 inches l.

Basal sheaths hirsute .... 8. *P. hirsutula.*

Basal sheaths glabrous, ovary smooth.

Petals not subulate.

Sepals greenish-yellow.

Leaf linear-lanceolate 9. *P. pruinosa.*

Leaf elliptical-lanceolate...................... 12. *P. Wilsonii.*

Sepals dark purple......... 11. *P. brachypetala.*

Petals subulate .... 14. *P. Morrisii.*

Basal sheaths glabrous, ovary warty......................... 13. *P. monophylla.*

Leaf 8 or 9 inches l. Sepals rough, about $\frac{3}{8}$ inch l. .... 19. *P. uncinata.*

2. Stem very short, leaf close to the root.

Stems clustered.

Flower-stalk shorter than leaf-stalk. Capsule covered with soft prickles....................... 18. *P. tribuloides.*

Flower-stalk shorter than leaf. Capsule smooth .... 20. *P. jamaicensis.*

Flower-stalks much longer than leaf (or about as long in *P. lanceola*). Capsule smooth.

Leaf roundish .... 15. *P. rotundifolia.*

Leaf not roundish.

Flower-stalk with one flower; lateral sepals connate nearly to apex .... 16. *P. corniculata.*

Flower - stalk few - flowered; lateral sepals free or not more than $\frac{1}{3}$-connate.

Flower-stalk much longer than leaf.

Sepals, apex obtuse; flower-stalk zig-zag...... 21. *P. delicatula.*

Sepals, apex tailed .... 22. *P. Helenæ.*

Flower-stalk about as long as leaf .... 23. *P. lanceola.*

Stems at intervals from creeping rhizome.
Leaf oval-roundish .... 17. *P. testifolia.*
Leaf linear-spathulate .... 24. *P. sertularioides.*

B. Sheaths (as in *Lepanthes*) with mouth dilated, margined, ciliate. Leaves ½ inch l. or less.
Flower-stalk much shorter than leaf; stems much shorter than leaf..... ... 25. *P. microlepanthes.*
Flower-stalk longer than leaf; stem not shorter than leaf.
Stem with more than one leaf........... 26. *P. foliata.*
Stem with only one leaf .... 27. *P. trilobata.*

In the sequence of the species we have followed the arrangement adopted by Prof. Cogniaux in Urban's Symbolæ Antillanæ vi. 391–7.

1. **P. longissima** *Lindl. Fol. Orch. Pleuroth.* 31 (1859); raceme solitary, many-flowered, two or three times as long as leaf, enclosed by large leathery spathe; sepals greenish-yellow, acute, lateral connate almost to apex.—*Griseb. Fl. Br. W. Ind.* 607; *Reichb. f. in Saund. Ref. Bot. t.* 141 (1872); *Cogn. in Symb. Ant. vi.* 397. P. racemiflora *Lodd. Bot. Cab. x. t.* 949 (1824); *Lindl. in Hook. Exot. Fl. t.* 123 (1825), non *in Fol. Orch.* Type in Lindley's Herb. at Kew.

On trees and rocks; in fl. March–June; *Bertero; Macfadyen*! St. Mary, *Purdie*! *McNab; Wilson*! Fairfield, *Wullschlaegel*, 1383! St. Mary, J.P. 2331, *Syme*! St. Ann, *Miss T. M. Barrett*! Claverty Cottage, *Moore*! Back Woods, Portland Gap, *Miss B. Gosset*! Mt. Moses, 2000 ft., *Harris*! Fl. Jam. 9018, 10,424, 8118 (fide Cogn.).—Guatemala.

Plant 1–1½ ft. high. *Stems* 4–17 cm. l., shorter or a little longer than the leaves, clustered, clothed below with 2 or 3 adpressed acute sheaths. *Leaf* 9–14 cm. l., 2–3 cm. br., sessile, oblong or oblong-elliptical, obtuse, tapering to base. *Raceme* to 35 cm. l., with 15–26 secund flowers; spathe 1–2 cm. l. *Bracts* 3–4 mm. l., slightly shorter than pedicels, scarious, sheathing, obliquely truncate and obtuse. *Pedicels* short, 5 mm. l. *Flowers* about ⅓ in. l., greenish-yellow. *Sepals* 7–9·5 mm. l., with mid-nerves prominent on back; lateral 4–4·3 mm. br., broadly ovately concave; median 3–3·3 mm. br., ovate, subacute, 3-nerved. *Petals* 6–8 mm. l., 2·1–2·5 mm. br., narrowly oval-oblong, broadest below the middle, to lanceolate, apex rounded, very minutely apiculate to acute, exceeding lip. *Lip* shorter than petals, when flattened barely exceeding 4 mm. l., limb 3–3·5 mm. l., 2–2·25 mm. br., with large basal auricles and short claw, roundish, 3-nerved, mid-nerve thickened in upper part. *Column* about 3 mm. l., slender; clinandrium membranous, ovate, blunt.

2. **P. racemiflora** *Lindl. Fol. Orch. Pleuroth.* 34 (1859); racemes usually more than one, but only one flowering, many-flowered, two or three times as long as leaf, enclosed by large leathery spathe; sepals purple, long acuminate, lateral ½-connate.—*Griseb. op. cit.* 607; *Cogn. op. cit.* 398. P. oblongifolia *Lindl. in Comp. Bot. Mag. ii.* 355 (1837). Epidendrum racemiflorum *Sw. Prodr.* 125 (1788). Dendrobium racemi-

florum *Sw. in Nov. Act. Upsal. vi.* 83 (1799) & *Fl. Ind. Occ.* 1543. Type in Herb. Mus. Brit.

On trees and rocks; in fl. Aug.-Nov., in fr. throughout the year; *Swartz*! *Macfadyen*! Mt. Moses, 2500 ft., J.P. 2073 (in part), *Syme*! Blue Mts., J.P. 2073 (in part), *Morris*! John Crow Peak, *Moore*! Vinegar Hill, *Sherring*! Cinchona, 5000 ft.; Morse's Gap; Old England; *Harris*! Fl. Jam. 7685, 7818.—Cuba.

Plant 5 to 12 in. high. *Stems* to 10 cm. l., 2 mm. br., with two developed internodes, clustered, terete at the base, compressed and angular above; sheaths, upper about 4 cm. l., loose, brown, membranous, longitudinally striate, widening considerably towards the mouth. *Leaf* generally about as long as the stem, blade 5·5–9 cm. l., 1·5–2·8 cm. br.; oval-oblong, narrowed at the base into a short stalk (1 cm. or less). *Racemes* 14–24 cm. l., with 10–12 flowers in two rows, lax; spathe 1·5–3 cm. l., compressed, obtuse. *Bracts* 4–7 mm. l., slightly shorter than the pedicels, scarious, ovate, obtuse, sheathing at base. *Flowers* about ⅜ in. l., magenta-coloured. *Sepals* 7–10 mm. l., connivent, concave, lanceolate, median 2 mm. br., lateral 4·5 mm. br. *Petals* 4 mm. l., 2 mm. br., less than half as long as the sepals, oval, narrowed at the base, apex round, crenulate, 3-nerved. *Lip* longer than the petals, clawed, limb 4 mm. l., 1·5 mm. br., concave, disk 3-keeled, elliptical, apex obtuse, triangular. *Column* scarcely equalling the petals; clinandrium shortly 3-fid. *Capsule* 7–9 mm. l., oblong.

3. **P. gelida** *Lindl. in Bot. Reg. xxvii. Misc.* 91 (1841); racemes two together, many-flowered, a little shorter than leaf, enclosed by large leathery spathe; sepals white, gibbous, lateral often ½-connate; lip undivided.—*Lindl. Fol. Orch. Pleuroth.* 4 (1859); *Griseb. op. cit.* 607; *Kränzl. in Xen. Orchid. iii.* 116, *t.* 267, *II.* (1900). *Cogn. op. cit.* 399. **P. univaginata** *Lindl. in Ann. & Mag. Nat. Hist. ser.* 3, *i.* 326 (1858) & *Fol. Orch. Pleuroth.* 4; *Griseb. loc. cit.*; *Cogn. op. cit.* 398.

On trees in rocky woodland; in fl. Dec.; Manchester, *Purdie*! Iron Face, Chester Vale, 3500 ft., *Harris*! Fl. Jam. 7836.—Cuba, Porto Rico.

Plant often 1½ ft. high. *Stems* clustered, about 3 mm. thick near the base; 2-jointed; sheath above the base of the stem somewhat compressed, membranous, blunt, to 8·5 cm. l. *Leaf* elliptical, 18–25 cm. l., including stalk (3–4 cm.), 4–5·5 cm. br. *Racemes* ·12–14 cm. l., with 16–20 secund flowers; spathe 1·5–2 cm. l. *Bracts* barely 2 mm. l., short, obtuse. *Flowers* nearly ⅓ in. l., pale yellow. *Sepals* 7 mm. l., erect, forming a cup, subelliptic, obtuse, with short hairs on the inside. *Petals* 3 5 mm. l., subovate. *Lip* 2·5 mm. l., widening above to a broad rounded apex, bearing two wing-like keels on the upper face which do not reach the apex, with two teeth in the middle of each side. *Column* less than 2 mm. l., with a minute tooth on each side at the apex. *Capsule* ellipsoidal, 3-ribbed.

Lindley says of *P. univaginata*, "very like the last" (*P. gelida*), "but distinguished by the state of the sepals and form of the lip." Examination of the type specimens in Lindley's herbarium and of other available material leads to the conclusion that we are dealing with only one species. The degree of union of the lateral sepals varies somewhat, while the hairs to which Lindley refers in his description "sepalis intus pilosis," are present to a very slight extent in the type specimen; the form of the lip is the same in both, the two median wing-like keels being present, though Lindley says of *univaginata*, "labello ecarinato."

Neither at Kew nor the British Museum are there any specimens named *P. gelida*, and Grisebach merely quotes Lindley's description, which was based on a cultivated plant obtained by Messrs. Loddiges from Jamaica.

The figure in *Xenia Orchidacea* (*iii. t.* 267, II.) represents the flower more widely open than obtains in any specimens we have seen.

4. **P. velaticaulis** *Reichb. f. in Linnæa xxii.* 824 (1849); racemes 1–5, with numerous flowers, shorter than leaf, or about as long, enclosed by a large leathery spathe; ovary terete, smooth; sepals keeled, lateral subfree; lip 3-lobed, lateral lobes emarginate.—*Lindl. Fol. Orch. Pleuroth.* 29 (1859); *Cogn. op. cit.* 400 & under. P. crassipes *Lindl.*, so far as concerns Jamaican specimen Fl. Jam. 7745. (Pl. 8, f. 17–22.)

On rocks and trees; in fl. Nov.; near Cinchona, 5000 ft., *Harris*! Fl. Jam. 7745.—Cuba.

Plant a foot or more high. *Stems* 13–21 cm. l., clustered, long, slender, angular, of two internodes, with a long thin tubular sheath at each node, 5 cm. l., with a shorter outer one at the base. *Leaf* oblanceolate-oblong, tapering into a long stalk, shorter than the stem, 10–12 cm. l., of which the stalk is about 2 cm. l., 2–2·5 cm. br. *Racemes* 6–10 cm. l. *Bracts* tubular below, dilated upwards, truncate, apiculate, 3–3·5 mm. l. *Pedicels* as long as the bracts. *Ovary* 3·5–4 mm. l., carrying the flower much above the bracts. *Sepals*, median ovate, subacute, 3-nerved, 4 mm. l., 2 mm. br.; lateral lanceolate, acute, 3-nerved, 4 mm., l., 1·3 mm. br. *Petals* linear-oblong, blunt, obscurely apiculate, 1-nerved, a little over 2 mm. l., ·7 mm. br. *Lip* broad, 3-lobed, 3-nerved, 1·7 mm. l., barely 1·5 mm. br., inserted by a short broad claw, above which is a transverse ridge, middle lobe shortly ovate, blunt; lateral lobes much shorter than the middle lobe, rounded, emarginate. *Column* short, broad, about 1 mm. l., without a membranous margin.

Reichenbach's description differs in that the sepals are oblong, obtuse, and the petals cuneate. This species is very near *P. crassipes* Lindl.

5. **P. alpestris** *Lindl. Gen. & Sp. Orch. Pleuroth.* 7 (1830); racemes, 1-several, with numerous flowers, not much longer than the small leaf, or about as long; sepals yellowish-green, shortly acute, connivent into a tube, the lateral connate nearly to apex. —*Griseb. op. cit.* 607; *Cogn. op. cit.* 401. P. nigroannulata *Cogn. in Fedde Repert. vi.* 306 (1909) *in part* (Fl. Jam. 7554) & *in Symb. Ant. vi.* 417. Epidendrum alpestre *Sw. Prodr.* 125 (1788). Dendrobium alpestre *Sw. in Nov. Act. Upsal. vi.* 84 (1799) & *Fl. Ind. Occ.* 1545. Type in Herb. Mus. Brit.

On trees; in fl. Nov.–Feb.; Blue Mts., *Swartz*! Clydesdale, 4000 ft., J.P. 29; 35; *Morris*! Cinchona, *C. Nicholls*! Mt. Hybla; Silver Hill Woodland, 4000 ft.; Hardware Gap; Woodcutters' Gap, 4000 ft.; Holly Mount, Mt. Diabolo, 2600 ft.; near Cinchona, 4000 ft.; *Harris*! Fl. Jam. 7554, 7613, 7829, 8908, 10,502.

Plant 4 or 5 in. high. *Stems* 2–3 cm. l., 1 mm. or less thick, clustered; somewhat compressed and angular when dry; a loose membranous sheath springs from a node just above the base and extends about half-way to the leaf. *Leaf* 3–4·5 cm. l., 7–13 mm. br., narrowly oval to oval-lanceolate, obtuse, sessile, at the base subacuminate. *Racemes* 3·5–5·5 cm. l., filiform. *Flowers* ¼ in. l., yellowish-green, lax. *Pedicels* scarcely longer than the bracts. *Bracts* 2 mm. l., small, acute, sheathing. *Sepals* 8 mm. l., median 1·75 to 2·5 mm. br., the united lateral 3 to 4 mm. br., erect, keeled, lanceolate. *Petals* yellowish-green tinged with purple at the base, 2·5–2·75 mm. l., ·75 to 1 mm. br., subovate, acute, one-third as long as the sepals. *Lip* deep purple, very small, 2·5 mm. l., shortly clawed, 2-keeled on the disk, keels running up near the apex, sub-panduriform with a blunt tooth on each side below middle, then constricted, and 2-auricled above the narrow claw. *Column* shorter than the petals, clinandrium crenulate, 2-denticulate, tinged with purple at the base. *Capsule* 8 mm. l., 2·5 mm. br., oblong, gibbous, with 6 muricate keels.

6. **P. multirostris** *Reichb. f. in Linnæa xli.* 49 (1877); raceme 1, with several flowers; sepals brownish-purple, curved, beak-like, the lateral connate nearly to apex, apex obtuse.—*Cogn. in Symb. Ant. vi.* 401.

In fl. Sept.

Stems clustered, about as long as the leaf, amply sheathed. *Leaves* oblong, acute, stalked. *Raceme* deflexed, with several flowers. *Bracts* minute, triangular. *Pedicels* about equalling the ovary. *Flowers* brownish-purple, curved, beak-like. *Sepals* linear-ligulate, lateral sepals connate nearly to apex, apex obtuse. *Petals* half as long as sepals, oblong, obtuse, minutely denticulate at apex, 3-nerved. *Lip* oblong, obtusely acute, obscurely 3-lobed at the middle, lateral lobes low, obtuse-angled and 3-nerved. *Column* thick, at apex membranous, denticulate.

We have not seem any specimen of this species, and it is solely on the authority of Reichenbach that it is included among the orchids of Jamaica.

7. **P. laxa** *Lindl. Gen. & Sp. Orch.* 7 (1830); racemes 1–3, sometimes with as many as six flowers on each, longer than the leaf; sepals deep purple or greenish striped with purple, acuminate, divaricate, lateral ½-connate.—*Lindl. Fol. Orch. Pleuroth.* 35; *Griseb. op. cit.* 607; *Cogn. op. cit.* 403. P. nigroannulata *Cogn. in Fedde Repert. vi.* 306 (1909) *in part* (Fl. Jam. 7536) & *in Symb. Ant. vi.* 417. Epidendrum laxum *Sw. Prodr.* 125 (1788). Dendrobium laxum *Sw. in Nov. Act. Upsal. vi.* 84 (1799) & *Fl. Ind. Occ.* 1547. Type in Herb. Mus. Brit.

On trees; in fl. Aug.–Nov.; in fr. Dec.; *Swartz*! Portland Gap, 5500 ft., J.P. 170; 2073 (in part); *Morris*! John Crow Peak, *Fawcett*! Pleasant Hill, 4500 ft; near Woodcutters' Gap, 4000 ft.; *Harris*! Fl. Jam. 7536.

Plant 3 to 5 in. high. *Stems* clustered, 2–5 cm. l., barely reaching 1 mm. in width, somewhat compressed and angular when dry, covered within 1 cm. or less of the leaf by three loose membranous keeled sheaths. *Leaf* 3–4 cm. l., 1–1·3 cm. br., oval-oblong. *Racemes* 5–7 cm. l. *Bracts* 2 mm. l., clasping, acute. *Flowers* deep purple, ⅓ in. l. *Sepals* 8 mm. l., lateral narrowly lanceolate, median linear-lanceolate. *Petals* 4 mm. l., purple or purple-streaked, narrowly lanceolate, half the length of the sepals. *Lip* slightly shorter than the petals, spotted with purple, shortly clawed, the lower part concave, obtusely lobed, the upper part ovate, recurved, with a pair of short, rounded erect lobes at base. 2-keeled on the disk. *Column* a little shorter than the petals, spotted with purple; clinandrium denticulate, anther-case dark purple.

8. **P. hirsutula** *Fawc. & Rendle in Journ. Bot. xlvii.* 3 (1909); basal sheaths with dark brown stiff short hairs; racemes 1–5 together, generally not more than one flowering at a time, with four or five flowers, about one-third the length of the leaf.—*Cogn. in Symb. Ant. vi.* 403. (Pl. 9, f. 15–19.) Type in Herb. Mus. Brit. & Jam. Herb.

On trunks of trees; in fl. Aug.; Holly Mount, Mt. Diabolo, 2600–3000 ft., *Harris*! Fl. Jam. 6681, 9890.

Plant 4½–7 in. high. *Rhizome* creeping, rooting. *Stems* 6–10 cm. l., clustered, longer than the leaves, two-edged, edges furrowed; sheaths three, loose. *Leaf* 5–7·5 cm. l., 1–1·3 cm. br., sessile, lanceolate, at the base clasping. *Raceme* 1·2–2·5 cm. l. *Bracts* 2 mm. l., sheathing, truncate. *Flowers* about ¼ in. l., very dark red. *Sepals* more or less erect, 3-nerved, somewhat fleshy; median 6·5 mm. l., lanceolate, somewhat acuminate, concave; lateral slightly shorter, narrowly triangular-lanceolate, acute, at the base connate, gibbous. *Petals* slightly exceeding 3 mm. l. by 1·5 mm. br., rhomboid, narrowed at the base, upper margins minutely toothed. *Lip* 2·75 mm. l., 1·25 mm. br., ovate, shortly clawed, margin undulate. *Column* 2·25 mm. l., slender, curved.

9. **P. pruinosa** *Lindl. in Bot. Reg. xxviii. Misc.* 75 (1842) stem longer than leaf; raceme one, with about six flowers, shorter than the leaf; leaf linear-lanceolate to narrowly oval; sepals greenish-white or pale yellow, lateral connate almost to apex; petals linear, somewhat acute.—*Lindl. Fol. Orch. Pleuroth.* 26; *Griseb. op. cit.* 608; *Cogn. op. cit.* 404.

On trees; in fl. April; *Wilson*; Fairfield, *Wullschlaegel*, 1358! Mt. Moses, 2500 ft., J.P. 2125, *Syme*! near Mabess River, 3500 ft.; Cascade, 2500 ft.; *Harris*! Fl. Jam. 7997 & 8111 (*fide* Cogniaux).—Hispaniola, Porto Rico, Guadeloupe, Martinique, Grenada, Trinidad, Guiana.

Plant 2 to 3 in. high. *Stems* to 5·5 cm. l., clustered, subterete, thin, sheathed at the base, and with one close, glabrous, obliquely truncate sheath below the middle. *Leaf* to 3·5 cm. l., 6 mm. br., sessile. *Spathe* 5–7 mm. l., closely compressed, lanceolate, subacute. *Bracts* sheathing, truncate, obtuse, glabrous, a little shorter than the pedicels. *Flowers* ⅛ in. l., pale yellow or greenish-white, minute, distant, very shortly stalked, cleistogamous (?). *Sepals* 3 mm. l., equal, broadly ovate, slightly 2-keeled; median ovate, subacute, slightly keeled. *Petals* a little shorter than the sepals. *Lip* half as long as sepals, erect, shortly clawed, limb triangular, somewhat acute, with 3 parallel keels on the disk. *Column* barely 1 mm. l., short, thick. *Capsule* 4 mm. l., elliptical, crowned by the persistent calyx.

10. **P. ruscifolia** *R. Br. in Ait. Hort. Kew. ed.* 2. *v.* 211 (1813); stem much longer than leaf; flower-stalks very short, in dense clusters, each with one flower; leaf 4 or 5 in. l.—*Hook. Exot. Fl. t.* 197; *Griseb. op. cit.* 608; *Cogn. op. cit.* 405. P. multicaulis *Poepp. & Endl. Nov. Gen. i.* 47, *t.* 82 (1835). Ruscus foliis &c. *Plum. Pl. Amer.* (*Burm.*) *t.* 176, *fig.* 2 (1755). Epidendrum ruscifolium *Jacq. Enum. Pl. Carib.* 29 (1760), *Select. Stirp. Amer.* 226. *t.* 133, *fig.* 3 (1763) & *Ed. pict. t.* 212; *Sw. Obs. Bot.* 331 (1791). Dendrobium ruscifolium *Sw. in Nov. Act. Upsal. vi.* 84 (1799).

On trees; in fl. July, Aug.; in fr. Aug.; *Swartz*! Moody's Gap, J.P. 2071, *Syme*! Mabess River, 2000 ft., *C. Nicholls*! *G. Nichols*! *Harris*!—Cuba, Hispaniola, Porto Rico, Martinique, Dominica, St. Vincent, Grenada, Trinidad, Margarita, B. Guiana, Ecuador, Peru, Bolivia.

Plant 7 to 13 in. l. Rhizome creeping, 1–1·5 mm. thick, with a dense covering of slender root fibres and from the upper surface a number of erect slender rounded clustered stems from 4 to 20 cm. l., with a slender adpressed cylindrical truncate bract (about 2 cm. l.) about or rather below the middle. *Leaf*, blade 11 cm. l., 2·5 cm. br., erect or suberect, stiff, many-nerved, glabrous, oblong-lanceolate, apex more or less acuminate, stalk about 1·5 cm. l. A dense cluster of inflorescences springs from the short conduplicate bract (6–7 mm. l.); each inflorescence contains several stalked flowers (stalks 5–6 mm. l.), the base of each stalk is surrounded by a thin hyaline bract. *Sepals* 7 mm. l., lanceolate, acuminate, lateral connate. *Petals* 3·5 mm. l., linear, acuminate, erect. *Lip* 1·25–1·5 mm. l., ·75–1 mm. br., lower portion nearly square in outline with raised convex sides and a broad central channel, apex bluntly triangular with a much thickened callus on the midrib. *Column* less than 1 mm. l., including the short rounded anther-cell. *Ovary* 2 mm. l. *Capsule* green, narrowly club-shaped, about 1 cm. l.

11. **P. brachypetala** *Griseb. in Cat. Cub.* 257 (1866); stems nearly as long as leaf; racemes 2–4 together, with two to five flowers, one-third to one-fourth the length of the leaf; leaf oval; sepals dark purple, lateral $\frac{2}{3}$-connate; petals cuneate, apiculate.—*Cogn. op. cit.* 406. (Pl. 8, f. 7–10.) Type: Cuba, Wright, 3349.

On trunks of trees amongst lichen and moss; in fl. Nov.-April; Mt. Moses, 3500 ft., J.P. 2277, *Syme*! above Mabess River, 3500–4000 ft.; Lancaster, 2500 ft.; near Cinchona, 4500 ft.; *Harris*! Fl. Jam. 7641, 7852, 9012, 10,501.—Cuba, Porto Rico.

Plant 2 to 3 in. high. *Stems* 2–3 cm. l., clustered, compressed, nearly covered with rather loose sheaths, the upper keeled, acute, the others blunt. *Leaf* 3–4 cm. l., 10–14 mm. br., sessile, oval to narrowly elliptical, base clasping. *Racemes* to 1 cm. l., with a prominent sheathing bract at base. *Flowers* dark purple. *Bracts* 2 mm. l., sheathing, loose, shortly acute. *Sepals* 4 mm. l., gibbous; median ovate, obtuse; lateral ovate-lanceolate, shortly acute. *Petals* half as long as sepals, ·8 mm. br. *Lip* shortly clawed, with two short lobes below the middle, elliptical above, apex rounded, incurved towards the column below, recurving above the middle. *Column* as long as petals, clavate, winged, with a distinct foot.

Specimens from Jamaica agree with Wright No. 3349 in Herb. Kew, which unfortunately has no flowers. Grisebach's description differs from our plant in petals being ovate and lip obovate.

12. **P. Wilsonii** *Lindl. in Ann. & Mag. Nat. Hist. ser.* 3, *i.* 326 (1858) (in part); stem longer than the small elliptical-lanceolate leaf; peduncles, one or (rarely) two or three, with one flower (or rarely two) on each, much shorter than leaf; petals oblanceolate, shortly acute.—*Lindl. Fol. Orch. Pleuroth.* 17 (in part); *Griseb. Fl. Br. W. Ind.* 608; *Cogn. op. cit.* 407. (Pl. 9, f. 8.) Type in Lindley's Herb. at Kew.

On trees and rocks; in fl. Aug.-Nov.; in fr. Dec.-May (*fide* Cogniaux); Bethabara, *Wilson*! and *Wullschlaegel*, 1079!—Porto Rico and Guadeloupe (*fide* Cogniaux).

Plant 2½ in. high, with creeping rhizome. *Secondary stem* 1·5–4 cm. l., slender, the lower half loosely sheathed. *Leaf* 1·5–3·3 cm. l., 5–8 mm. br., sessile, margined. *Peduncles* 3–5 mm. l. *Bracts* about 1 mm. l., sheathing, truncate, obtuse. *Flowers* greenish-yellow, striped violet inside (*fide* Cogniaux), about ⅛ in. l. *Sepals* about 4 mm. l., 3-nerved; median narrow, triangular, acute, concave, 1·2 mm. br.; lateral triangular, falcate, slightly narrower. *Petals* 2 mm. l., ·5 mm. br. *Lip*, lower portion somewhat broadly ovate above a short broad claw, indistinctly 3-lobed, middle lobe narrow, oblong, blunt, about 2 mm. l.; lower portion about two-thirds of whole length; margin papillose. *Capsule* obovoid, narrowly 3-keeled, 8–9 mm. l.

Lindley included two plants in his description of *P. Wilsonii*—one from Cuba (Wright, No. 668), and the other from Jamaica, collected by Wilson. The two plants differ remarkably in the conformation of the lip, and must be regarded as distinct species. We retain the name *P. Wilsonii* for the Jamaican plant, as Lindley's description of the lip evidently refers to it, and not to the Cuban plant, which we have described, in Journ. Bot. xlvii. 129 (1909), under the name *P. confusa*. The lip only of *P. Wilsonii* is represented in Pl. 9, f. 8. A portion of the plant *P. confusa* is shown in f. 5, a petal in f. 6, and the lip in f. 7.

Cogniaux's description of Porto Rico and Guadeloupe specimens differs somewhat from ours, *e.g.*, the median sepal is oblong-subspathulate, obtuse, the petals are narrowly ovate, 1·5–1·75 mm. l., ·75 mm. br., the peduncles are 6–14 mm. l.

13. **P. monophylla** *comb. nov.*; stem shorter than the leaf; racemes, two or three together (only one flowering), with two flowers, shorter than the leaf; leaf elliptical-lanceolate; ovary warty; sepals pale-green, lateral united near base and somewhat gibbous; petals lanceolate, acute.—P. emarginata *Lindl. Fol. Orch. Pleuroth.* 25 (1859); *Griseb. op. cit.* 608; *Cogn. op. cit.* 408. Epidendrum (?) monophyllum *Hook. Exot. Fl. t.* 109 (1825).

Without locality, *Wiles*.

Plant about 2 in. high. *Stems* about 2 cm. l., clustered, erect, slender, covered for the most part by sheathing bracts. *Leaf* 3–4 cm. l., 1·2 cm. br., obtuse. *Pedicels* about one-third as long as sepals, rather swollen upwards. *Bract* scarcely exceeding the pedicel, sheathing, with ovate acute mouth. *Sepals* 7 or 8 mm. l., 1-nerved, slightly keeled, lanceolate, acute. *Petals* minute, about one-third as long as sepals, whitish with a red central line, 1-nerved. *Lip* small, about as long as the petals, pink or deep purple, lanceolate, parallel with the petals and enclosed by them, nearly plane, slightly serrated at the margin, near the base above having two indistinct tubercles. *Column* shorter than the petals, straight, semicylindrical at the base, above at the base of the anther expanding into a sort of hood formed by the concave broadly winged and toothed margin, the base deep purple. *Ovary* very short, purple.

No specimen seen; described from figure and description in Hooker's Exotic Flora.

14. **P. Morrisii** *Fawc. & Rendle in Journ. Bot. xlvii.* 3 (1909); stem about as long as the narrowly lanceolate, somewhat falcate, subacuminate sessile leaf; peduncles two together, 1-flowered, very short; petals subulate; lip entire.—*Cogn. op. cit.* 408. (Pl. 9, f. 1–4.) Type in Herb. Mus. Brit. and Jam. Herb.

Without locality, *Morris*, 28!

Plant 3 or 4 in. high. *Stems* 3–5 cm. l., filiform, clustered, terete, with two dark-brown, adpressed sheaths at the base, the upper sheath 7–9 mm. l. *Leaf* 4–5 cm. l., 5–8 mm. br. *Sepals* 3·5 mm. l., broadly elliptical; median 2 mm. br., 3-nerved; lateral somewhat broader, connate, 2-nerved, at apex 2-toothed. *Petals* 2·5 mm. l. *Lip* 2·4 mm. l., 1·6 mm. br., shortly clawed, articulated to foot of column, obtuse, concave, with a thickened margin. *Column* 1·5 mm. l., short; clinandrium with a long narrow process in front.

The species is named after Sir Daniel Morris, formerly Director of Public Gardens and Plantations in Jamaica.

15. **P. rotundifolia** *Rolfe in Kew Bulletin*, 1895, 191; stems almost none, clustered; peduncle much longer than the roundish leaf, solitary, few-flowered; sepals straw-coloured, lateral connate nearly to apex.—*Cogn. op. cit.* 409. (Pl. 8, f. 11–16.) Type in Herb. Kew.

Without locality, *Morris*!

Plant very small, 1½–2 in. high, acaulescent. *Leaves* roundish-obovate, marginate, fleshy, 6–8 mm. l., 5–7 mm. br.; petiole short, 2–4 mm. l. *Scape* slender, with about six flowers, 4–5 cm. l. *Bracts* tubular at the base, triangular-ovate, acute, small, 1 mm. l. *Pedicels* 2 mm. l. *Flowers* straw-coloured, about ⅛ in. l. *Sepals* oval; median acute, concave, 3-nerved, 4 mm. l., 1·8 mm. br., with light red-purple nerves; lateral 2-nerved, 4·5 mm. l., 2·5 mm. br. *Petals* spathulate-oblong, subobtuse, 1·4 mm. l., straw-coloured with mid-nerve purple-red. *Lip* subrecurved, undivided, oblong, obtuse, 3-nerved, 1·7 mm. l., ·7 mm. br., basal three-fourths purple-red, upper fourth straw-coloured. *Column* winged, with acute apex, 1·25 mm. l.

16. **P. corniculata** *Lindl. in Bot. Reg. xxviii. Misc.* 83 (1842) (excl. syn.); stems very short, clustered; peduncle longer than the oval-oblong leaf, solitary, with only one flower; lateral sepals connate almost to apex.—*Lindl. Fol. Orch. Pleuroth.* 42; *Griseb. op. cit.* 609; *Cogn. op. cit.* 412. Epidendrum corniculatum *Sw. Prodr.* 123 (1788). Dendrobium corniculatum *Sw. in Nov. Act. Upsal. vi.* 83 (1799) & *Fl. Ind. Occ.* 1537. Type in Herb. Mus. Brit.

On trunks of trees; in fl. Sept.–April; *Swartz*! near Rodney Hall, *Purdie*! Beaufort, *Wullschlaegel*, 1081 (in part)! *Prior*! lowlands to 1000 ft.; St. Mary; J.P. 2403; *Syme*! Hopeton, Westmoreland, 1400 ft., *Harris*! Fl. Jam. 9781.—Cuba.

Plant 1½–2 in. high. *Stems* closely clustered, barely exceeding ·5 cm. l., enveloped in loose overlapping brown scarious blunt sheaths. *Leaf* 1–2 cm. l., 5–7 mm. br., margined, with a short stalk (3–8 mm. l.). *Peduncle* 2·5–3·5 cm. l., solitary, longer than the leaf, filiform, with one flower, with a scarious sheathing bract above the middle. *Flower* yellowish, not open (cleistogamous?), about ⅕ in., slightly curved. *Sepals* 5 mm. l., lanceolate-acuminate; median 3-nerved, median nerve keeled; lateral connate almost to the apex, 2-keeled, gibbous at the base. *Petals* 2·75 mm. l., erect, suboblanceolate, 3-nerved, a little more than half as long as the sepals. *Lip* yellow, 2·5 mm. l., subrhomboid, obtusely acuminate above, 3-nerved. *Column*, including the anther, 2 mm. l., erect, a little shorter than the petals; clinandrium, with a lateral tooth and prolonged anteriorly on both sides into a slightly curved tooth. *Capsule* 6 mm. l., oblong.

17. **P. testifolia** *Lindl. in Ann. & Mag. Nat. Hist. ser.* 3, *i.* 328 (1858); stems very short from a creeping rhizome; leaf oval or roundish oval, 7–9-nerved, sessile at the apex of the stalk-like stem, minutely 3-toothed at the obtuse apex; sepals purple, oblong, 3-nerved, covered with white hairs, median obtuse and minutely apiculate, lateral connate above the middle, gibbous at base, connivent above over the lip and column, concave, narrowing to a shortly apiculate apex.—*Lindl. Fol. Orch. Pleuroth.* 43; *Griseb. op. cit.* 609; *Cogn. op. cit.* 414. Epidendrum testæfolium *Sw. Prodr.* 122 (1788). Cymbidium testæfolium *Sw. in Nov. Act. Upsal. vi.* 71 (1799) & *Fl. Ind. Occ.* 146.

In fl. Sept.–Feb.; *Wilson*! *Purdie*! *March*! Bethabara, *Wullschlaegel*, 1082! J.P. 2122, *Syme*! Dollwood; Lancaster, 2200–2500 ft.; near Mabess River, 4000 ft.; near John Crow Peak; *Harris*! Fl. Jam. 7548, 7551, 7787, 10,484.—Cuba, Martinique, Venezuela.

Plant creeping, branch with leaf to 1¼ in. l. *Rhizome* long, creeping, rooting at the nodes, ringed, 1 mm.—or a little more—thick; sheaths about equal to internodes, membranous, obtuse, keeled. *Secondary stems* reduced, distant along rhizome, usually at every third node, 1 cm. or less apart, about 5 mm. l., enveloped in a pair of overlapping truncate sheaths. *Leaf* 20–30 mm. l., 10–20 mm. br. *Flower* purple, nearly ¼ in. l., single, subsessile, subtended by a sheath; sheath ultimately about 6 mm. l., persistent, accrescent, conduplicate, with a lacerate-winged keel. *Ovary* covered with white hairs. *Sepals*, median 5·5 mm. l., 1·5 mm. br., lateral 6 mm. l. *Petals* 2·5–3 mm. l., erect, not quite half the length of the sepals, spathulate, crenulate at upper margin. *Lip* including the narrow claw (·75 mm. l.), 2·75–3 mm. l., broadly oblong-pandurate, shortly auriculate at the base, apex rounded, 3-nerved, margin with long hairs. *Column* slightly shorter than the petals; clinandrium membranous, curled-crenulate.

18. **P. tribuloides** *Lindl. Gen. & Sp. Orch.* 6 (1830); stems very short, clustered; peduncle shorter than leaf-stalk, one or a few together, 1-flowered; capsule covered with soft prickles.—*Lindl. Fol. Orch. Pleuroth.* 39; *Griseb. op. cit.* 609; *Kränzl. in Xen. Orch. iii.* 131, *t.* 275, *f. III. &* 14–19; *Cogn. op. cit.* 415. Epidendrum tribuloides *Sw. Prodr.* 123 (1788). Dendrobium tribuloides *Sw. in Nov. Act. Upsal. vi.* 83 (1799), *Fl. Ind. Occ.* 1535 & *Adnot. Bot. t.* 1, *f.* 4. Type in Herb. Mus. Brit.

On trees in moist woods; in fl. Aug.–Jan.; in fr. Oct.–Jan.; *Swartz*! *Wilson*! *Macfadyen*! St. Mary, *Purdie*! Beaufort, *Wullschlaegel*, 1080! St. Mary, 2000 ft., J.P. 2402, *Syme*! Mabess River, 2000 ft., *G. Nichols*! Slope of John Crow Peak, 200 ft. above Mabess River, *Fawcett*! Union Hill, near Moneague, *Lady Blake*! Trelawney, *Miss T. M. Barrett*! Mt. Moses; Mabess River; Mt. Airy, 2000 ft.; Belvedere, Hanover, 500 ft.; near Christiana, 2600 ft.; *Harris*! Fl. Jam. 7546, 7648, 10,402.—Cuba, Costa Rica, Mexico.

Plant 1 or 2 in. high. *Stems* about 2 mm. l., closely clustered on a short densely rooted rhizome, and enveloped in scarious sheaths. *Leaf* 3–5 cm. l., 8–10 mm. br., narrowly elliptical to elliptical-oblong, tapering into a long stalk, minutely emarginate. *Peduncle* closely surrounded by short blunt scarious sheaths. *Ovary* densely covered with long papillæ.

Flowers about ¼ in. l., brownish-red. *Sepals* papillose, 6 mm. l.; median nearly 2 mm. br. at base, broadly linear-tapering; lateral 2·5 mm. br. together, 2-nerved, about ⅓-connate, free part acuminate. *Petals* about 2 mm. l., one-third as long as sepals, oval-oblong. *Lip* about as long as petals, short-clawed; limb narrowing from a broader base to a short linear recurved fleshy shortly pubescent tip. *Column* nearly as long as petals, narrowly margined, margins forming a slender tooth at each side near apex. *Capsule* about ·5 cm. l., broadly ellipsoidal.

19. **P. uncinata** *Fawc. in Journ. Bot. xxxiii.* 12 (1895); stem about as long as the leaf; peduncles much shorter than the leaf, three or four together, 1-flowered; leaf 8 or 9 in. l.; sepals and ovary rough.—*Cogn. op. cit.* 416. (Pl. 10.)

On trees and peaty banks; in fl. Feb., June; fr. July; Greenhill, near Newcastle; Ridge below Vinegar Hill, 3900 ft.; *Harris*! Hardware Gap, 4000 ft., *G. Nichols*! Fl. Jam. 10,092.

Plant about 15 in. high. *Stems* 12–24 cm. l., clothed with 5-7 loose scarious flaccid sheaths, terete below, compressed and angular above, striate. *Leaves* 15–23 cm. l., 2·5–3 cm. br., oblong-lanceolate or lanceolate, with numerous prominent nerves. *Peduncles* 3–4 cm. l., rough; pedicels about 6 mm. l.; floral bracts 1·3 cm. l., reaching nearly to top of ovary, sheathing at the base, gradually tapering to an obtuse apiculate apex; spathaceous bract about 2 cm. l. at base of peduncles. *Sepals* purplish, surface rough on inner and outer sides, tomentose on outer side; median 1·7 cm. l., 5 mm. br., oblong, tapering to a subacute apex, 5-nerved; lateral 1·6 cm. l., 5·5 mm. br., free or shortly connate at base, oblong-elliptical, tapering to a subacute apex, 4-nerved. *Petals* about 8 mm. l., 2 mm. br., half as long as the lateral sepals, oblanceolate, apex acute, 3-nerved, purplish above, orange-coloured at base. *Lip* slightly exceeding 7 mm. l., nearly as long as the petals, about 3 mm. br. above, purplish, indistinctly 3-lobed, 3-nerved; middle lobe roundish, margin with minute denticulations; lateral lobes conduplicate, uncinate, each forming a sharp curved hook pointing upwards, with a thickened margin above the hook. *Column* nearly 7 mm. l., narrowly winged in front, with a short foot. *Ovary* about 8 mm. l., about 3·5 mm. br. above, rough. *Capsule* 2·2 cm. l., 1·1 cm. br., ellipsoidal, surface rough and tomentose, with 6 furrows.

20. **P. jamaicensis** *Rolfe in Journ. Bot. xlvii.* 122 (1909); plant 1–1½ in. high; stems clustered, shorter than the leaf; peduncle one-half or one-third as long as the leaf, with one or a few flowers; sepals reddish-purple, lateral very shortly connate at base; petals linear-lanceolate, acuminate.—*Cogn. op. cit.* 422.

Without locality, *Morris*! Flowered at Kew in Sept. 1886 and on subsequent occasions.

Plant 1–1½ in. high. *Secondary stems* clustered, 1–2 cm. l., shorter than the leaf; sheaths tubular, apiculate, veined. *Leaf* shortly stalked, elliptical, obtuse, leathery, 1·5–2·5 cm. l. *Racemes* shortly stalked, with one or a few flowers, 1–1·5 cm. l. *Bracts* sheathing, shortly acuminate, 2 mm. l. *Flowers* light reddish-purple with a darker lip, scarcely ⅓ in. l. *Sepals* somewhat spreading, ovate-lanceolate, acute, 7–8 mm. l. *Petals* a little shorter than the sepals. *Lip* linear, obtuse, recurved, 2 mm. l. *Column* slender, curved, 1 mm. l.

21. **P. delicatula** *Lindl. Fol. Orch. Pleuroth.* 38 (1859); stems almost none, clustered; peduncle flexuose, generally much

longer than leaf, one flower developed at a time, but pedicels of former flowers present (to 16); lateral sepals ⅓-connate.—*Griseb. op. cit.* 608; *Cogn. op. cit.* 423.

On trees in moist woods; in fl. Oct.-April; *Purdie*! near Mabess River, 3500–4000 ft.; near Troy, 2500 ft.; near Christiana, 3000 ft.; near John Crow Peak; *Harris*! Ellen Aire, Moy Hall, *Sullivan*! slope of John Crow Peak, 100 ft. above Mabess River, *Fawcett*! Fl. Jam. 7617, 7826, 8869, 9007, 10,399, 10,476, 10,478.

Plant (without the raceme) ½–¾ in. high. *Stems* about 3 mm. l., very short, closely clustered, filiform, enveloped by the membranous sheath. *Leaf* 1–2 cm. l., including the stalk, oblanceolate, margined, narrowed at the base into a slender stalk. *Racemes* 2·5–4 cm. l., generally much longer than the leaf, but occasionally shorter, clustered, flowering solitary, filiform, flexuose, with several (to 16) flowers, laxly secund or in two ranks. *Bracts* about 1 mm. l., short, half the length of the pedicels, sheathing, obtuse, shortly apiculate. *Pedicels* about 2 mm. l. *Flowers* about ⅛ in., greenish-yellow streaked with crimson or purple. *Sepals* 3·75 mm. l., suberect, acuminate above and somewhat thick, obtuse; lateral narrowly ovate, about ⅓-connate; median oblong or lanceolate-oblong, keeled. *Petals* about 2 mm. l., half as long as the sepals, cuneate, fimbriate-serrulate above, acute. *Lip* about 3 mm. l., a little shorter than the sepals, 3-lobed; middle lobe oblong or oblong-oblanceolate, somewhat thick; much exceeding the short broad incurved serrulate lateral lobes. *Column* a little shorter than the petals, upper margin dilated, serrulate; clinandrium 2 or 3-dentate. *Capsule* 6 mm. l., obovate-elliptical, glabrous.

22. **P. Helenæ** *Fawc. & Rendle in Journ. Bot. xlvii.* 4 (1909); stems very short, clustered; peduncle filiform, much longer than the leaf, few-flowered; sepals caudate, lateral free; lip nearly half as long as sepals.—*Cogn. op. cit.* 424. (Pl. 9, f. 9–14.) Type in Herb. N. York Bot. Gard. and in Herb. Mus. Brit.

On trunks of trees; in fl. July; in fr. Aug.; Mt. Moses, 3500 ft., J.P. 2279, *Syme*! Mabess River, 3500 ft., *Harris*!

Plant 1½–2 in. high. *Stems* very short, clustered; sheaths tubular below, acute. *Leaf*, including stalk, 10–15 mm. l., 2–3 mm. br., oblanceolate-elliptical or oblanceolate-linear, narrowed at the base into the stalk. *Racemes* 1·5–5 cm. l., branched. *Bracts* sheathing, acute or apiculate. *Flowers* ⅛ in. l., pale green. *Sepals* about 3·25 mm. l., pale greenish, 1-nerved, spreading, caudate, glabrous; median ovate, 1 mm. br.; lateral lanceolate, free, narrower than median. *Petals* two-thirds as long as sepals, 2·6 mm. l., ·75 mm. br., pale greenish, ovate, fimbriate on the upper margin, caudate at the apex. *Lip* half as long as the sepals when spread out, 1·5 mm. l., ·75 mm. br., pale greenish marked with crimson, 3-lobed, conduplicate, lateral lobes broadly rounded, fimbriate on the upper margin, middle lobe ovate, obtuse. *Column* lined with crimson, including the conical anther, as long as the lip, slender, thicker above; clinandrium 3-toothed. *Capsule* obovoid-ellipsoidal, 3–4 mm. l., 2 mm. br.

The figures in the plate are from a detailed drawing by Miss H. A. Wood, after whom the plant is named.

23. **P. lanceola** *Spreng. Syst. iii.* 731 (1826); stems very short, clustered; peduncle about as long as the leaf, clustered, only one flowering at a time, 2–4-flowered; lateral sepals connate at base.—*Lindl. Fol. Orch. Pleuroth.* 40; *Griseb. op. cit.* 609; *Cogn. op. cit.* 425. Epidendrum lanceola *Sw. Prodr.* 123 (1788).

Dendrobium lanceola *Sw. in Nov. Act. Upsal. vi.* 83 (1799), *Fl. Ind. Occ.* 1539 & *Adnot. Bot. t.* 2, *f.* 5. Type in Herb. Mus. Brit.

On trees and stones; in fl. and fr. Oct.–Feb.; mountains, *Swartz*! St. George, J.P. 171, 231, *Morris*! Mt. Moses, 3800 ft., J.P. 2091, *Syme*! St. George, Portland, 2500 ft.; near Mabess River, 3000–4000 ft.; near John Crow Peak; *Harris*! Fl. Jam. 7467, 7520, 7742, 10,479.

Plant 1½ to 2½ in. high. *Stems* generally 1–1½ cm. l., sometimes shorter, enveloped in scarious sheaths, filiform. *Leaf* 3–4 cm. l., 3–3·5 mm. br., narrowing gradually below into a distinct stalk, linear-lanceolate, shortly apiculate, margins revolute when dry, midrib prominent on the under side. *Pedicels* ·5 cm. l. *Bracts* about 1·5 mm. l., sheathing, truncate. *Flowers* nearly ¼ in. l., orange or scarlet. *Sepals* 5–6·5 mm. l., lanceolate, obtuse, 3-nerved, keeled, spreading at apex, lateral connate at the base, gibbous. *Petals* 2·5 mm. l., nearly half as long as the sepals, oblong-spathulate, with 1–2 nerves, prominent. *Lip*, as long as the petals, oblanceolate, 3-nerved, reflexed about the middle, with undulate margin there. *Column* scarcely shorter than the petals, winged; clinandrium, margin crenulate. *Capsule* ·5 cm. l., ellipsoidal to obovoid.

24. **P. sertularioides** *Spreng. Syst. iii.* 731 (1826); stems very short, from creeping rhizome; leaf narrowly oval, oval-oblong or linear-spathulate, narrowed into the stalk; sepals straw-coloured or greenish, membranous, lanceolate, acuminate, lateral shortly connate, slightly gibbous.—*Lindl. Fol. Orch. Pleuroth.* 42; *Griseb. op. cit.* 609; *Cogn. op. cit.* 426. Epidendrum sertularioides *Sw. Prodr.* 122 (1788). Dendrobium sertularioides *Sw. in Nov. Act. Upsal. vi.* 83 (1799) & *Fl. Ind. Occ.* 1541. Type in Herb. Mus. Brit.

On trees; in fl. and fr. Jan., Feb.; mountains, *Swartz*! Beaufort, *Wullschlaegel*, 1081 (in part)! below Vinegar Hill Gap, Portland, J.P. 2039, *Campbell*! near Mabess River, *Fawcett*! Belvedere, Hanover, 500 ft.; Troy, 1600 ft.; *Harris*! Walderston, *Mrs. Longstaff*! Fl. Jam. 7569, 10,446, 10,487.—Cuba, Trinidad.

Plant barely an inch high. *Rhizome* less than 1 mm. br., slender, creeping, ringed, rooting at the nodes, covered with membranous sheaths. *Stems* less than 5 mm. l., from 5 to 10 mm. apart, generally springing from alternate nodes or every third node of the rhizome, enclosed by sheaths. *Leaf* 1·5–2·5 cm. l. *Peduncles* solitary, capillary, scarcely as long as the leaf, 1-rarely 2-flowered, with a short, truncate, 2-dentate, spathaceous bract at the base and above the middle, sometimes with 3 bracts; occasionally the peduncles of two preceding flowering seasons persist. *Bracts* barely exceeding 1 mm. l. *Flowers* about ⅛ in. l., straw-coloured or light green. *Sepals* about 4 mm. l., 1 mm. br., 1-nerved. *Petals* 3 mm. l., ·7 mm. br., linear-lanceolate, acuminate, 1-nerved. *Lip* 2·4–2·75 mm. l., ·4–·75 mm. br., sessile, with minute square auricles at base, linear-lanceolate, margin undulate, denticulate about midway, recurved above. *Column* 1·5 mm. l., with a membranous wing. *Capsule* ellipsoidal, 3-keeled, about 9 mm. l.

25. **P. microlepanthes** *Griseb. op. cit.* 610; plant not more than ½ in. high; stems clustered, shorter than the leaf; sheaths with mouth dilated, margined and ciliate; peduncle half as long as leaf; sepals yellow in drying, lateral connate nearly to apex.—*Cogn. op. cit.* 430.

On trees; in fl. Nov.; *Macfadyen*! *Wilson*! *March*! Mabess River, *Harris*. Fl. Jam. 7468 (*fide* Cogniaux).

Stems reduced, glabrous, 2–5 mm. l.; sheaths 2–3, tubular, with a glabrate deltoid mouth. *Leaf* margined, elliptical, or elliptical-roundish, somewhat pointed, tapering into the stalk, 8–6 mm. l., 4–6 mm. br. *Racemes* stalked, clustered, with 2–5 flowers. *Flowers* yellow and purple. *Sepals* ovate, acute, erect, scarcely 1 mm. l., yellow in drying; lateral connate nearly to apex. *Petals* minute, spathulate, half as long as the breadth of the sepals. *Lip* oblong, obtuse, a little shorter than the sepals.

26. **P. foliata** *Griseb. op. cit.* 610; stems clustered, with more than one leaf, much longer than leaf; sheath with mouth dilated, margined and ciliate; peduncle longer than leaf; sepals, lateral connate to apex; petals acute or acuminate; lip undivided.—*Cogn. op. cit.* 431. Type in Herb. Kew.

Without locality, *Wilson*!—Cuba.

A small *plant* 2 to 3 in. high. *Stems* 3–7 cm. l., puberulous at the angles, often somewhat branched or distantly 2-several-leaved. *Leaf* margined, elliptical or obovate, apex somewhat rounded and sometimes minutely apiculate, narrowed into the stalk (2–3 mm. l.), 12–8 mm. l., 5 mm. br. *Racemes* stalked, one or sometimes two, distantly 4–8-flowered, 15–25 mm. l. *Pedicels* 2–3 mm. l. *Bracts* acuminate, 1–1·5 mm. l. *Flowers* yellowish. *Sepals* oblong-lanceolate, erect, 2–3 mm. l., yellow in drying. *Petals* ovate, acute, half as long as the sepals. *Lip* linear, obtuse, puberulous above, a little longer than the petals, dark purple at apex.

27. **P. trilobata** *Fawc. & Rendle in Journ. Bot. xlvii.* 4 (1909); stems clustered; about as long as leaf; sheaths with mouth dilated, margined and ciliate; peduncle much longer than leaf; sepals, lateral ½-connate; petals very obtuse; lip 3-lobed.—*Cogn. op. cit.* 431. (Pl. 8, f. 1–6.) Types in Herb. Mus. Brit., in Jam. Herb. & N. York Bot. Gard. Herb.

On trunks of trees; in fl. Feb. and Sept.; in fr. Sept.; between Newhaven Gap and Vinegar Hill, 4000 ft., *Harris*! Fl. Jam. 10,084. John Crow Peak, *Britton*, 267!

Plant minute, ½ to ¾ in. high. *Stem* 7 mm. l. *Leaf* 7 mm. l., nearly 3 mm. br., elliptical, narrowed at the base into a thin stalk. *Sheaths* 2 or 3, covering the stem. *Scape* 2 cm. l., one or few-flowered, solitary or two, filiform, flexuose. *Bracts* ·6 mm. l., spathaceous, lanceolate, acute. *Sepals*, median 4–4·3 mm. l., 1·5 mm. br., lateral somewhat shorter and much narrower; median elliptical, apex caudate, 3-nerved, lateral 1-nerved. *Petals* 1·6 mm. l., ·7 mm. br., oblong, narrowed at the base, 1-nerved, about one-third as long as the sepals. *Lip* 1·5–1·7 mm. l., nearly as broad, 3-nerved, middle lobe oblong, sometimes crenulate at the apex, a little longer than the rounded lateral lobes. *Column* a little over 1 mm. l.

23. LEPANTHES Sw.

Small or dwarf epiphytic herbs. Stems clustered, thin, erect, covered throughout the length with sheaths, and bearing at the apex a single leaf; sheaths tubular, enlarging towards the margined oblique mouth, and marked with several vertical ridges. Leaf sessile or subsessile, rigid, elliptical or roundish elliptical, margined, minutely 3-toothed at the apex. Racemes axillary, solitary or clustered; flowers few or several, somewhat close together, in two ranks, generally minute. Sepals spreading or erect, more or less ovate, subequal, lateral more or less connate. Petals minute, the short claw adnate to the base of the column, the limb generally much wider than long. Lip adnate at the base or above the base of the column, 2-lobed, the lobes erect, parallel to the column, cuneate, with the upper edges thickened, forming flattened surfaces embracing the column; or sometimes simple, with or without two posterior lobes. Column short; anther terminal, opercular, generally obovate; pollinia 2, waxy, pear-shaped, pedicellate. Capsule obovoid.

Species about 50, on the mountains of tropical America and the West Indies.

§ 1. *Diplocheilus.* Lip dividing near the base into two erect, diverging lobes of which the upper edges thicken into more or less expanded lanceolate surfaces, embracing the column.

A. Breadth of petals shorter than length of sepals.
  I. Breadth of petals longer than upper surfaces of lip.
    a. Posterior lobe of petal subequal to anterior.
      1. Sepals acuminate, lateral ⅓—or less—connate.
        Sepals ciliolate, lateral connate only at base .... 1. *L. pulchella.*
        Sepals glabrous, lateral ⅓-connate 7. *L. divaricata.*
      2. Sepals acuminate, glabrous, lateral ½-connate; petals rhomboidal ...... 9. *L. obtusipetala.*
      3. Sepals acute or obtuse, ½–⅔-connate.
        Sepals ciliolate, lateral ⅔-connate 2. *L. sanguinea.*
        Sepals glabrous, lateral ½-connate.
          Petals deltoid or obversely deltoid.
            Plants 1½–2½ inches high, petals obversely deltoid.
              Sepals acute, crimson ... 3. *L. Wullschlaegelii.*
              Sepals obtuse, yellow ... 4. *L. Woodiana.*
            Plants 4–7 inches high; sepals acute, petals deltoid 10. *L. obtusa.*
          Petals lanceolate .... 5. *L. Loddigesiana.*
      4. Sepals connate nearly to apex ...... 6. *L. rotundata.*
    b. Posterior lobe of petal much larger than anterior .... 8. *L. ovalis.*
  II. Breadth of petals shorter than upper surfaces of lip .... 11. *L. arcuata.*

B. Breadth of petals about as long as length of sepals .... 12. *L. cochlearifolia.*

§ 2. *Haplocheilus*. Lip not dividing near the base, but expanding variously above the petals.

Lip simple in outline.
  Petals 2-lobed .... 13. *L. bilabiata.*
  Petals subulate .... 14. *L. elliptica.*
  Petals small, rounded.
    Sepals free nearly to base .... 15. *L. brevipetala.*
    Sepals ½-connate .... 16. *L. Harrisii.*
Lip prolonged posteriorly into 2 lobes.
  Lip-lobes linear.
    Anterior part of lip rounded ...... 17. *L. tridentata.*
    Anterior part of lip square .... 18. *L. quadrata.*
  Lip-lobes triangular........................ 19. *L. concolor.*

§ 1. **Diplocheilus.** Lip dividing near the base into two erect diverging lobes, of which the upper edges thicken into more or less expanded lanceolate surfaces, embracing the column.

1. **L. pulchella** *Sw. in Nov. Act. Upsal. vi.* 86, *t.* 5, *figs.* 6, *c, d, e* (1799); sepals yellow, lateral with a crimson spot at base, median crimson along midrib; ciliolate, caudate-acuminate, lateral connate only at base; posterior lobe of petal subequal to anterior, breadth of petals shorter than length of sepals, longer than upper surfaces of lip.—*Sw. Fl. Ind. Occ.* 1559 & *Adnot. Bot. t.* 2, *fig.* 3; *Griseb. Fl. Br. W. Ind.* 610; *Fawc. & Rendle in Trans. Linn. Soc. vii.* 7, *t. I. f.* 15 (1904). Epidendrum pulchellum *Sw. Prodr.* 125 (1788). Type in Herb. Mus. Brit.

In fl. throughout the year; top of Catherine Peak, *Swartz*! Moody's Gap, J.P. 2031, *Syme*! Mabess River, 3000 ft.; John Crow Peak; Newhaven, 4000 ft.; Newhaven Gap, 5600 ft.; Gap to Vinegar Hill, 4000 ft.; Morse's Gap; ridge above Morse's Gap, 5300 ft.; *Harris*! John Crow Peak, *Fawcett*! Moy Hall, *Sullivan*! Fl. Jam. 7761, 7830, 10,085.

Plant less than an inch high. *Stems* 1 cm. or less in length, about as long as the leaves; mouth of the sheaths acute, ciliolate, ridges glabrous. *Leaves*, including the short stalk, 10–14 mm. l., 5–6 mm. br., oval, acute. *Racemes* 1·5–2 cm. l., exceeding the leaf, generally 2–5-clustered, rarely solitary. *Flowers* about ¼ in. l., several, three or four appearing at the same time. *Sepals* 6–8 mm. l., 2–3 mm. br., ovate, long-acuminate, ciliolate, lateral connate only at the base. *Petals* crimson, inclining to yellow at the ciliolate margin; apex obscure, indicated by an inconspicuous notch or tooth; lobes subequal, subtriangular, varying somewhat in shape, especially in breadth, more or less bluntly and asymmetrically triangular, the outer edges of the two forming an almost continuous line, 2·2–4 mm. br. *Lip* crimson, minutely ciliolate, 1–1·5 mm. br. *Column* to base of clinandrium 1 mm. l., crimson; anther subglobose, emarginate. *Capsule* (unripe), 5 mm. l., 6-keeled. (The flowers of some specimens are only about half the above measurement.)

Swartz's figure in Adnot. Bot. *l. c.* represents a larger plant than any that we have seen, having stems to 2 cm. and leaves to nearly 2 cm. l., 1·8 mm. br.

2. **L. sanguinea** *Hook. in Bot. Mag.* 4112 (1844); sepals magenta-crimson, ciliolate, subacute, lateral ⅔-connate; posterior

lobe of petal subequal to anterior, breadth of petals shorter than length of sepals, longer than upper surfaces of lip.—*Griseb. op. cit.* 611 ; *Fawc. & Rendle op. cit.* 9, *t. I. f.* 16, 17.

In fl. throughout the year; near Mabess River, 4000 ft.; near John Crow Peak; Mt. Diabolo, 2500 ft.; *Harris* ! Fl. Jam. 7616, 7747, 7825.

Plant ½ to 1 in. high. *Stems* 1 cm. or less l., shorter than the leaves; mouth of the sheaths acute, ciliolate, ridges muriculate. *Leaves* 1·2–1·8 cm. l., 5–8 mm. br., subsessile, oval to elliptical, shortly apiculate. *Racemes* 4–6 mm. l., ⅓ or ¼ as long as leaves, several clustered, few-flowered. *Bracts* acute, more or less hirsutulous on the dorsal nerve and margin. *Sepals* 4 mm. l., ovate, median 2·5–3 mm. br. *Petals* ·7 mm. l., 2·5–3 mm. br., rose-coloured and crimson, somewhat equally extended on both sides, lobes oblong or oblong-elliptical, ciliolate, apex forming an almost straight line or slightly retuse. *Lip* deep crimson, upper margins elliptical to elliptical-lanceolate. *Column* short, dilated above; anther obovately orbicular, emarginate.

3. L. **Wullschlaegelii** *Fawc. & Rendle in Journ. Bot. xlvii.* 126 (1909); plant 1½–2½ in. high; leaves green; bracts ciliolate on keel and margins; sepals crimson, glabrous, acute, lateral ½-connate; petals obversely deltoid, apex abruptly apiculate, lobes subequal, glabrous, breadth of petals shorter than length of sepals, longer than upper surfaces of lip.—Type in Herb. Reg. Monac.

Beaufort, *Wullschlaegel*, 1078 !

Plant 1½–2½ in. high. *Stem* longer than leaf, 2·5–6 cm. l., slender, sheaths long, slender, ciliolate on the ridges and at the ovate acuminate mouth. *Leaves* oval, shortly acuminate, 2–2·5 cm. l., about 1 cm. br. *Racemes* shorter than the leaf; flowers many, in two ranks. *Bracts* clasping, broadly ovate, 1–1·2 mm. l. *Sepals* ovate, lateral, nearly 2 mm. l., 1·75 mm. br.; median nearly 2 mm. l., 1·2 mm. br. *Petals* 1·3–1·4 mm. br. *Lip*, upper surfaces lanceolate, ·6 mm. l. *Column* long, ·8 mm. l. *Capsule* small, obovoid, 2·5 mm. l.

Closely resembles *L. divaricata* in habit, but the structure of the flower brings it near to *L. obtusa* and *L. Woodiana*.

4. L. **Woodiana** *Fawc. & Rendle op. cit.* 6 ; plant 1½ in. high; bracts glabrous; sepals transparent yellow, glabrous, obtuse, lateral ½-connate; petals obversely deltoid, apex short, obtuse, lobes subequal, glabrous, breadth of petals shorter than length of sepals, longer than upper surfaces of lip.—(Pl. 11, f. 12–17.)

On trunks of trees; in fl. Feb.; Holly Mount, Mt. Diabolo, 2600 ft., *Harris* !

Stem to 3·5 cm. l., generally longer than the leaf; sheaths very acute, muriculate on the ridges. *Leaf* generally 8–14 mm. l., elliptical, subsessile. *Racemes* several, clustered, shorter than the leaf. *Bracts* somewhat acute. *Flowers* $\frac{1}{12}$ in. l., yellow, several. *Sepals* broadly ovate, median 2 mm. l., lateral barely as long as median. *Petals* slightly over 1 mm. br., a deeper shade of yellow than the sepals, shaded with crimson. *Lip* ·5 mm. l., crimson, upper margins elliptical to elliptical-lanceolate. *Column* deep pink, short, dilated above; anther case white.

The specific name is given in honour of Miss H. A. Wood.

5. **L. Loddigesiana** *Reichb. f. Xen. Orch. i.* 145 (1856); sepals yellow and purple, glabrous, acute, lateral ½-connate; petals lanceolate, downy, lobes subequal, anterior obtuse, posterior acute, breadth of petals shorter than length of sepals, longer than upper surfaces of lip.—*Fawc. & Rendle in Trans. Linn. Soc. vii.* 8 (1904). L. tridentata *Lindl. Bot. Reg. t.* 1762 (1835) (non *Swartz*); *Griseb. op. cit.* 610, "a form with reduced stems."

Highest parts of mountains.

Plant 1–3 in. high. *Stems* varying from 4 mm. to 5 cm. l., generally shorter than the leaves; sheaths smooth. *Leaves* 3·7 cm. l., oval, with shortly acute apex, tapering at the base into an obscure petiole. *Racemes* about half as long again as the leaves, solitary, 1–2-flowered. *Sepals* equal, ovate, median purple with a yellow edge, lateral yellow, slightly tinged with purple, the surface frosted with brilliantly glittering tubercles. *Petals* half as long as the sepals, deep purple. *Lip* downy, yellowish, lower part tinged with purple. *Column* short, with two small lateral horns.

We have not seen this plant. The description is compiled from the account and figure in the "Botanical Register," where it is assigned by Lindley to *L. tridentata* Sw. It differs widely from that species in the form of the petals and lip, which resemble somewhat those of *L. pulchella.*

6. **L. rotundata** *Griseb. op. cit.* 610; sepals, lateral connate nearly to apex; petals obversely deltoid, breadth of petals shorter than length of sepals, longer than upper surface of lip.—*Fawc. & Rendle op. cit.* 6.

Without locality, *Macfadyen.*

Plant about 2 in. high. *Stems* much longer than the leaf; sheaths with a subulate minutely ciliate limb. *Leaf* 12–16 mm. l., 8 mm. br., elliptical. *Racemes* clustered, sometimes solitary, longer than the leaf. *Flowers* about ⅛ in. l. *Sepals* deltoid, lateral cohering below the 2-dentate apex. *Petals* much smaller than the sepals. *Lip*, lobes diverging, spathulate, adnate to the middle of the column.

We have not seen this plant.

7. **L. divaricata** *Fawc. & Rendle op. cit.* 11, *t. II.* 27–29; sepals light yellow shaded with crimson down the centre, glabrous, acuminate, lateral ⅓-connate; posterior lobe of petals subequal to anterior, breadth of petals shorter than length of sepals, longer than upper surfaces of lip.

In fl. Aug.–April; Blue Mt. Peak, J.P. 14 (in part); Cinchona, J.P. 16 (in part); *Morris*! Moy Hall, *Sullivan*! John Crow Peak, 5500 ft.; Newhaven, 4000 ft.; ridge from Newhaven Gap to Vinegar Hill, 4000 ft.; near Woodcutters' Gap, 4000 ft.; ridge above Morse's Gap, 5800 ft.; Hardware Gap, 4500 ft.; Blue Mt. Peak, 7000 ft.; *Harris*! Portland Gap; Morse's Gap, 4800 ft.; John Crow Peak; *Fawcett*! Retreat, J.P. 2031, *Miss T. M. Barrett*! Fl. Jam. 7758, 7832, 7833, 10,082, 10,127, 10,459.

Plant 3–8 in. high. *Stems* 2·5 cm. l., rarely longer (to 15 cm.), much longer than the leaves; mouth and ridges of the sheaths minutely ciliolate. *Leaves*, including the short stalk, generally 2–3 cm., rarely to 4 cm. l., 8–13 mm. br., often purple beneath, elliptical, acuminate. *Racemes* 2–3·5 cm. l., several, clustered, sometimes shorter, sometimes longer than the leaves; flowers several, close together. *Sepals* very variable in size, 4–6 mm. l., 2–4 mm. br., ovate, acuminate; lateral ⅓-connate, apices diverging. *Petals* 2·5–3·5 mm. br., apex obscure, extended nearly equally on both sides; lobes ·75–1·5 mm. br., linear to linear-oblong, obtuse; anterior half orange, posterior half crimson, the crimson colour sometimes extending nearly to the edge. *Lip* crimson, with more or less orange shading, upper surfaces lanceolate, anterior acute, posterior obtuse. *Column* crimson; anther light-purplish. *Capsules* 5 mm. l., 4 mm. br., obovoid, with 6 ridges.

The colour of the sepals varies with age; when the flower opens, it is a transparent light yellow slightly shaded with crimson down the centre; as the flower grows older the crimson shading extends almost to the edges showing very little of the yellow. When the flower opens, the tips of both lobes of the petals overlap, then they straighten, but in the old flower converge again; when the lobe-tips converge, the apex of the petal appears distinctly acute, but when the petals are straight, the apex is not evident.

Var. **minor** *Fawc. & Rendle loc. cit.*

In fl. Sept., Moy Hall, *Sullivan*!

Plant smaller. *Stems* to 3 cm. l. *Leaves* broadly elliptical, very shortly acuminate, rarely exceeding 1·5 cm. l., 1 cm. br. *Racemes* to 2 cm. l., slightly longer than the leaf. *Sepal*, median 3 mm. l., barely 2 mm. br.; lateral 3·2 mm. l. *Petals* less than ·5 mm. l., 2·5 mm. br. Upper surfaces of lip-lobes lanceolate, acute, a little over 2 mm. l.

8. **L. ovalis** *comb. nov.*; stem many times (3–5) longer than leaf; leaf 4–7 cm. l.; sepals tawny yellow, sometimes flushed with crimson, deltoid, acute, lateral ⅓-connate; petals, posterior lobe much larger, elliptical, obtuse, anterior lanceolate, acute; breadth of petals shorter than length of sepals, longer than upper surfaces of lip; lip, upper surfaces scarcely longer than the anterior lobes of the petals, lanceolate-subulate, apex pointing abruptly upwards.—L. concinna *Sw. in Nov. Act. Upsal. vi.* 85 (1799); *Fl. Ind. Occ.* 1557 & *Adnot. Bot. t.* 2, *f.* 4; *Griseb. op. cit.* 611; *Fawc. & Rendle op. cit.* 10, *t. II. f.* 22–25. Epidendrum ovale *Sw. Prodr.* 125 (1788).

In fl. Oct.–Feb.; summits of the mountains, *Swartz*! Cinchona, J.P. 16 (in part), *Morris*! Mt. Moses, J.P. 2268, *Syme*! Green Hill Wood, 3500 ft.; Mabess River, 3500–4000 ft.; John Crow Peak; north side below John Crow Peak, 5000 ft.; below Vinegar Hill, 3900 ft.; Morse's Gap, 4800 ft., and ridge above the Gap, 5500 ft.; Holly Mount, Mt. Diabolo, 2600 ft., 3000 ft.; *Harris*! John Crow Peak, *Fawcett*! Greenhill, 3000 ft. *Moore*! Fl. Jam. 7755, 7767, 7782, 8909, 9782, 9892. 10,091.—Cuba.

Plant sometimes as much as 10 in. high. *Rhizome* short, woody. *Secondary stems*, sometimes as much as 20 cm. l., barely 1 mm. br., slender, wiry, erect or suberect; sheaths thin, minutely ciliolate on the acute mouth and ridges; by the perishing of the sheaths, the older stems often become bare, exposing the cylindrical internodes, which often have a reddish tinge. *Leaves*, generally 4–7 cm. l., 1·5–2·8 cm. br., elliptical, acuminate, shortly stalked. *Racemes* to 5 cm. l., shorter than the leaves; besides the flowering peduncle, a number (sometimes as many as 15) of dried peduncles of former flowering seasons often form a dense cluster in the leaf-axil. *Bracts* glabrous, slightly acuminate. *Flowers* 8–16, in two close ranks in the upper 6–12 mm. *Sepals* deltoid, acute, 2·5–4 mm. l.; median 2·5–3·75 mm. br.; lateral 3–4·5 mm. br. *Petals* 2·5–3 mm br., posterior lobes 1·5–1·6 mm. × 1–1·3 mm.; orange, shaded on the inner side with crimson or purple; anterior lobes 1·4–1·5 × ·75 mm., yellow. *Lip* crimson or purple; upper surfaces ·75–1·1 mm. l. *Column* short, crimson, dilated above; anther obovate, apex retuse. *Capsule* 4 mm. l., 3 mm. br., obovate, gibbous at the apex, the pedicel elongating considerably in the fruit (7 mm. in our specimen).

9. **L. obtusipetala** *sp. nov.*; plant 4–7 in. high; stem many times (2–3) longer than leaves; leaf 3–5 cm. l.; bracts glabrous; sepals deltoid, acuminate, lateral ½-connate; petals rhomboidal, apex mucronate, lobes subequal, somewhat square, breadth of petals shorter than length of sepals, longer than upper surfaces of lip; lip, upper surfaces longer than anterior lobes of petals, narrowly oblanceolate, tapering both anteriorly and posteriorly to an acute point.—L. concinna *Sw.* var. obtusipetala *Fawc. & Rendle op. cit.* 11 (1904). Types in Herb. Mus. Brit. and Jam. Herb.

In fl. June; J.P. 14 (in part) *Morris*! Portland Gap, 5550 ft., *Fawcett*!

Plant much like *L. ovalis*, differing in some details given above, and also as follows: 4–7 in. high. *Stems* 7–14 cm. l. *Leaves* 1·3–2 cm. br. *Racemes* 1–2 cm. l., much shorter than the leaves, few (3–5) in a cluster, and most of them with one flower open at the same time, pedicels of 2–4 older flowers remaining on each peduncle. *Sepals*, median 4·5–5 mm. l., 3·7–4 mm. br., lateral 5 mm. l., together 4·5 mm. br. *Petals* 1·2 mm. l., 2·3–2·5 mm. br., posterior lobe a little larger than the anterior. *Lip*, upper surfaces of lobes 1·5–2 mm. l. (The flowers on some specimens measure scarcely more than half the above.)

On reconsideration, we are inclined to regard this as a distinct species.

10. **L. obtusa** *Fawc. & Rendle op. cit.* 11, *t. II. f.* 26; plant 4–5 in. high; stems (1½–2 times) longer than leaves; leaf 3·5–5·5 cm. l.; bracts glabrescent; sepals crimson, glabrous, deltoid, acute, lateral ½-connate; petals somewhat deltoid, apex obtuse, surface minutely pubescent, lobes subequal, posterior somewhat rhomboidal, anterior obtusely triangular, breadth of petals shorter than length of sepals, longer than upper surfaces of lip; lip, upper surfaces a little longer than the anterior lobes of petals, posterior half oblong, anterior tapering to an acute apex.

In fl. Nov.–April; Blue Mt. Peak, J.P. 14 (in part), *Morris*! Greenhill, 3000 ft., *Moore*! Portland Gap Road, *Fawcett*! Newhaven Gap, 4000 ft.; between Morse's Gap and Vinegar Hill, 4000 ft.; Newhaven, 4000 ft.; ridge from Newhaven Gap to Vinegar Hill, 4000 ft.; *Harris*! Fl. Jam. 7834, 10,081.

Plant much like *L. ovalis* but smaller. *Stems* slender, 10–12 cm. l., much longer than the leaves. *Leaves* 1·2–2 cm. br., narrowly elliptical, acuminate, bright violet on lower side. *Racemes* 2 cm. l., scarcely half as long as the leaves, few (3–6) in a cluster, of which two or three have one (sometimes two) flowers open at same time. *Bracts* sheathing, shortly acuminate, about 2 mm. l. *Sepals*, median 5–6 mm. l. and br.; lateral,

5·5–6 mm. l., 6·3–6·5 mm. br. below the point of union. *Petals* 3·5 mm. br., crimson. *Lip* crimson. *Column* short, dilated above. *Capsule* 5 mm. l., broadly obovoid. (The flowers on some specimens scarcely measure more than half the measurements given above.)

11. **L. arcuata** *Fawc. & Rendle in Journ. Bot. xlviii.* 6 (1909); sepals magenta-crimson, lateral ⅓-connate; breadth of petals shorter than length of sepals, and shorter than upper surfaces of lip.—(Pl. 11, f. 1–7.) Types in Herb. Mus. Brit. and Jam. Herb.

On trunks of trees; in fl. and fr., Feb.; Holly Mount, Mt. Diabolo, 2600 ft., *Harris*!

Plant 1 to 1½ in. high. *Stem* 1·5–2·5 cm. l., longer than the leaf; sheaths ciliolate on the mouth and ridges. *Leaf* 1·2–1·7 cm. l., narrowly-elliptical to roundish-elliptical. *Racemes* with several flowers, clustered, shorter than the leaf. *Bract* shortly acute, glabrescent. *Flowers* about $\frac{1}{10}$ in. l. *Sepals*, median 2·3 mm. l., ovate-lanceolate, acuminate; lateral 2 mm. l., broadly ovate, acuminate, 1-nerved. *Petals* brick-orange-yellow, barely 1 mm. br., 3-lobed, lobes subequal, obtusely triangular. *Lip* brick-orange-yellow tinged with crimson, 1·5 mm. l., upper surfaces of lateral lobes much longer than the petals and column, 1 mm. l., linear-lanceolate, subfalcate. *Column* short, ·6 mm. l., of a rather lighter shade of magenta-crimson than the sepals; anther white.

This species is near *L. tridentata* in appearance, but differs in the lip, the short column, and in other respects. *Petals* in breadth barely 1 mm., while those of *L. tridentata* are 1·5 mm. broad. *Lip* 1·5 mm. l., the lateral lobes nearly twice as long as the column, being 1 mm. l., whereas the lip of *L. tridentata* is just over 1 mm. l., the lateral lobes being barely 1 mm. l. The column is ·6 mm. l. as contrasted with that of *L. tridentata*, which is 1 mm. l.

12. **L. cochlearifolia** *Sw. in Nov. Act. Upsal. vi.* 86, *t.* 5, *figs.* 6, *a*, *b* (1799); stem much longer than leaf; leaf roundish, 1·5–2·5 cm. l.; sepals purple, lateral ⅓-connate; breadth of petals about as long as length of sepals.—*Sw. Fl. Ind. Occ.* 1563 & *Adnot. Bot. t.* 2, *fig.* 1; *Griseb. op. cit.* 611; *Fawc. & Rendle in Trans. Linn. Soc. vii.* 9, *t. II. f.* 18–21. Epidendrum cochlearifolium *Sw. Prodr.* 126 (1788). Type in Herb. Mus. Brit.

In fl. Sept.–April; summits of the Blue Mts., *Swartz*! Swift River, J.P. 468 (2679); Mt. Moses, 3500 ft., J.P. 2373; *Syme*! Browns Town, *Miss T. M. Barrett*! John Crow Peak, slopes of spur, 100 to 300 ft. above Mabess River, *Fawcett*! near Mabess River, 3500–4000 ft.; near John Crow Peak; *Harris*! Fl. Jam. 7788, 7828, 9015.

Plant less than 3 in. high. *Stems* 2–6 cm. l.: mouth of the sheath broadly open, apex acute; margin and ridges covered with numerous stiffish white hairs. *Leaves* 1–1·8 cm. br., sometimes tinged with purple, varying in shape from orbicular to broadly elliptical, shortly apiculate, narrowing at the base into a short stalk, the marginal and three principal veins uniting below the apex. *Racemes* generally less than 1 cm. l., less than half as long as the leaves, generally several clustered, lax, with few flowers. *Sepals* 5–6 mm. l., 3–3·5 mm. br., ovate, acuminate. *Petals* brick-orange-yellow, tinged with crimson towards the centre, with a short blunt apex and a short rhomboidal body ·65 mm. l., shortly toothed at the lower corners and drawn out at the upper into longer tapering upcurving horns; 6 mm. br., including the horns. *Lip* purplish-crimson; upper surfaces lanceolate. *Column* short, dilated under the apex; anther obovately rectangular, emarginate.

§ 2. **Haplocheilus.** Lip not dividing near the base, but expanding variously above the petals.

13. **L. bilabiata** *Fawc. & Rendle op. cit.* 4, *t. I. f.* 1–4; sepals, median dark crimson-purple, lateral reddish, $\frac{2}{3}$-connate; petals 2-lobed; lip suborbicular, cordate.

In fl. Sept.-Feb., above Mabess River, 3000 and 4000 ft., *Fawcett*! *Harris*! near Hardware Gap, 4300 ft; ridge below Vinegar Hill, 3800 ft.; *Harris*! Fl. Jam. 7786, 10,095.

Plant $\frac{3}{4}$ to $1\frac{1}{2}$ in. high. *Stems* 1–2·5 cm. l., rarely shorter, thin, usually exceeding the leaf; sheaths with acuminate mouth ciliolate, ridges muriculate. *Leaf* narrowly elliptical, narrowing at the base into a short stalk 2–3 mm. l., blade 10–17 mm. l., 4–7 mm. br., obscurely apiculate. *Racemes* often several clustered, half as long as the leaf, lax, with a few flowers. *Bracts* subacute, almost glabrous. *Sepals*, median 3 mm. l. and br., concave and shaped like an open basket, mucronate; lateral 2·5 mm. l., forming an anterior bifid lip, like the dorsal but scarcely as wide, apices acute. *Petals* about 3 mm. br., apex obscure, the anterior lobe whitish-yellow, shortly subulate; the posterior lobe larger, dark crimson-purple. *Lip* about 2 mm. l., and somewhat broader when flattened; whitish-yellow, concave, 2- or 3-denticulate at apex, enclosing the column. *Column* short, dilated above; anther obovate, deeply emarginate. *Capsule* subglobose, gibbous, with six conspicuous keels, markedly asymmetrical.

14. **L. elliptica** *Fawc. & Rendle in Journ. Bot. xlviii.* 5 (1909); sepals crimson, lateral $\frac{1}{3}$-connate; petals subulate; lip elliptical, apiculate.—(Pl. 11, f. 18–21.)

On trunks of trees; in fl. Sept.; Holly Mount, Mt. Diabolo, 2700 ft. *Harris*! Fl. Jam. 9895a.

Plant 1 to $1\frac{1}{2}$ in. high. *Stem* 1–3 cm. l., exceeding the leaf; the large spreading hirsute sheaths and stems are much like those of *L. cochlearifolia*. *Leaf* 1–2 cm. l., ·5–1·1 cm. br., sessile, elliptical, narrowing at the base. *Racemes* 2–several clustered or solitary, shorter than the leaf, with few flowers, bracteoles ciliolate at length glabrate. *Flowers* about $\frac{1}{4}$ in. l., crimson. *Sepals* 3·5 mm. l., the median slightly over 2 mm. br., triangular-ovate, shortly acuminate, lateral similar to median. *Petals* only developed anteriorly into a subulate lobe reaching in bud to about the top of the anther. *Lip* small, simple, springing from the column. *Column* (without anther) 1 mm. l.

15. **L. brevipetala** *Fawc. & Rendle loc. cit.*; sepals crimson, lateral shortly connate; petals very short, roundish-ovate, obtuse; lip large, simple, semi-circular, cordate.—(Pl. 11, f. 22, 23.) Types in Herb. Mus. Brit. and in Jam. Herb.

On trees; in fl. and fr. Dec.; J.P. 17, *Morris*! Silver Hill, Woodland, 4000 ft., *Harris*! Fl. Jam. 7539.

Plant 2 to $3\frac{1}{2}$ in. high. *Stem* 3–7 cm. l., robust for the genus, exceeding the leaf; sheaths large, spreading, much like those of *L. coch-*

laurifolia, ferruginous-hirsute, mouth broadly open with acute apex. *Leaves* 2·5–4 cm. l., ·7–1 cm. br., narrowly elliptical, obtuse. *Racemes* generally several clustered, often a little longer than the leaf. *Flowers* very small, $\frac{1}{18}$ in. l., crimson, several. *Sepals*, median 1·4 mm. l., broadly ovate, obtuse, concave; lateral 1·2 mm. l., subacute. *Petals* ·3 mm. l., simple. *Lip* 1 mm. l., a little more than 1 mm. br., concave, strongly 3-nerved. *Column* short, about ·5 mm. l. *Capsule* 4 mm. l., ellipsoidal, with three narrow wings.

16. L. **Harrisii** *Fawc. & Rendle loc. cit.*; sepals dark crimson-purple, lateral $\frac{1}{2}$-connate; petals short, shortly elliptical, rounded at apex; lip ovate-roundish, subcordate, rounded at apex.—(Pl. 11, f. 8–11.)

On trees; in fl. and fr. Feb.; Holly Mount, Mt. Diabolo, 2600 ft., *Harris*!

Plant about $1\frac{1}{2}$ in. high. *Stem* 3–4 cm. l., longer than the leaf; sheaths robust, ferruginous-hirtellous on the broad mouth and ridges. *Leaf*, blade 1·5–2 cm. l., 7–9 mm. br., elliptical. *Racemes* few-flowered, longer than the leaf; bracts acute, glabrous. *Flowers* $\frac{1}{10}$ in. l., dark crimson-purple, less caducous than usual. *Sepals* 2·5 mm. l., median long-ovate, 3-nerved; lateral lanceolate, 1-nerved. *Petals* ·7 mm. l., like those of *L. brevipetala*, but twice as long. *Lip* 1·3 mm. l., strongly 3-nerved. *Column* short, ·5 mm. l. *Capsule* ellipsoidal, 6-keeled.

17. L. **tridentata** *Sw. in Nov. Act. Upsal. vi.* 86 (1799); sepals yellowish, lateral $\frac{1}{2}$-connate; petals subrhomboid but extended transversely, truncate at the base, with the upper margins slightly concave, the posterior lobes embracing the column and meeting behind it; lip anteriorly rounded and suddenly mucronate, posteriorly with two erect awl-shaped lobes lying close to the column and of equal length to it.—*Sw. Fl. Ind. Occ. iii.* 1561 & *Adnot. Bot. t.* 2, *f.* 2; *Griseb op. cit.* 610, *excl. syn.*; *Fawc. & Rendle in Trans. Linn. Soc. vii.* 6, *t. I. f.* 8–11. Epidendrum tridentatum *Sw. Prodr.* 125 (1788). Type in Herb. Mus. Brit.

In fl. throughout the year; Blue Mts., *Swartz*! near Mabess River, 4000 ft.; Morse's Gap, 5000 ft.; near John Crow Peak, 5200 ft.; *Harris*! near John Crow Peak; near Newhaven Gap; *Fawcett*! Moy Hall, *Sullivan*!—Cuba.

Plant $1\frac{1}{2}$–3 in. high. *Stems* 2–7 cm. l., very much longer than the leaves; sheath margins ciliolate, ridges muriculate. *Leaves* somewhat elliptical, and tapering at base and apex, 1·5–2·4 cm. l., 5–7 mm. br. *Racemes* a little shorter than the leaf, 2–4-clustered, rarely solitary. *Flowers* numerous, in two ranks, closely crowded in the upper half or third of the raceme. *Sepals* 2·5 mm. l., ovate-acuminate; median 1 mm. br.; lateral 1·5 mm. br. at the point of union. *Petals* yellowish, suberect. *Lip* crimson. *Column* 1 mm. l., pink, scarcely dilated at the apex; anther white, roundly obcordate. *Capsule* 5 mm. l., subglobose, with prominent deep-purple keels.

Swartz describes the flower as purple-yellow.

18. L. **quadrata** *Fawc. & Rendle op. cit.* 7, *t. I. f.* 12–14; sepals purplish-crimson, lateral $\frac{1}{3}$-connate; petals with a short

blunt apex, the lobes of equal length, posterior oblanceolate, converging, anterior linear; lip consisting of a large anterior subquadrate portion attached by a very short insertion on the column-base, with a shallowly notched broadly rounded apex, and produced behind into a pair of bluntly tapering narrow lobes about equal in length to the anterior portion.

In fl. Nov.–April; near John Crow Peak; near Mabess River, 3500 ft., 4000 ft.; Holly Mount, Mt. Diabolo, 2700 ft.; *Harris*! Fl. Jam. 7746, 7824, 7827, 9016, 9898.

Plant about 1½ in. high. *Stems* to 2·5 cm. l., longer than the leaves; mouth of the sheaths spreading, minutely ciliolate; ridges muriculate. *Leaves* 1–1·6 cm. l., 4–9 mm. br., oval to elliptical. *Racemes* generally several clustered, sometimes solitary, not half as long as the leaves. *Bracts* glabrous. *Flowers* few, distant. *Sepals* 3 mm. l., ovate-acuminate, ventricose, margin and nerves ciliolate, lateral a little smaller. *Petals* 1·5 mm. br. *Column* short, dilated above; anther ovoid, rectangular, emarginate. *Capsule* obovoid.

19. L. **concolor** *Fawc. & Rendle op. cit.* 5, *t. I. f.* 5–7; sepals pale yellow, lateral ½-connate; petals obversely deltoid, at the apex unequally concave; lip anteriorly shallowly retuse with a short apiculus, lobes thick, extended backwards and embracing the column.

In fl. Aug.–Feb.; Moy Hall, *Sullivan*! Morse's Gap and Blue Mountains ridge, *Fawcett*! ridge above Morse's Gap, 5400 ft.; Newhaven Gap, 5600 ft.; Rose Hill, 4000 ft.; Newhaven, 4000 ft.; ridge from Newhaven Gap to Vinegar Hill, 4000 ft.; John Crow Peak; above Mabess River; *Harris*! Fl. Jam. 7831, 10,083, 10,465, 10,468.

Plant ½–1½ in. high. *Stems* generally 2–3 cm. l., much longer than the leaf; sheaths conspicuously ridged, the mouth (sharply acute) and the ridges minutely ciliolate. *Leaf*, blade generally 1 to 1·5 cm. l., 5–8 mm. br., oval to roundish oval. *Racemes* 6–8 mm. l., generally shorter than the leaf, 1–3-clustered, lax, with a few flowers, one or two open at a time. *Bracts* glabrate. *Flowers* about ⅛ in. l., pale yellow. *Sepals* 3 mm. l., ovate, glabrous, lateral each barely 1·5 mm. br. at union, median 2·5 mm. br. *Petals* a little over 1 mm. l. and slightly broader, pale yellow. *Lip* 1·5 mm. br., lobes 1 mm. l. *Column* short, dilated above. *Capsule* 4 mm. l., 3 mm. br., obliquely obovoid, 6-keeled.

The plant from Rose Hill differs from the other specimens in having the racemes longer than the leaf; they reach nearly 2 cm. The flowers, however, resemble those of specimens from other localities.

24. BRACHIONIDIUM Lindl.

Epiphytic herbs, forming a long prostrate rhizome, which often branches, and is more or less covered with scarious sheaths. Secondary stems with a solitary leaf, very short below the leaf, covered by one or two scarious sheaths. Leaf somewhat leathery, rather small, contracted at the base into a short petiole. Peduncle solitary, enclosed with the petiole in a scarious sheath with one flower and generally also an aborted bract. Sepals produced at the apex into long tails, lateral connate nearly to apex. Petals similar to sepals. Lip small, jointed to foot of column. Column short, broad, erect, with a foot; clinandrium 3-lobed. Anther opercular, incumbent, somewhat flat, cordate, indistinctly 2-celled; pollinia 6, elongated, clavate, in two bundles. Capsule obovoid.

Species 5 or 6, natives of the Andes of Bolivia and Colombia, of Mt. Roraima in British Guiana, and of the West Indies.

B. Sherringii *Rolfe in Kew Bulletin*, 1893, 4.—B. parvum *Cogn. in Fedde Repert. vi.* 307 (1909), as regards the Jamaican plant. (Pl. 12.)

Amongst mosses on banks, or on trees; in fl. Oct.–Dec.; below Mount Moses, 3000 ft., J.P. 2361 *Syme*! *Harris*! Regale, below Hardware Gap, *A. Moore*! Fl. Jam. 7802, 10,474.—Grenada, Martinique, St. Kitts.

Plant 1–1½ in. high from the creeping or climbing rhizome. *Secondary stem* very short, 6–12 mm. l., enclosed by scarious truncate apiculate sheaths. *Leaf* oblanceolate, or oblong-lanceolate, narrowed into the petiole, 3-denticulate at apex, 7-nerved; blade 1·5–3 cm. l., 4–12 mm. br. *Peduncle* slender, longer than the leaf, 1·5–3·5 cm., springing from a scarious bract, bearing an acuminate sheathing bract at or below the middle, 3–4·5 mm. l., and a third at the apex, 2–3 mm. l., tubular, spreading, enclosing the short thick pedicel of the flower and a slender pedicel 4 mm. l., bearing an aborted bract. *Flowers* claret-coloured, sometimes tinged with tawny-yellow on the sepals, perianth 1 in. l. *Sepals*, median oval, acuminate, caudate, 1·2 cm. l. without tail, 3–4 mm. br.; tail reaching nearly 2 cm. l.; lateral connate, tail 2-fid at apex, similar to median but broader, 4–5 mm. br., tail 1·5 cm. l. *Petals* as long as the sepals, subfalcate, ovate, caudate, margin minutely ciliolate, 1–1·1 cm. l., 2·5–4 mm. br., tail 1·5 cm. l. *Lip*, in the fresh flower, apparently simple, horseshoe-shaped, embracing the column, attached by a rather broad claw to the foot of the column, purple; when spread out under the microscope, it is evidently 3-lobed, with a short, slender claw, 1·5 mm. l., middle lobe transversely oblong-elliptical, at apex mucronate, 3-nerved, 1 mm. br., lateral lobes oblong, incurved over the disk; breadth of lip at lobes 3 mm. *Column* broad, a little over 1 mm. l.

25. ISOCHILUS R. Br.

Stems erect from a creeping rhizome, covered with the sheaths of leaves, no pseudobulbs. Leaves in two ranks, spreading, linear or lanceolate, somewhat rigidly herbaceous, generally obtuse or retuse. Flowers medium-sized, rose-coloured, lake-magenta or crimson, in a terminal dense one-sided raceme, with very short stalks. Bracts concave, much shorter than the flower. Sepals equal, erect, keeled-concave, at the base somewhat saccate. Petals slightly shorter than the sepals, flat. Lip equal to the petals, and with them attached to the base of the column, contracted below the middle, and slightly sigmoid-flexuose, dilated towards the middle. Column erect, somewhat long, semiterete, not winged, at the apex at the sides of the stigma produced into two erect points, with a short foot; clinandrium little prominent, with a short posterior tooth, bearing the anther. Anther terminal, opercular, convex, distinctly 2-celled; cells divided again into two cells by an imperfect longitudinal septum; pollinia 4, waxy, two in each cell, equal, ovoid-oblong, compressed parallel, erect, the appendage granular-viscous, laminiform, ascending from the base, incumbent on the pollinia. Capsule globose or ovoid, ribs only slightly prominent.

Species about 13, natives of warmer America from the West Indies and Mexico to Brazil and Peru.

I. linearis *R. Br. in Ait. Hort. Kew. ed.* 2, *v.* 209 (1813); *Bot. Reg. t.* 745; *Lindl. Gen. & Sp. Orch.* 112; *Griseb. Fl. Br. W. Ind.* 623. Helleborine tenuifolia repens *Plum. Pl. Amer.* (*Burm.*) *t.* 182, *f.* 1. Epidendrum lineare *Jacq. Stirp. Amer.* 221, *t.* 131, *f.* 1 (1763), *Ed. pict. t.* 201. Cymbidium lineare *Sw. in Nov. Act. Upsal. vi.* 72 (1799). (Pl. 13, f. 8–13.)

On trees and rocks; in fl. Sept.–March, in fr. Nov.–March; *Shakespear*! *Swartz*! *Macfadyen*! Petersfield, St. Mary, *McNab*! Radnor, *Purdie*! Fairfield, *Wullschlaegel*, 1063! *March*! *Prior*! Blue Mts. J.P. 32, *Morris*! Catherine Peak, *Eggers*, 3527; near Browns Town, *Miss T. M. Barrett*! Unity Valley, St. Ann, *Miss A. Roper*! Cinchona, 4500 ft., *Fawcett*! *Harris*! Tweedside, south St. Andrew, *Harris*! Fl. Jam. 7752, 7843, 10,460.—Cuba, Hispaniola, Porto Rico, St. Vincent, Guadeloupe, Dominica, Martinique, Trinidad, Costa Rica, Nicaragua, Guatemala, Mexico, Venezuela, Colombia, B. Guiana, Brazil, Argentine, Bolivia, Peru.

Plant 1 ft.–2 ft. 4 in. high. *Stems* 2–6 dm. l. *Leaves* 3·5–6·5 cm. l., 3–6 mm. br.; sheaths 1–2·5 cm. l. *Racemes* 1·5–3·5 cm. l. *Bracts* coloured like the flowers, obovate, 7–10 mm. l. *Flowers* about ⅓ in. l. *Ovary* 2-winged. *Sepals* 2–3 nerves on each side of median curve, lateral with a strong keel, which is decurrent on ovary, and forms a mucro at apex, oblong, tapering near apex, shortly connate at base, about 10 mm. l., 3 mm. br.; median not keeled, oblong, abruptly acute, a little shorter and broader than the lateral. *Petals* 5-nerved, oblanceolate-elliptical, apex obtuse, somewhat oblique at base, a little shorter and narrower than the median sepal. *Lip* 7-nerved, oblong, expanding in middle, tapering near the apex, and slightly tapering to base, curved from base inwards to embrace column and then outwards, 8–10 mm. l., 1–1·25 mm. br. *Column* ·5 cm. l. *Capsule* 6-ribbed, two ribs prominent above, 7–9 mm. l., about 3 mm. br.

26. OCTADESMIA Benth.

Epiphytic herbs, with simple leafy stems. Leaves linear-oblong or broadly lanceolate, in two ranks, somewhat rigid but not fleshy, with numerous fine prominent nerves, jointed with, and deciduous from, the sheath. Peduncle terminal, simple, or slightly branched, the branches laxly few-flowered. Flowers medium-sized, shortly stalked. Bracts short, sheathing at the base. Sepals, all of about the same length, somewhat spreading, lateral somewhat falcate, connected under the lip into a slightly prominent chin. Petals somewhat like the lateral sepals. Lip suberect from the base of the column, free, oblong or roundish,

3-lobed, lateral lobes slightly prominent. Column somewhat long, slightly incurved, concave anteriorly, with a short foot; clinandrium obscurely 3-lobed. Anther terminal, opercular, incumbent, semi-globose, with two cells which are divided again. Pollinia 8, four in each cell, connected at the apices by a small amount of viscid substance. Capsule oblong or elongate, spindle-shaped, contracted into a short beak.

Species 2, natives of West Indies and Brazil.

Plant 1–2 ft. high .... 1. *O. montana.*
Plant reaching to 6 ft. high.................... 2. *O. elata.*

1. **O. montana** *Benth. in Journ. Linn. Soc. xviii.* 311 (1881) & *in Benth. & Hook. f. Gen. Pl. iii.* 525; leaves oblong-lanceolate; lip about as broad as long.—*Cogn. in Fl. Bras. iii.* pt. 5, 19 & *in Symb. Ant. iv.* 172. Epidendrum montanum *Sw. Prodr.* 121 (1788). Cymbidium montanum *Sw. in Nov. Act. Upsal. vi.* 72 (1799) & *Fl. Ind. Occ.* 1469. Octomeria serratifolia *Hook. in Bot. Mag. t.* 2823 (1828). Bletia montana *Reichb. f. in Walp. Ann. Bot. vi.* 445 (1862). Tetramicra montana *Griseb. Fl. Brit. W. Ind.* 622 (1864). (Pl. 17, f. 13–16.) Type in Herb. Mus. Brit.

On trees and rocky banks; in fl. Nov.-March; Blue Mts., *Swartz*! *Wiles*! *Macfadyen*! *Wilson*! *Jenman*! Blue Mts., 5000–5500 ft, J.P. 27, *Morris*! near Gordon Town, *Watt*! Morse's Gap, 5000 ft., *Fawcett*! *Harris*! Ridge from Newhaven Gap to Vinegar Hill, 4000 ft.; Hardware Gap, 4000 ft.; below Sir John's Peak, 6000 ft.; *Harris*! Fl. Jam. 7550, 10,090, 10,128, 10,436.—Hispaniola, Porto Rico.

Plant 15–24 in. high. *Stem* 2–6 dm. l., covered below with sheaths, leafy above. *Sheaths* striate and with brown spots and minute warts, 2·5–3 cm. l., free portion short. *Leaves* 4–11 cm. l., 7–20 mm. br. *Panicle* (sometimes a raceme) 4–20 cm. l. *Bracts* minutely spotted and warty, free portion triangular, acute or acuminate; sterile 1–1·7 cm. l.; floral ·5–1 cm. l. *Ovary* and *pedicel* each 4·5 mm. l. *Flowers* creamy-white, with a delicate scent of violets, perianth over ½ in. l. *Sepals* 5-nerved, oblong-lanceolate, median 1·4–1·8 cm. l., 4·5 mm. br., apex obtuse; lateral rather shorter, apex thickened and shortly apiculate. *Petals* 3-nerved, linear-oblong, tapering to apiculate apex, 1·2–1·5 cm. l., 2·5–3·5 mm. br. *Lip* very shortly clawed, 5-nerved with numerous lateral nerves branching off, roundish in outline, disk 2-lamellate below, with minute warty calli along the three nerves of the terminal lobe, lamellæ 2-lobed; lateral lobes short, obtuse; terminal lobe much larger, minutely crenulate, apiculate, 9–10 mm. l., and about as broad. *Column* narrowly 2-winged, 6–6·5 mm. l. *Capsule* ellipsoidal, 2·5 cm. l.

2. **O. elata** *Benth. & Hook. f. Gen. Pl. iii.* 525 (1883); leaves broadly lanceolate; lip longer than broad.—Type in Herb. Kew.

On trees, in rocky soil, and in peaty mould in forest; in fl. March; Mt. Moses, 3500–4000 ft., *Syme*! *Harris*!

Plant robust, erect, upright, reaching to 6 ft. high. *Stem* 15 or 16 dm. l., below bearing the fibrous remains of the leaf-sheaths, leafy above. *Leaves*, passing above into spathaceous bracts, blade 8–11 cm. l., 2–3 cm. br. *Panicle* 2 dm. l. *Bracts*, sterile 4–2 cm. l., floral about 1 cm. l. *Ovary* (fertilised) and *pedicel*, each generally about 1 cm. l. *Flowers*, perianth over ½ in. l. *Sepals* 7-nerved, narrowly oblong, slightly apiculate;

median 1·75 cm. l., 4 mm. br.; lateral keeled and slightly hooded towards apex, about 1·5 cm. l., barely 4 mm. br. *Petals* 5-nerved, narrowly oblong, slightly keeled and hooded, 1·5 cm. l., 3·5 mm. br. *Lip* sessile, 7-nerved with lateral nerves, obovate-oblong in outline, 1 cm. l., 7 mm. br., disk 2-lamellate below, lamellæ and prominent middle nerve extending as prominent lines towards apex, lateral lobes rounded, terminal lobe much larger, at apex 3-lobulate, apiculate. *Column* about 7 mm. l.

27. CŒLIA Lindl.

Epiphytic herbs. Stems with several leaves, increasing at the base to form fleshy pseudobulbs. Leaves long, narrow, plaited, veiny. Scapes at the base of the pseudobulbs, short, simple, with imbricate subspathaceous sheaths at the base. Flowers medium-sized, densely racemose, shortly pedicellate. Bracts membranous or paleaceous, narrow, longer than the flowers. Sepals, all of about the same length, erect from the base, spreading above; median free, concave; lateral broader, connate at the base with the foot of the column, forming a short or long chin. Petals somewhat similar to the median sepal. Lip jointed at foot of column, narrow, incumbent at the base, spreading above, undivided, contracted into a flexuose claw. Column short, rather broad, acute, margined, produced at the base into a foot. Anther terminal, opercular, incumbent, semi-globose, perfectly 2-celled; pollinia 8, waxy, four in each cell, ovoid, cohering at the apex by a sparse viscid substance. Capsule prominently winged.

Species about 6, natives of the West Indies, Central America and Mexico.

C. triptera *G. Don ex Steud. Nom. ed.* 2, *i.* 394 (1840).—*Griseb. Fl. Br. W. Ind.* 629. C. Bauerana *Lindl. Gen. & Sp. Orch.* 36 (1830), *in Bauer Ill. Orch. Pl. Gen. t.* 3 & *Bot. Reg. xxviii. t.* 36 (1842). Epidendrum tripterum *Sm. Ic. Pict. t.* 14 (1793). Cymbidium tripterum *Sw. in Nov. Act. Upsal. vi.* 70 (1799). *Salisb. Ic. ined. vi. fol.* 571. (Pl. 13, f. 1–7.) Type in Herb. Smith in Linn. Soc.

On rocky banks; in fl. Jan.–Feb., in fr. March; J.P. 486, *Morris*! near Browns Town, *Miss T. M. Barrett*! Chester Vale, 3000 ft.; Silver Hill Woodland; *Harris*! Fl. Jam. 7614.—Cuba, Mexico.

Plant 1 ft. and more high. *Pseudobulbs* ovate-oblong, 3–5 cm. l. *Leaves* 3 or 4 from apex of pseudobulb, linear-lanceolate, long-tapering to base where they are articulated with closed sheaths, 2–4 dm. l. (incl. sheath), 1–2 cm. br. *Scape* 10–14 cm. l. *Bracts* linear-lanceolate, long-acuminate, 2–2·5 cm. l. *Flowers* white, perianth a little over ¼ in. l. *Ovary* with nine wings, three very broad, about 1 cm. l. *Sepals* 5-nerved, ovate, obtuse, about 7·5 mm. l., and 3 mm. br. *Petals* oblong, concave, obtuse, about 7 mm. l., nearly 2 mm. br. *Lip* with seven branching veins, oblong below, expanding to form an obscure lobe on each side about the middle, tapering to an obtuse apex, with a callus along the upper margin as far as the lobes, about 6 mm. l., more than 3 mm. br. at lobes. *Column* scarcely 2 mm. l. *Capsule* ellipsoidal, about 1 cm. l.

28. SERAPHYTA Fisch. & Mey.

Epiphytic herb. Stems leafy, sheathed at the base, scarcely fleshy. Leaves leathery, elliptical, oblong or lanceolate, in two ranks along the stem, with sheaths tightly adpressed. Panicle terminal, laxly and thinly branched, with small stalked flowers laxly racemose along the branches. Bracts minute. Sepals nearly equal, free, acute, somewhat spreading. Petals narrowly linear, nearly as long as the sepals, spreading. Claw of the lip as long as the column, with which it is connate forming a cup; limb broadly ovate-cordate, undivided, spreading; disc with a callus at the mouth of the cup. Column short, the wings forming a cup with the claw of the lip, foot wanting; clinandrium short, lateral lobes roundish, median obsolete. Anther terminal between the lateral lobes, opercular, incumbent, subglobose, thick and fleshy at the back, as if appendaged, 2-celled anteriorly, each cell divided again into two cells; pollinia 4, waxy, ovoid-globose, distinct, without an appendage, connected in pairs by a somewhat copious viscid substance elastically extensible into two threads, and often attached to the rostellum. Capsule broadly ellipsoidal, somewhat beaked, the ribs scarcely prominent.

Species 1, native of the West Indies, Mexico, Colombia and Surinam.

S. diffusa *Pfitz. in Engl. & Prantl Nat. Pflanzenf. ii.* 6, 142 (1889).—S. multiflora *Fisch. & Mey. in Bull. Sc. Acad. Petersb. vii.* 24 (1840). Epidendrum diffusum *Sw. Prodr.* 121 (1788) & *Fl. Ind. Occ.* 1503; *Lindl. Gen. & Sp. Orch.* 102 & *Fol. Orch. Epidendr.* 88; *Hook. in Bot. Mag. t.* 3565; *Griseb. Fl. Br. W. Ind.* 619. (Pl. 14, f. 1–3.)

On trees; in fl. Sept.–Jan., in fr. Dec., Jan., July; *Swartz*! *Macfadyen*! Manchester and St. Mary, *Purdie*! Fairfield, *Wullschlaegel*, 1052! *March*! Moneague, *Prior*! Blue Mts., 2000–5000 ft. J.P. 30, *Morris*! Blue Mts., *Mrs. Maclaverty*! near Browns Town, *Miss T. M. Barrett*! Cinchona, *Plaxton*! Arntully, *Miss Sabonadière*! Ramble, Hanover; near Gordon Town; *Fawcett*! Ramble, Claremont, *Fawcett & Harris*! Walderston, *Mrs. Longstaff*! Cold Spring; Belvedere, Hanover, 500 ft.; Lancaster; Clydesdale, 3500 ft.; Maryland; Troy, 1600 ft.; *Harris*! Lyndhurst, near Green Vale, *Harris & Britton*! Fl. Jam. 7625, 7656, 7770, 10,394, 10,447, 10,623.—Cuba, Mexico, Colombia, Surinam.

Plant 7–27 in. high. *Stem* 3–30 cm. l. *Leaves* 3–6 cm. l., 1·5–3·5 cm. br. *Panicle* 1–3·5 cm. l. *Bracts* deltoid, acute, 1–2 mm. l. *Pedicels* 1–3 mm. l. *Flowers* yellowish-green, drying red, perianth about ⅓ in. l. *Sepals* 3-nerved, linear-lanceolate, subacute, 7·5 mm. l., lateral somewhat oblique, 2·25 mm. br., median 1·75 mm. br. *Petals* 1-nerved, obtuse, 7 mm. l., ·65 mm. br. *Lip* as long as lateral sepals, median nerve prominent below and forming a mucro at apex, limb about 5 mm. l., 3·5 mm. br. *Column* 2·5 mm. l. *Capsule* 13–18 mm. l.

29. ARPOPHYLLUM Llav. & Lex.

Epiphytic herb. Stems erect from a creeping rhizome, with one leaf at the apex, and covered below the leaf with ample sheaths. Leaf leathery, very long, folded together at the base, but not closed into a sheath. Peduncle terminal with one or two sheaths above the leaf; spike long, dense, cylindrical, with numerous medium-sized shortly stalked or stalkless flowers. Bracts very small. Sepals of nearly equal length, free among themselves, spreading, the lateral a little broader than the median, concave at the base, adnate to the foot of the column. Lip attached to the foot of the column, concave and somewhat saccate quite at the base, above the swelling contracted and incumbent, then erect; crenulate on the margin above. Column erect, slightly bent, not winged, produced at the base into a short foot; clinandrium little dilated, rostellum shorter. Anther opercular, incumbent, subglobose, 2-celled; pollinia 8, four in each cell, obovoid, acute or shortly acuminate, cohering into two groups by a sparse viscid substance. Capsule ellipsoidal, 3-keeled, 3-ribbed.

Species 5, natives of Jamaica, Mexico and Guatemala.

A. giganteum *Hartweg ex Lindl. in Ann. & Mag. Nat. Hist. iv.* 384 (1840); *Reichenb. f. in Walp. Ann. vi.* 448. (Pl. 14, f. 4–9)

On rocks; in fl. March-May; Old England, 4000 ft., J.P. 26, *Morris*! *Harris*! Fl. Jam. 7652.—Mexico, Guatemala.

Plant 2 ft. high and over. *Rhizome* about 8 mm. thick, covered with sheaths, sending out very long stout roots. *Stem* 14–25 cm. l., compressed, of three or four long joints; sheaths short at base, increasing above to 11 cm. l. *Leaves* ligulate, flat above, keeled, 4–5 dm. l., 2 to nearly 3 cm. br. *Peduncle* 10–15 cm. l. to base of spike; spathe at base 9–13 cm. l. *Spike* 14–15 cm. l. *Bracts* triangular, acuminate, sterile, lowest clasping, 13–15 mm. l., higher 2·5–5 mm. l., floral 1·5 mm. l. *Flowers* light purple, stalkless, perianth ⅛ in. l. *Sepals*, lateral with 3 nerves branching from base, ovate-oblong, obtuse, slightly oblique, 6 mm. l., about 3 mm. br. at base; median 3-nerved, oblong, obtuse, 5·5 mm. l., 2 mm. br. *Petals*, 1 median nerve with a few pinnate nerves, oblanceolate-linear, apex rounded, margin erose above, about 5·5 mm. l., 1·6 mm. br. *Lip* about 5·5 mm. l., 7-nerved, not divided, hooded and erose above. *Column* 3·5–4 mm. l. *Capsule* 7–8 mm. l.

30. EPIDENDRUM Sw.

Epiphytic herbs. Stems leafy, sometimes fleshy, or at length thickened into a pseudobulb, sometimes thinner, sometimes branching. Leaves leathery or more rarely grass-like. Peduncle terminal on a leafy stem, simple or paniculate. Flowers medium-sized or somewhat small, racemose on a simple rachis or the branches of a panicle, shortly stalked, with small or narrow bracts, more rarely crowded on a very short, sessile raceme with imbricate bracts. Sepals free, equal, spreading or reflexed, or more rarely somewhat erect. Petals somewhat similar to the sepals, or more rarely much narrower. Claw of the lip erect, adpressed to the column and more or less connate with it into a tube; lateral lobes obsolete or more rarely dilated embracing the column; blade spreading, undivided or 3-fid; disc with various calli. Column generally narrow, connate with the claw of the lip even to the apex, sometimes free, semiterete or more rarely 2-winged or 2-auriculate; clinandrium generally short. Anther terminal, opercular, incumbent, convex or semiglobose, distinctly 2-celled, each cell again divided longitudinally; pollinia 4, waxy, ovate or broad, equal, compressed from the side, in one series, connected in each cell by a granular-viscous appendage. Capsule ovoid or oblong, the ribs often very prominent, or sometimes winged.

Species about 750, widely dispersed through tropical America, a few extending to the southern States of North America.

I. Column connate with lip beyond its middle or even as far as the apex (but only to about the middle in *E. cochleatum* and *E. spondiadum*).

1. With pseudobulbs.
    - Lip 3-lobed, with fringed lobes .... 1. *E. ciliare.*
    - Lip entire.
        - Lip greenish-white with purple streaks ... 2. *E. fragrans.*
        - Lip purple .... 3. *E. cochleatum.*
    - Lip indistinctly 2-lobed, cream-colour blotched with reddish-purple .... 4. *E. spondiadum.*
2. Without pseudobulbs.
    - a. Floral bracts inconspicuous, much shorter than the flowers.
        - Sterile bracts none, or a few similar to floral bracts.
            - Flowers more than 1½ inch l., apparently solitary.
                - Lateral lobes of lip semi-ovate .... 5. *E. nocturnum.*
                - Lateral lobes of lip linear................. 6. *E. angustilobum.*
            - Flowers less than ¾ inch l.
                - Flowers umbellate .... 7. *E. difforme.*
                - Flowers in a raceme or panicle.
                    - Lip undivided .... 8. *E. jamaicense.*
                    - Lip 3-lobed with terminal lobe 2-partite.
                        - Lip longer than sepals, lateral lobes oblong, apex of lobes eroded... 9. *E. verrucosum.*
                        - Lip not longer than sepals, lateral lobes rounded, apex of lobes entire................................ 10. *E. patens.*
        - Sterile bracts spathaceous.
            - Lateral lobes of lip entire.
                - Flowers in a panicle; lateral lobes of lip trapezoid-roundish .... 11. *E. nutans.*
                - Flowers in a crowded raceme, almost like a head; lateral lobes of lip roundish, almost obsolete........... 12. *E. anceps.*
            - Lateral lobes of lip fringed .... 13. *E. rivulare.*

b. Floral bracts large, broad, concave, spathe-like, enclosing the ovary or even the flower.
  Lip undivided.
    Leaves small, not more than 4 inches l.
      Leaves oblong-elliptical, 2 inches l.; spike longer than leaves .... 14. *E. rigidum.*
      Leaves linear-ligulate or linear-lanceolate.
        Spike with few flowers.
          Leaves linear, 3–4 inches l. Flowers ¼ inch l. .... 15. *E. ramosum.*
          Leaves linear-lanceolate, ¾ inch l. Flowers ⅛ inch l. .... 16. *E. strobiliferum.*
        Flowers solitary .... 17. *E. repens.*
    Leaves large; 6–8 inches l. .... 18. *E. imbricatum.*
  Lip 3-lobed .... 19. *E. bifarium.*

II. Column free, or only connate below the middle with the lip (connate about half-way in *E. teretifolium*).
1. With pseudobulb.
  a. Lip undivided.
    Flowers solitary .... 20. *E. polybulbon.*
    Flowers racemose or paniculate.
      Lip somewhat similar to sepals and petals 21. *E. Ottonis.*
      Lip long-clawed, broadly ovate .... 22. *E. subaquilum.*
      Lip very short-clawed, elliptical, acute, ventricose-concave .... 23. *E. bletioides.*
  b. Lip 3-lobed.
    Lip quite free.
      Lateral lobes rounded, median lobe rounded and not larger .... 24. *E. parvilobum.*
      Lateral lobes narrow, middle lobe differing in form, and much larger.
        Ovary rugose .... 25. *E. monticolum.*
        Ovary smooth .... 26. *E. belvederense.*
    Lip attached to column about one-third of its length .... 27. *E. angustifolium.*
2. Without pseudobulb.
  Stem with one leaf.
    Flower-stalk longer than leaf, with one flower .... 28. *E. brachyglossum.*
    Leaf serrulate, spike of flowers shorter than leaf .... 29. *E. serrulatum.*
  Stem with a few semi-terete fleshy leaves.
    Stem short with a few minute flowers clustered at apex .... 30. *E. globosum.*
    Stem long with one flower at apex .... 31. *E. teretifolium.*

1. **E. ciliare** *L. Syst. ed.* 10, 1246 (1759); pseudobulbs long, thicker above, of several joints, 2-leaved; lip white, 3-lobed, lateral lobes with long fringes, median lobe subulate-filiform; column connate with lip as far as apex.—*Jacq. Sel. Stirp. Amer.* 224, *t.* 179, *f.* 89, *Ed. pict. t.* 209; *Bot. Reg. t.* 784; *Lindl. Fol. Orch. Epid.* 29; *Griseb. Fl. Br. W. Ind.* 615; *Cogn. in Fl. Bras. iii. pt.* 5, 70; *Salisb. Ic. ined. vi. fol.* 574. Helleborine graminea &c. *Plum. Pl. Amer.* (*Burm.*) *t.* 179, *f.* 2, & *Ic. ined. iii. t.* 143.

On trees; in fl. July–Dec.; *Shepherd* in Herb. Kew!—Porto Rico, S. Thomas, Tortola, St. Cruz, St. Martin, St. Bartholomew, Saba, St. Eustatius, St. Kitts, Antigua, Guadeloupe, Dominica, Martinique, St. Lucia, St. Vincent, Bequia, Barbados, Grenada, Trinidad, Venezuela, B. Guiana, French Guiana, Colombia, Brazil, Guatemala, Costa Rica, Mexico, B. Honduras.

Plant 1½ ft. high and more. *Pseudobulbs* 1–2 dm. l. *Leaves* leathery, oblong-ligulate to elliptical-oblong, obtuse, 1–2 dm. l., 2·5-4·5 cm. br. *Raceme* with few or many flowers above, as long as the leaves, sometimes longer, clothed below with compressed keeled sterile bracts (2·5–5 cm. l.). *Bracts*, floral large, spreading, broadly lanceolate, acute, 2·5–5 cm. l. *Pedicels* about 1 cm. l. *Flowers* large, sweet-smelling, pale yellowish-green with white lip. *Sepals* narrowly linear-lanceolate, long acuminate, many-nerved, 4–5·5 cm. l., 3–5 mm. br. *Petals* like the sepals but slightly smaller, and somewhat falcate. *Lip*, lateral lobes 1–2 cm. l., 2–3 mm. br., median lobe 3–3·5 cm. l. *Column* white, dilated above, 13–15 mm. l. *Capsule* oblong-spindle-shaped, 4–5 cm. l., 1·5–2 cm. br.; beak 2–2·5 cm. l.

The sole authority for including this species rests on a specimen from Shepherd "from Jamaica" in the Kew Herbarium. If truly native, it is remarkable that so striking a plant should have escaped the notice of so many collectors from Swartz onwards.

2. E. **fragrans** *Sw. Prodr.* 123 (1788); pseudobulb spindle-shaped or ovoid-lanceolate, 1-leaved; lip creamy-white with purple lines, on upper side of flower, undivided, ovate-roundish, acuminate; column connate with lip beyond the middle.—*Sw. Fl. Ind. Occ.* 1487 & *Ic. ined. t.* 46; *Andrews Bot. Rep. x. t.* 645; *Bot. Mag. t.* 1669; *Lindl. Fol. Orch. Epid.* 39; *Griseb. op. cit.* 615; *Cogn. op. cit.* 83 & *in Symb. Ant. iv.* 174; *Salisb. Ic. ined. vi. fol.* 575.

On trees and rocks; in fl. all the year; *Swartz*! *Bertero*, *Distin*! *Lane*! Fairfield, *Wullschlaegel*, 1066! Moneague, *Prior*! J.P. 11, *Morris*! Ramble, Hanover; Bath; *Fawcett*! Silver Hill, 4000 ft.; Dolphin Head, 1600 ft.; *Harris*! near Radnor, *Miss B. Gosset*! Fl. Jam. 7859, 10,441.—Cuba, Florida, Hispaniola, Dominica, Martinique, St. Vincent, Grenada, Trinidad, Venezuela, B. Guiana, Surinam, Brazil, Panama, Colombia, Ecuador, Nicaragua, Guatemala.

Plant 12–18 in. high. *Pseudobulb* 6–13 cm. l., with a few, rather large scarious sheaths, soon decaying. *Leaves* leathery, ligulate-oblong, obtuse, 2–3·5 dm. l., 2–4 cm. br. *Raceme* 6–10 cm. l., with few or several flowers, much shorter than the leaf; spathe at base broadly ovate, compressed, about 4 cm. l. (3–6). *Bracts* minute, triangular or triangular-lanceolate, acute or shortly acuminate, 3–6 mm. l. *Flowers* fragrant, shortly stalked, perianth 1 in. or more l. *Sepals* lanceolate, long-acuminate, 5–7 nerved, yellowish-white with greenish tint outside, 25–35 mm. l., 3·5–5 mm. br. *Petals* oval-lanceolate, long acuminate, 7-nerved, yellowish-white, 22–28 mm. l., 7–10 mm. br. *Lip* 11–13-nerved, 16–18 mm. l. (to junction with column), 10–13 mm. br. *Column* greenish-white, short, 5–8 mm. l., thicker above, at the apex three obtuse lobes with a long tooth on the inner side of the median lobe; anther indistinctly 2-lobed above. *Capsule* ellipsoidal, 2·5–3 cm. l.

3. E. **cochleatum** *L. Sp. Pl. ed.* 2, 1351 (1763), *excl. syn. Sloane*; pseudobulb with one to three small joints at base and apex, ovoid-lanceolate or cylindrical, with two or sometimes

three leaves; lip purple, whitish round column, with conspicuous purple radiating veins, rarely yellow, on the upper side of flower, undivided, shell-shaped, cordate, broadly roundish, apiculate, with two calli at base; column connate with lip to about the middle.—*Jacq. Ic. Pl. Rar. iii. t.* 605; *Andrews Bot. Rep. t.* 13; *Bot. Mag. t.* 572; *Griseb. op. cit.* 616; *Cogn. in Symb. Ant. iv.* 175. Viscum caryophylloides &c. *Catesby Nat. Hist. Carol. ii.* 88 *t.* 88. Helleborine cochleato flore *Plum. Pl. Amer.* (*Burm.*) 9, *t.* 185, *f.* 2.

On trees and rocks; in fl. throughout year; *Swartz*! Fairfield, *Wullschlaegel*, 1064 bis! Newcastle Road, *Prior*! St. Ann (lip yellow), *Steer*! J.P. 10, 2153, *Morris*! *Syme*! Green River, *G. Nichols*! Silver Hill; Chester Vale, 3000 ft.; Vinegar Hill, 4000 ft.; *Harris*! near Radnor, *Miss B. Gosset*! Moneague, *Mrs. Balston*! Fl. Jam. 7676, 8000, 10,461. —Cuba, Bahamas, Florida, Hispaniola, Porto Rico, Costa Rica, Nicaragua, B. Honduras, Guatemala, Mexico, Venezuela.

Plant 1–2 ft. high. *Pseudobulb* 9–16 cm. l., with a few rather large scarious sheaths, soon decaying. *Leaves* oblong-lanceolate, acute, 2–3 dm. l., 2–3·5 cm. br. *Raceme* 2–5 dm. l., with few or several flowers, longer than the leaf; spathe at base 7–14 cm. l. *Bracts* small, triangular-lanceolate, acuminate, 5–12 mm. l. *Flowers* stalked, perianth 1 in. and more l., lip about ½ in. l. *Pedicels* 1–2 cm. l. *Ovary* 3-cornered, about 1 cm. l. *Sepals* linear-lanceolate from a broad base, acute or acuminate, pale green or greenish-white with one or two purple spots at base, pendulous, 2·5–4·2 cm. l., 3–5 mm. br. *Petals* narrowly oblanceolate, otherwise like sepals, but shorter and narrower. *Lip* 12–20 mm. l. *Column* purple below, white above, short, about 6 mm. l., thicker above; clinandrium with three obtuse lobes. *Capsule* ellipsoidal, with three broad wings, recurved, about 4 cm. l.

4. **E. spondiadum** *Reichb. f. in Bot. Zeit. x.* 731 (1852); pseudobulb with one or two small joints at base, cylindrical, sometimes slightly larger at base, with one leaf; lip reddish-purple with cream-coloured or yellowish-green margin, on the upper side of flower, cordate, broadly ovate with obscure lateral lobes, apex apiculate, two calli at base running up to a circular callus on the lower part of the disk, and level with top of column; column connate with lip to about the middle.—*Hook. f. in Bot. Mag. t.* 7273. E. variegatum *Hart in Gard. Chron. xxvi.* 11 (1886) (non *Hook.*).

On trees and rocks; in fl. and fr. Nov.–Feb.; Coldspring; Whitfield Hall; J.P. 245; *Morris*! Clydesdale, *Harris*! Fl. Jam. 7866.—Costa Rica.

Plant 12–16 in. high. *Pseudobulbs* clustered along a creeping rhizome, 10–17 cm. l., with two somewhat large scarious sheaths, soon decaying. *Leaves* ligulate, obtuse, 2–3 dm. l., 3–3·5 cm. br. *Raceme* 8–11 cm. l., with a few or several flowers, much shorter than the leaf, spathe at base 3·5–4·5 cm. l. *Bracts* small, triangular, obtuse, 3–4 mm. l. *Flowers* shortly stalked, perianth more than ½ in. l., lip about ½ in. br., but not so long. *Pedicels* 4–6 mm. l. *Ovary* 3-winged, about 1 cm. l. *Sepals* somewhat fleshy, oblong, with an acute apex, creamy-white or greenish-yellow tinted with reddish-purple, 13–15 mm. l., 3·5–4 mm. br. *Petals* 7-nerved, somewhat fleshy, oblanceolate, acute, creamy-white or greenish-yellow with a reddish-purple blotch in the centre, 11–15 mm. l., 4–4·5 mm. br. *Lip* about 11 mm. l. from junction with column. *Column* short, about 5 mm.

l.; clinandrium with three lobes, lateral lobes obtuse, retuse or slightly 2-lobed, median lobe denticulate. *Capsule* ellipsoidal, with three broad wings, recurved, 3·5–4·5 cm. l.

5. **E. nocturnum** *Jacq. Enum. Pl. Carib.* 29 (1760); without pseudobulbs; bracts very much shorter than flowers; flowers solitary, perianth $1\frac{3}{4}$ in. l.; ovary about 5 cm. l.; lip deeply 3-lobed, lateral lobes semi-ovate, obtuse, entire, about 2 cm. l., 5–7 mm. br., the median lobe about 2·5 cm. l., 1 mm. br., resembling petals; column connate with lip nearly to apex.—*Jacq. Sel. Stirp. Amer.* 225, *t.* 139, *Ed. pict. t.* 210; *Sw. Obs. Bot.* 327; *Lindl. Gen. & Sp. Orch.* 105 & *Fol. Orch. Epid.* 81; *Hook. in Bot. Mag. t.* 3298; *Griseb. op. cit.* 619; *Cogn. in Fl. Bras. iii. pt.* 5, 134, *t.* 37; *Ames Orch. S. Fla.* 16, *t.* 7. (Pl. 15, f. 2.)

On trees, stones and rocky land; in fl. Aug.-Nov., in fr. Sept.-Dec.; *Macfadyen*! Moneague, *Prior*! *March*! J.P. 3, *Morris*! Moody's Gap, J.P. 2005, *Syme*! St. Ann and Trelawney, *Miss T. M. Barrett*! Cedar Valley, *Moore*! Lancaster, 2500 ft., *Harris*! Fl. Jam. 7565.—Cuba, Bahamas, Florida, Porto Rico, Guadeloupe, Dominica, Martinique, St. Lucia, St. Vincent, Grenada, Trinidad, Nicaragua, Honduras, Guatemala, Mexico, Venezuela, B. Guiana, Surinam, Brazil, Peru.

Plant 14–15 in. high. *Rhizome* short, with numerous stout roots. *Stems* 10–40 cm. l., 4–5 mm. br., several, erect, long, stout, covered with the leaf-sheaths, the blades of which are fallen below, compressed and leafy above. *Leaves* 7–12 cm. l., generally from 2–3 cm. br., in two ranks, leathery, linear-oblong to oblong, apex rounded or slightly emarginate, at the base clasping and sheathing; sheaths slightly keeled. *Bracts* 3–7 mm. l., broadly ovate, acute, at base clasping. *Flowers* greenish-white or cream-coloured, sessile, five or six or more, in a terminal panicle, the rachis of which is scarcely developed, and generally remains enclosed by the leaf-sheaths, only one or two flowers generally appearing at a time. *Ovary* about 5 cm. l., as long as, or longer than, perianth. *Sepals* 4–4·5 cm. l., greenish-white or cream-coloured, narrowly lanceolate, acuminate, many-nerved, lateral 5 mm. br., slightly oblique, median slightly narrower. *Petals* as long as the sepals, 1–1·5 mm. br., whitish or cream-coloured, linear, long-acuminate. *Lip* as long as, or a little shorter than, the petals; disk with two long broad parallel lamellæ. *Column* 15–18 mm. l., much dilated above, clinandrium dentate. *Capsule* 6 cm. l., about 1·5 cm. br., oblong, spindle-shaped, much attenuated above and at the base.

6. **E. angustilobum** *Fawc. & Rendle in Journ. Bot.* xlvii. 124 (1909); without pseudobulbs; bracts very much shorter than flowers; flowers solitary, large, perianth $1\frac{3}{4}$ in. l.; ovary very long, 12·5 cm. l.; lip deeply 3-lobed, lateral lobes linear, acuminate, 7 mm. l., barely 1 mm. br., the median lobes three times as long, linear, long-acuminate, 2·2 cm. l., 1 mm. br.; column auricled, connate above the middle with the lip, about three-quarters of its length. (Pl. 15, f. 1.)

In fl. Oct.; Rose Hill, 4000 ft., *Harris*! Fl. Jam. 10,485.

Plant over 2 ft. high. *Stem* high, erect, stout, terete at the base, bare above the base, compressed above, covered with the sheaths of the leaves, with two leaves at the apex, 5 dm. l., 4–10 mm. br. *Leaves* oval or elliptical-oblong, leathery, obtuse, many-nerved, at the base clasping and long-sheathing, blade about 11 cm. l., 4–5 cm. br., sheaths compressed, slightly keeled. *Panicle*, rachis scarcely protruded and covered by sheaths. *Bracts* sheathing at the base, ovate-oblong, acute or somewhat acuminate, 7–8 mm. l. *Flower* cream-coloured. *Sepals* narrowly oblanceolate, acuminate, many-nerved, 4·2–4·3 cm. l. *Petals* as long as the sepals, linear, long-acuminate. *Lip* a little shorter than the sepals; disk with two narrow parallel lamellæ at the base. *Column* 2 cm. l., above very dilated; clinandrium with three posterior ovate acuminate lobes, and three anterior, the middle acuminate, and the lateral obtuse; anther purplish.

7. **E. difforme** *Jacq. Enum. Pl. Carib.* 29 (1760); without pseudobulbs; bracts much shorter than flowers; flowers umbellate, perianth about ½ in. l.; lip kidney-shaped, emarginate, margin undulate, with two calli on disk at base; column connate with lip to apex.—*Jacq. Sel. Stirp. Amer.* 223, *t.* 136, *Ed. pict. t.* 206; *Cogn. in Fl. Bras. iii. pt.* 5, 139. E. umbellatum *Sw. Prodr.* 121 (1788) & *Fl. Ind. Occ.* 1501; *Bot. Mag. t.* 2030; *Lindl. Fol. Orch. Epid.* 79; *Griseb. op. cit.* 618. (Pl. 15, f. 3.) Type in Herb. Mus. Brit.

On trees; in fl. Aug.–April; mountains, *Swartz*! *Purdie*! Nazareth, *Wullschlaegel*, 1395! *Prior*! *Eggers*; J.P. 18, 1516, *Morris*! lowlands, J.P. 524, *Syme*! Cedar Hurst, *Fawcett*! Cinchona, *Isaacs*! *Harris*! Keith Hill, 2200 ft., *Thompson*! Clydesdale, 4000 ft.; Lancaster 3000 ft.; Belvedere, Hanover, 500 ft.; Silver Hill, 4000 ft.; *Harris*! Mahogany Vale Ridge, *Miss B. Gosset*! Fl. Jam. 7629, 7651, 7865, 9022, 10,455, 10,498.—Cuba, Florida, Hispaniola, Porto Rico, St. Eustatius, St. Kitts, Guadeloupe, Dominica, Martinique, St. Lucia, St. Vincent, Grenada, Trinidad, Guatemala, Mexico, Costa Rica, Venezuela, B. Guiana, Surinam, Brazil.

Plant 2–10 in. high. *Stems* 4–20 cm. l., few or several, clustered, covered with leaf-sheaths. *Leaves* leathery, elliptical-oblong or ligulate, obtuse, obliquely retuse, clasping at base, blade 2·5–7·5 cm. l., 1–2·5 cm. br. *Bracts* linear, 7–12 mm. l. *Flowers* pale green or whitish. *Pedicels* 1–2·5 cm. l. *Ovary* 1–2·5 cm. l. *Sepals* with seven incomplete nerves prominent on inside, elliptical, median apex thickened, subacute, 15 mm. l., about 5·5 mm. br.; lateral somewhat oblique, apex keeled at back, keel forming an acute point, 15–16 mm. l., 6 mm. br. *Petals* 3-nerved, linear-lanceolate, 14 mm. l., 2·5 mm. br. *Lip* with numerous nerves, 13–14 mm. l., about 18 mm. br. *Column* short, about 1 cm. l. (to apex of clinandrium), thicker above; clinandrium with denticulate margin. *Capsule* ellipsoidal, shortly and thickly beaked, 2·5 cm. l., 1·5 cm. br.

8. **E. jamaicense** *Lindl. Fol. Orch. Epid.* 82 (1853); without pseudobulbs; bracts shorter than flowers; flowers in a raceme, perianth ⅔ in. l.; lip undivided, roundish or broader than long, undulate, subentire, inconspicuously mucronate, two calli on disk at base; column connate with lip to apex.—*Griseb. op. cit.* 618.

On trees; in fl. Sept.; Manchester, Dunrobin Castle, *Purdie*! Cedar Hurst, Rose Hill, J.P. 481, *Syme*! near Cedar Hurst, 2409, *Morris*! near Bath, *Harris*!

Plant to 1 ft. high. *Stem* 2–3 dm. l., covered with leaf-sheaths, leafy above. *Leaves* oblong-lanceolate, very acute, longer than the raceme, blade 4–11 cm. l., 1–2 cm. br. *Bracts* triangular-lanceolate, acuminate, keeled, 12–5 mm. l. *Pedicels* 11–12 mm. l. *Ovary* rather longer than pedicel. *Sepals* oblanceolate, 12–16 mm. l. *Petals* narrowly linear, slightly shorter than sepals. *Lip*, limb 8 mm. l., 12 mm. br. *Column* short, branching upwards, 9 mm. l.; clinandrium hooded, undivided.

9. E. **verrucosum** *Sw. in Nov. Act. Upsal. vi.* 68 (1799); without pseudobulbs; bracts much shorter than flowers; flowers in a diffuse panicle, sepals and petals ⅓ in. l.; lip much longer than sepals, 3-lobed with median lobe 2-lobed, lateral lobes and median lobes somewhat oblong with upper margins more or less eroded, three calli on disk at base; column connate with lip to apex.—*Sw. Fl. Ind. Occ.* 1497; *Lindl. Fol. Orch. Epid.* 92. (Pl. 15, f. 11.)

On trees, rocks and dry banks; in fl. Dec., Jan.; *Swartz*! *Wright*! *Bancroft*! Christiana, *Purdie*! *Lane*! *March*! Fairfield, *Wullschlaegel*, 1089! Manchester, J.P., 2008, 2040, *Syme*! *Morris*! Blue Mts., J.P. 2048, *Morris*! Cinchona; Clydesdale; *Harris*! Fl. Jam. 7542, 7819.—Colombia.

Plant 1–4 ft. high. *Stems* 1–6 dm. l., covered with leaf-sheaths. *Leaves* lanceolate or linear-lanceolate, acute, blade 8–14 cm. l., 13–30 mm. br.; sheaths covered with minute warts. *Bracts* triangular, acuminate, 2·5–4 mm. l. *Flowers* creamy-white or light yellow. *Pedicels* about 1 cm. l. *Ovary* about 1 cm. l. *Sepals* with seven to nine nerves, prominent on inside, elliptical, apex thickened, obtuse, 9 mm. l., 4 mm. br., lateral rather narrower and slightly oblique. *Petals* with three to five branching nerves, linear-oblanceolate, 9 mm. l., 2·25 mm. br. *Lip* with numerous fine nerves, about 12 mm. l. *Column* short, about 5 mm. l. (to apex of clinandrium); clinandrium with four broad truncate lobes; anther slightly 2-lobed above. *Capsule* spindle-shaped, beaked, 2·5 cm. l.

10. E. **patens** *Sw. in Nov. Act. Upsal. vi.* 68 (1799); without pseudobulbs; bracts much shorter than flowers; flowers in a panicle; perianth nearly ⅓ in. l.; lip as long as median sepal, 3-lobed with median lobe 2-lobed, lateral lobes roundish, median lobes oblong, margins of lobes entire, three calli at base of median lobe; column connate with lip to apex.—*Sw. Fl. Ind. Occ.* 1495; non *Lindl. in Paxt. Flow. Gard. ii. Glean.* 119, *f.* 198, nec *Fol. Orch. Epid.* 75; *Griseb. op. cit.* 617 (in part).

Stems 20–30 cm. l. *Leaves*, blade 5–10 cm. l., oblong or oblong-lanceolate; sheaths smooth. *Bracts* linear-lanceolate. *Flowers* pale yellow, shortly stalked. *Sepals* oblanceolate-oblong, acute, median 1–1·1 cm. l., lateral ·9–1 cm. l. *Petals* oblanceolate to elliptical, 1 cm. l. *Lip* 1–1·1 cm. l. *Column* stout, concave in front, widening upwards, with a dorsal keel; clinandrium 3-lobed, median lobe broad, truncate, crenulate.

We have not seen any specimen from Jamaica, but a specimen in Herb. Mus. Brit. labelled in Solander's hand "Epidendrum patens fl. ind. occid., Barthol." agrees with Swartz's description. Swartz assigns the species to Jamaica and the Carribean Islands and cites Fahlberg as the collector. Samuel Fahlberg, a Swede, visited St. Bartholomew in 1785. The specimens referred by Lindley to this species, belong to the section *Amphiglottum*, characterised by a closely sheathed peduncle, and have flowers 1½ in. across.

11. **E. nutans** *Sw. Prod.* 121 (1788); without pseudobulbs; floral bracts much shorter than the flowers; panicle sheathed at base by two large spathaceous bracts; lip cordate, 3-lobed, median lobe 2-lobed, lateral lobes subrotund or subrhomboid, disk with three lamellæ, the lateral forming at the base two auriculate calli; column connate with lip to apex.—*Sw. Fl. Ind. Occ.* 1499; *Bot. Reg. t.* 17; *Hook. Exot. Fl. i. t.* 50; *Lindl. Fol. Orch. Epid.* 56; *Griseb. op. cit.* 616. (Pl. 15, f. 5.) Type in Herb. Mus. Brit.

On rocks and trees; in fl. Aug.–Feb.; *Swartz*! Springfield, *Wullschlaegel*, 1090! Moneague, *Prior*! *Macfayden*! *Jenman*! J.P. 2007, *Syme*! J.P. 527, *Morris*! Cedar Hurst, *Fawcett*! Trelawney, *Miss T. M. Barrett*! Walderston, *Mrs. Longstaff*! Mt. Moses; Woodcutters' Gap, 4000 ft.; Chester Vale, 3000 ft.; Wallenford, 2000 ft.; Hollymount, near Ewarton, 2500 ft., *Harris*! Fl. Jam. 7624, 7864, 8907.—Trinidad, Venezuela, Brazil.

Plant 1½–2 ft. high. *Stem* 4–5 dm. l., 7–10 mm. br., erect, stout, covered with leaf sheaths, leafy above. *Leaves* leathery, oblong to oblong-lanceolate, acute to subobtuse, clasping and sheathing, blade 12–20 cm. l., 3–5·5 cm. br. *Panicle* terminal, large, longer than the leaves, lax, many flowered, subnutant; spathaceous bracts much smaller than the leaves and passing above into the floral bracts, lower floral bracts linear, decreasing upwards, the uppermost 3–4 mm. l., triangular, acuminate, very much shorter than the pedicels. *Flowers* between ½ and ¾ in. l., greenish, or whitish-green, pedicels long; ovary and pedicel together 2–3 cm. l. *Sepals* 1·5–1·8 cm. l., 3–5 mm. br., oblanceolate, distinctly 5-nerved, median obtuse, lateral acute, rather broader and longer. *Petals* a little shorter than the sepals, 3-nerved, subspathulate-linear, obtuse. *Lip* as long as or longer than the sepals. *Column* 1·2–1·3 cm. l., erect, thick, club-shaped; clinandrium with a tooth on each side anteriorly.

Var. tridentatum *var. nov.*; E. tridentatum *Fawc. in Journ. Bot. xxxiii.* 12 (1895); leaves narrower and more acute, sterile bract small or none, middle lobe of lip with three triangular lobules.

In fl. April, Blue Mts., *Harris*!

Var. obtusifolium *var. nov.*; stems reaching to 4½ ft. in length; leaves oblong to oblong-elliptical, obtuse, broader than in type; flowers smaller, generally dark-purple lip and column and the same colour on back of sepals; sepals broader than in type, 1·3 cm. l. or less, 5–6 mm. br.; petals same length, barely 2 mm. br.; lip as long as the sepals, and as broad as long.

On rocks in woodlands, in fl. Oct.–Dec., Mandeville, *Lady Blake*! Near Browns Town, *Miss T. M. Barrett*! near Chapelton, *Craig*! Troy, 2000 ft., *Harris*! Fl. Jam. 10,420; Walderston, *Mrs. Longstaff*!

12. **E. anceps** *Jacq. Sel. Stirp. Amer.* 224, *t.* 138 (1763), *Ed. pict. t.* 208; without pseudobulbs; sheaths on long flower-stalk below, flowers crowded above almost into a head; floral bracts much shorter than the flowers; lip broadly kidney-shaped-cordate, 2-lobed at apex; column connate with lip nearly to apex.—E. fuscatum *Smith Spicil.* 21, *t.* 23 (1791);

Bot. Mag. t. 2844 ; *Lindl. Fol. Orch. Epid.* 65 ; *Griseb. op. cit.* 617 ; *Salisb. Ic. ined. vi. fol.* 574. (Pl. 15, f. 10.)

On trees and rocks; in fl. March–Nov.; *Bancroft*! Moneague, *Prior*! *March*! J.P. 19, 2401, *Morris*! *Syme*! Trelawney, *Griffith*! Bath, *G. Nichols*! Iron River, 1500 ft.; Cascade, 3000 ft.; Holly Mount, Mt. Diabolo, 2800 ft.; *Harris*! Fl. Jam. 9021, 10,428.—Cuba, Florida, Hispaniola, Porto Rico, St. Martin, Saba, St. Eustatius, St. Kitts, Antigua, Guadeloupe, Dominica, Martinique, St. Lucia, St. Vincent, Barbados, Grenada, Trinidad, Mexico, Honduras, Nicaragua, Colombia, Venezuela, B. Guiana, Surinam, Brazil.

Plant 1½ to 3 ft. high. *Stem* to 8 dm. l., compressed. *Leaves* oblong-elliptical or sometimes oblong-lanceolate, obtuse, sometimes subacute, sessile, sheathing at base; limb 7–17 cm. l., 2·5–4·5 cm. br. *Peduncle* 2–5 dm. l.; sheaths compressed, keeled, oblong-triangular above, imbricating, 2–6 cm. l.; raceme short, dense, sometimes branching, 1·5 sometimes to 7 cm. l. *Bracts* lanceolate, acuminate, 3–6 mm. l. *Flowers* fleshy, light-brown, greenish-brown, or tawny yellow. *Ovary* slender, with pedicel about 1 cm. l. *Sepals* obovate-elliptical, somewhat acute, light brown, lateral slightly broader, 6–8 mm. l., 3–4 mm. br.; when dry, with 3–5 prominent nerves. *Petals* spathulate-linear, light brown, 1-nerved when dry, 5·5–6·5 mm. l., 1–15 mm. br. *Lip* greenish-brown, with numerous veins when dry; limb 5–6·5 mm. l., 6–9 mm. br. *Column* 4 mm. l., greenish brown.

13. **E. rivulare** *Lindl. Ann. & Mag. Nat. Hist. ser.* 3, *i.* 330 (1858); without pseudobulbs; stems long, slender; leaves linear-lanceolate; base of peduncle clothed with closely-sheathing, subfoliaceous sterile bracts; flowers few above, long-stalked, lax, somewhat corymbose; floral bracts much shorter than the flowers; lip 3-lobed, lateral lobes large, fringed; median lobe linear, 3-toothed at dilated apex; disk with two calli at base; column connate with lip to apex.—*Griseb. op. cit.* 616. (Pl. 15, f. 7.)

On trees; in fl. Aug.–March; Dolphin Head, *Purdie*! Green River, J.P. 485, *Morris*! near Browns Town, *Miss T. M. Barrett*! Christiana, *Tomlinson*! St. Ann, *Stewart*! between Cinchona and Old England; Holly Mount, Mt. Diabolo; *Harris*! Fl. Jam. 10,421, 10,429, 10,449.—Cuba, Hispaniola.

Plant 1½–2 ft. and over. *Stem* to 6 dm. and over, covered by the leaf-sheaths. *Leaves* 6–12 cm. l., 8–11 mm. br. *Peduncle* 7–12 cm. l., sterile bracts in two rows, compressed, tapering upwards, keeled above, imbricating, to 5 cm. l. *Raceme* short, rarely branching, 2–3 cm. l. *Bracts* lanceolate, acute, 10–4 mm. l. *Flowers* yellowish-green speckled with red, lip white or light yellow, column tipped with red, perianth nearly ¾ in. l. *Ovary* slender, 12–13 mm. l. *Pedicel* about 15 mm. l. *Sepals* with 7-nerves prominent on outside, oblong-lanceolate, acute, 1·5–1·6 cm. l., 3·5 mm. br. *Petals* 1-nerved, linear, as long as sepals. *Lip* about 1 cm. l. above attachment to column. *Column* thicker above, 7–8 mm. l.; clinandrium with two shallow lobes at each side. *Capsule* ellipsoidal, tapering to base, and forming a beak at apex, about 3 cm. l., 1 cm. br.

14. **E. rigidum** *Jacq. Enum. Pl. Carib.* 29 (1760); without pseudobulbs; stems erect from a creeping rhizome, compressed;

leaves leathery, ligulate-oblong or oblong-elliptical, obtuse, bracts ample, broadly triangular, obtuse, keeled, keel decurrent, clasping, as long as ovary and enclosing it; lip on the upper side of the flower, leathery, a little longer than the sepals, undivided, roundish-ovate, cordate, apex obtuse, disk with two calli at base; column connate with lip to apex.—*Jacq. Sel. Stirp. Amer.* 222, *t.* 134, *Ed. pict. t.* 204; *Sw. Prodr.* 121 & *Fl. Ind Occ.* 1507; *Hook. Ic. Pl. t.* 314; *Lindl. Fol. Orch. Epid.* 76; *Griseb. op. cit.* 618; *Cogn. in Fl. Bras. iii. pt.* 5, 171.

On trees; in fl. Aug.-Nov., in fr. Nov.-Jan.; *Macfadyen*! St. Thomas in Vale, *Purdie*! New Carmel, *Wullschlaegel*, 1086! *March*! St. Mary, J.P. 2013, *Syme*! *Morris*! Trelawney, *Miss T. M. Barrett*! Ramble, St. Ann, *Fawcett & Harris*! Belvedere, Hanover; near Cinchona; *Harris*! Fl. Jam. 7246, 7623, 7754.—Cuba, Florida, Hispaniola, Porto Rico, Guadeloupe, Dominica, Martinique, St. Lucia, St. Vincent, Grenada, Tobago, Trinidad, Costa Rica, B. Honduras, Mexico, Venezuela, B. Guiana, French Guiana, Surinam, Brazil, Colombia, Bolivia.

Plant 4–10 in. high. *Stems* somewhat stout, simple, covered by the leaf-sheaths, 5–14 cm. l. *Leaves*, blade 2–6·5 cm. l., 1·2–2 cm. br.; sheaths 1–2 cm. l. *Spike* 3-11 cm. l., longer than the leaves, simple, very compressed, with two winged edges. *Bracts* 10–15 mm. l. *Flowers* few or several, green, perianth about ¼ in. l. *Sepals* 6–7-nerved, leathery, oval, apex with thickened margin, somewhat acute, 6 mm. l., 3 mm. br., lateral slightly oblique. *Petals* 3-nerved, linear, obtuse, as long as the sepals, 1 mm. br. *Column* short, thick, 3 mm. l.; clinandrium denticulate on margin. *Capsule* ellipsoidal, with a short and thick beak, 16–20 mm. l., 8–9 mm. br.

15. **E. ramosum** *Jacq. Enum. Pl. Carib.* 29 (1760); without pseudobulbs; spike with few flowers; bracts ample, elliptical-ovate, obtuse, sometimes acute, slightly keeled at apex, hooded, as long as, or scarcely longer than the ovary and enclosing it; leaves leathery, linear-ligulate, apex obliquely emarginate; lip on the lower side of the flower, slightly shorter than the sepals, leathery, with numerous nerves, undivided, ovate, cordate, acute, with two calli at base of disk which unite above and form a rib running to apex; column connate with lip nearly to apex.—*Jacq. Sel. Stirp. Amer.* 221, *t.* 132, *Ed. pict. t.* 202; *Sw. Prodr.* 120 & *Fl. Ind. Occ.* 1505; *Lindl. Fol. Orch. Epid.* 76; *Griseb. op. cit.* 618; *Cogn. in. Fl. Bras. iii. pt.* 5, 172. (Pl. 15, f. 12.)

On trees and rocks; in fl. Nov.-March; mountains, *Swartz*! *Bancroft*! *Macfadyen*! St. Ann, *McNab*! Portland Gap, *Purdie*! Bethabara, *Wullschlaegel*, 1054! *March*! mountains, J.P. 28; lowlands, 400–800 ft. J.P. 2012, *Morris*! Mt. Moses, 2500 ft. J.P. 526, *Syme*! Cinchona, *Clute*! Newcastle, *Hart*! Belle Vue, Port Royal Mts., *Miss B. Gosset*! Cinchona, 5000 ft.; Moody's Gap, 3000 ft.; Clydesdale, 4500 ft.; above Mabess River, 3500 ft.; Newhaven Gap to Vinegar Hill, 4000 ft.; *Harris*! Fl. Jam. 7535, 7799, 7881, 9897, 10,089.—Cuba, Hispaniola, Porto Rico, Guadeloupe, Dominica, Martinique, St. Lucia, St. Vincent, Grenada, Trinidad, Mexico, Guatemala, Costa Rica, Ecuador, Peru, Venezuela, B. Guiana, Surinam, French Guiana, Brazil.

Plant 1–3 ft. long. *Stems* to 1 m. l., much branched. *Leaves* sheathing at the base, 2–9 cm. l., 3–13 mm. br. *Spike* terminal on the branches, shorter than the leaves, or as long, laxly few- or several-flowered, 2–3 cm. l. *Bracts* becoming scarious and white, 9–11 mm. l. *Ovary* with a cavity above, within its walls, below the lip. *Flowers* greenish, perianth about ¼ in. l. *Sepals* 5–9-nerved, lateral ovate-oblong, slightly oblique at base, keeled outside near apex, point of keel 3-denticulate, acute, 8 mm. l., 2·75 mm. br.; median oblong, slightly keeled outside at apex, keel ending in a small tooth, 8 mm. l., 2·25 mm. br. *Petals* 5-nerved, linear, somewhat acute or obtuse, shorter than the sepals, 7–8 mm. l., 1 mm. br. *Lip* 3·5 mm. br. *Column* 2–2·5 mm. l., short, thick, thicker upwards; clinandrium, lateral lobes very broadly ovate, median lobe very short, 2-denticulate. *Capsule* ellipsoidal-ovoid, with a short, thick beak, 10–12 mm. l., 6–8 mm. br.

16. **E. strobiliferum** *Reichb. f. in Nederl. Kruidk. Arch. iv.* 333 (1859); without pseudobulbs; spike with few flowers; bracts ample, somewhat leathery, scarious, strongly nerved, plicate-hooded, broadly ovate, obtuse, keeled, clasping, nearly as long as the flowers, somewhat densely imbricate in two ranks; leaves in two ranks, fleshy-leathery, small, shortly ligulate-oblong or linear-lanceolate, at the apex obtuse, or subtruncate or sometimes obliquely emarginate, at the base half-clasping and shortly sheathing; lip on the upper side of the flower, leathery, about as long as the sepals, undivided, broadly ovate-cordate, acute, sub-acuminate, disc with several, somewhat prominent nerves; column connate with lip nearly to apex.—*Griseb. op. cit.* 618; *Cogn. op. cit.* 174, *t.* 25, *f.* 2; *Ames Orch. i.* 37, *t.* 10.

On trees; in fl. towards end of year, in fr. Jan.; *Swartz*! *Masson*! New Carmel, *Wullschlaegel*, 1085! Wag Water and Ginger Rivers, J.P. 2014 (in part), *Syme*! Belvedere, Hanover; Ramble, St. Ann; *Fawcett*! *Harris*! Fl. Jam. 7245a, 7519, 7543.—Florida, Cuba, St. Cruz, Saba, Dominica, Martinique, St. Vincent, Grenada, Trinidad, Venezuela, B. Guiana, French Guiana, Dutch Guiana, Brazil.

Plant 1–6 in. high. *Stems* clustered, branching, generally leafy to the base, 3–15 cm. l. *Leaves*, blade 1·5–2 cm. l., 4–6 mm. br.; sheaths 5–9 mm. l. *Spike* very short, 5–8 mm. l., shorter than the leaves, with 3–6 flowers. *Bracts* 5–6 mm. l. *Ovary* with a cavity below the lip. *Flowers*, perianth about ⅛ in. l. *Sepals* with seven or eight prominent nerves, lateral oblong-lanceolate, acute, about 4 mm. l., 1·75 mm. br.; median oblong, tapering slightly to a somewhat blunt point, as long as the lateral, 1·25 mm. br. *Petals* with three to five prominent nerves, linear, a little shorter than the sepals, ·5 mm. br. *Column* 1·5 mm. l., short, thick, thicker about the middle; clinandrium with two teeth on each side, one long and one short, and two short teeth behind. *Capsule* ellipsoidal, with short thick beak, 8–9 mm. l., 5–6 mm. br.

17. **E. repens** *Cogn. in Fedde Repert. vii.* 122 (1909); without pseudobulbs; bracts two or three, with a solitary flower, ample, triangular-ovate to roundish, obtuse, slightly keeled, overlapping, clasping, as long as or longer than ovary and enclosing it; leaves somewhat rigid, small, narrowly ligulate-oblong, apex rounded and obliquely and slightly emarginate, at the base roundish-subcordate and shortly sheathing; lip erect on the

upper side of the flower, slightly shorter than the sepals, leathery, with numerous nerves prominent on the outside and sometimes slightly on the inside, undivided, ovate, cordate, acute; column connate with lip nearly to apex.

On trees and rocks; in fl. Dec.-Feb.; *Macfadyen*! Newcastle, J.P. 2014 (in part), *Morris*! near Woodcutters' Gap; near Newcastle, 4000 ft.; ridge from Newhaven Gap to Vinegar Hill, 4000 ft.; *Harris*! Morse's Gap, *G. Nichols*! Fl. Jam. 7561, 10,086.—Hispaniola.

Plant 1-1½ ft. long or longer. *Stems* long, creeping, slender, much and irregularly branched, with many leaves even to the base, 3-4 dm. l. or longer. *Leaves*, blade 1·5-2 cm. l., 4-6 mm. br.; sheaths 6-8 mm. l. *Bracts* 5-6 mm. l. *Ovary* with a cavity above, within its walls, below the lip. *Flowers* terminal at the apex of branches, solitary; perianth scarcely ⅓ in. l. *Sepals* leathery, with seven to nine prominent nerves, ligulate-lanceolate, shortly acuminate, 7-8 mm. l., 2 mm. br., median sepal slightly longer. *Petals* 3-5-nerved, narrowly linear, very acute, a little shorter than lateral sepals, about ·7 mm. br. *Column* 2·5 mm. l., very short, thick, thicker upwards; clinandrium 3-lobed, lateral lobes triangular, acute, longer than median. *Capsule* erect, ovoid-subglobose, with short beak, 11-12 mm. l., 7-8 mm. br.

18. E. **imbricatum** *Lindl. Gen. & Sp. Orch.* 110 (1831); without pseudobulbs; bracts ample, spathe-like, leathery, broadly ovate, obtuse, concealing the flower in bud, keeled above; leaves large, leathery, ligulate-oblong, obtuse; lip undivided, leathery, ovate, obtuse, obsoletely apiculate, cordate, strongly 1-ribbed; column connate with lip to apex.—*Lindl. Fol. Orch. Epid.* 78; *Cogn. in Fl. Bras. iii. pt.* 5, 170, *t.* 14, *f.* 2. (Pl. 15, f. 9.)

On trees; in fl. and fr. Nov.-April; *Macfadyen*! below Morse's Gap, J.P. 244, *Morris*! near Mabess River, 4000 ft; near Vinegar Hill, Portland, 3500 ft; Catherine Peak, 5000 ft.; Thomson's Gap; *Harris*! Fl. Jam. 6540, 7577, 7807, 7867, 9092.—Brazil, Costa Rica.

Plant 15 in. high and more. *Stem* 40 cm. l. and more, long, stout, branching, covered even to the apex with truncate, adpressed, sheathing bases, from which in the lower part the blades have dropped, succeeded near the base by simple sheaths. *Leaves*, blade 9-18 cm. l., 1-2 cm. br. *Spikes* several-flowered, much shorter than the leaves. *Bracts* 2·1-2·5 cm. l. *Flowers* white, sessile, about ⅓ in. l. *Sepals* 12-14 mm. l., 3·5 mm. br., leathery, equal, narrowly oblong, subacute to acute, keeled above. *Petals* 11-13 mm. l., 2-2·5 mm. br., a little shorter and much narrower than the sepals, linear-lanceolate to oblanceolate, acute. *Lip* a little shorter than the petals, 8-10 mm. l., 6 mm. br. *Column* 4-5 mm. l., short, thick; clinandrium obtusely auricled. *Capsule* 2·5 cm. l., 12 mm. br., ovoid, constricted at apex, slightly 6-ribbed.

19. E. **bifarium** *Sw. in Nov. Act. Upsal. vi.* 68 (1799); without pseudobulbs; spike long, 2-edged, many-flowered; bracts ample, spathe-like, compressed, keeled, triangular, sheathing, acute, about as long as the ovary; flowers pale greenish shaded with reddish-brown; petals setaceous, white; lip 3-lobed; column connate with lip nearly to apex.—*Sw. Fl. Ind. Occ.* 1509 & *Ic. ined. t.* 42; *Lindl. Fol. Orch. Epid.* 78; *Griseb. op. cit.* 618.

E. anceps *Sw. Prodr.* 121 (1788) (non *Jacq.*). (Pl. 15, f. 6.) Type in Herb. Mus. Brit.

On trees; in fl. Nov.–April; mountains, *Swartz!* *Macfadyen!* *March!* J.P. 150, *Morris!* Clydesdale, 4000 ft.; Cedar Hurst, 2000 ft.; Wallenford, 3000 ft.; ridge below Vinegar Hill, 3800 ft.; *Harris!* Fl. Jam. 7558, 7753, 9024, 10,095a.

Plant 6–13 in. high. *Stem* 5–15 cm. l., leafy, covered with leaf-sheaths, sometimes decayed below, branching below, throwing out very long slender roots. *Leaves* fleshy, linear-lanceolate, cordate, clasping, somewhat falcate, 3–6 cm. l., 4–8 mm. br. *Spikes* much longer than the leaves, 10–25 cm. l. *Bracts* 5–10 mm. l. *Flowers*, perianth about ¼ in. l. *Sepals* 3-nerved, oblong-elliptical, tapering to the keeled apex, lateral slightly falcate, 6–7 mm. l., 3 mm. br.; median as long, but a little narrower. *Petals* 5 mm. l. *Lip* 5–6 mm. br., greenish-yellow, lobes somewhat equal, broader than long, lateral somewhat roundish, with undulate margin, median 2-lobed, apiculate; disk with two calli at base. *Column* short, thick, laterally compressed, thicker upwards, 3–4 mm. l.; clinandrium, margin undulate. *Capsule* spindle-shaped, long-tapering at base and apex.

20. **E. polybulbon** *Sw. Prodr.* 124 (1788); growing in masses with long creeping rhizome bearing many pseudobulbs each with two leaves and one white or pale yellow flower; lip undivided; column free.—*Sw. Fl. Ind. Occ.* 1491; *Hook. Exot. Fl. t.* 112; *Bot. Mag. t.* 4067. Dinema polybulbon *Lindl. Gen. & Sp. Orch.* 111 (1831). Type in Herb. Mus. Brit.

On trees and rocks; in fl. Oct.–Feb.; mountains, *Swartz!* *Macfadyen!* Manchester, *Purdie!* Bethabara, *Wullschlaegel*, 1083! *March!* Mt. Moses, 3000 ft. J.P. 2060, *Syme!* Bath, *Fawcett!* Silver Hill, 4000 ft.; Lancaster, 2500 ft.; Holly Mount, Mt. Diabolo, 2400 ft.; above Mabess River, 3000 ft.; below Vinegar Hill, 3800 ft.; near Christiana, 3000 ft.; *Harris!* Fl. Jam. 7538, 7838, 8870, 9796, 10,098, 10,400, 10,401; near Belle Vue, Port Royal Mts., *Miss B. Gosset!*—Cuba, Mexico, Guatemala.

Plant 2 or 3 in. high. *Rhizome* jointed, covered with remains of decayed sheaths. *Pseudobulb* ellipsoidal to cylindrical, with a short joint at apex, rising directly from the rhizome or on branches of it, 1·5–2·5 cm. l., covered at first with two scarious sheaths which soon decay. *Leaves* linear-oblong to oblong-elliptical, obtuse, obliquely emarginate, with numerous nerves, half-clasping at base, 2–5 cm. l., 7–14 mm. br. *Ovary* cylindrical, with stalk about 3 cm. l.; stalk with 3 bracts. *Bracts* sheathing below, shortly acuminate, about 1 cm. l. *Flowers*, perianth a little more than ½ in. l. *Sepals* 5-nerved, lanceolate-linear, very acute, spreading, 15–17 mm. l., 2·5 mm. br. *Petals* like the sepals but narrower. *Lip* white, subroundish-cordate, clawed, as long as the sepals, 10–11 mm. br. *Column* short, 2-angled and with two projecting teeth in front, and two long, subulate, erect teeth, one on each side, 4–5 mm. l. *Capsule* cylindrical, narrow.

21. **E. Ottonis** *Reichb. f. in Hamburg. Gartenz. xiv.* 213 (1858); pseudobulbs with one long narrow leaf and flowering stalk nearly as long as leaf; flowers cream-coloured; lip undivided, somewhat similar to sepals and petals; column quite free.—*Griseb. op. cit.* 613; *Cogn. in Symb. Ant. iv.* 173. (Pl. 15, f. 8.)

On trees; in fl. and fr. Jan.; Belvedere, Hanover, 500 ft., *Harris!* Fl. Jam. 7540, 7678.—Cuba, Porto Rico, Trinidad, Venezuela.

Plant 6–8 in. high. *Pseudobulbs* on short many-jointed clustered branches of a creeping rhizome, covered with three or four scarious sheaths, of which one is longer than pseudobulb, ellipsoidal, 2–2·5 cm. l. *Leaves* linear-ligulate, obtuse, mucronate, 9–15 cm. l., 6–9 mm. br. *Peduncle* many-jointed, covered with bracts, 7–10 cm. l. *Bracts*, sterile and floral similar, folded together and compressed, linear-lanceolate, acuminate, 12–15 mm. l. *Ovary* smooth, 8–10 mm. l. *Pedicel* 2 mm. l. *Flowers*, perianth about ⅓ in. l. *Sepals* 5-nerved, lanceolate, shortly acuminate, 8–9 mm. l., 2–3 mm. br., lateral slightly falcate. *Petals* oblanceolate, acute, 3-nerved, 6–6·5 mm. l., 2–2·5 mm. br. *Lip*, margin undulate, nearly as long as petals, 5·5–6 mm. l., 2 mm. br. *Column* 4 mm. l., thicker upwards, slightly winged at the sides; clinandrium 5-toothed; anthers 3, the lateral anthers apparently somewhat aborted, obtuse at apex and sides. *Capsule* narrowly ellipsoidal, with three shallow keels, very shortly beaked, 14–15 mm. l., 5 mm. br.

22. **E. subaquilum** *Lindl. in Bot. Reg. xxxii. sub. t.* 64 (1846); pseudobulbs with one leaf and panicle longer than, or about equal to, leaf; lip undivided, broadly ovate above claw; column almost free.—*Lindl. Fol. Orch. Epid.* 5. E. Harrisii *Fawc. in Symb. Ant. i.* 291 (1899).

On trees; in fl. Jan., Feb., in fr. March; near Bath, J.P. 4, *Morris*! Bridge Hill, above Gordon Town, 1500 ft., *Harris*! Fl. Jam. 5583, 7645, 7858, 10,422.—Mexico or Guatemala.

Plant 8–12 in. high. *Pseudobulbs* densely clustered, with three or four scarious sheaths, soon splitting into ribbons, of which two are longer than the pseudobulb, ovoid, 2–2·5 cm. l., about 1 cm. br. *Leaves* linear-ligulate, obtuse, obliquely emarginate, 8–19 cm. l., 3·5–6 mm. br. *Scape* slender, paniculate, 16–27 cm. l. *Bracts* sterile, small, few, distant, sheathing, adpressed, triangular, acute or obtuse, 4–6 mm. l.; floral smaller, spreading, subulate. *Ovary* about as long as the pedicel, about 4 mm. l. *Flowers* greenish-yellow striped with reddish-brown, perianth ¼–⅓ in. l. *Sepals* 3-nerved, linear-lanceolate, somewhat acute, 8–9 mm. l., 2 mm. br., median slightly narrower. *Petals* 1-nerved, linear, 7–8 mm. l., less than 1 mm. br. *Lip* embracing the column, disk with three calli, as long as the sepals. *Column* 4–5 mm. l., slightly thicker above, 2-edged; clinandrium with two short rounded lobes. *Capsule* ellipsoidal, 10–11 mm. l., 5–6 mm. br.

23. **E. bletioides** *Griseb. Fl. Br. W. Ind.* 615 (1864); pseudobulbs with two (?) leaves and a scape longer than leaf; lip undivided, concave, elliptical, pointed, with two linear crests below the middle, claw very short; column free.

Hanover, *Purdie*! near Browns Town, *Miss T. M. Barrett* (in Herb. Kew.)!

Plant 1–1½ ft. high. *Pseudobulbs* ovate-lanceolate, 3·5 cm. l. *Leaves* fleshy, oblong, apex rounded, 12–15 cm. l., about 2·5 cm. br. *Scape* with flowers distant in a spreading panicle, or racemose. *Bracts*, sterile distant, short, adpressed-sheathing, blunt, about 1 cm. l.; floral short, broadly clasping, acute, much shorter than the pedicel, 3 mm. l. or less. *Flowers*, perianth ½ in. l. *Sepals* oblong-lanceolate, acute, 12–13 mm. l. *Petals* oblong to slightly oblanceolate, acute, as long as sepals. *Column* obovate, sharply 3-dentate at the broad summit, 7 mm. l. *Capsule* (immature) spindle-shaped, keeled, 2·5 cm. l.

24. **E. parvilobum** *Fawc. & Rendle in Journ. Bot. xlvii.* 123 (1909); pseudobulbs with one leaf and panicle much longer than leaf; lip 3-lobed, lateral lobes obovate, somewhat roundish, middle lobe small, semicircular, minutely apiculate, disk with three thick raised lines and near the base a large callus; column quite free. —(Pl. 16, f. 5–7.)

On trees; in fl. May; Dolphin Head, 1600 ft., *Harris*! Fl. Jam. 10,439.
Plant 2 ft. high. *Pseudobulbs* 3–5 cm. l., lanceolate-ovate, covered by two scarious sheaths. *Leaf* linear-ligulate, obtuse, conduplicate at base, with many prominent nerves, 22 cm. l., 1·5–2 cm. br. *Panicle* laxly few-flowered. *Bracts*, sterile sheathing, small, obtuse, 5–10 mm. l.; floral deltoid, half-clasping, shortly acute, 1·5–2 mm. l. *Pedicels* 7 mm. l. *Sepals* 5-nerved, oval, obtuse; median 10 mm. l., 4 mm. br.; lateral 8 mm. l., 4 mm. br. *Petals* reticulate-3-nerved, oblanceolate-oblong, 8 mm. l., 2·5 mm. br. *Lip* 8 mm. l., 10 mm. br. *Column* small, 3 mm. l., wider at apex, winged; clinandrium with minute teeth.

25. **E. monticolum** *Fawc. & Rendle op. cit.* 124; pseudobulbs with one leaf or sometimes two, and scape as long as, or shorter than, the leaf; ovary rugose; lip shortly and broadly clawed, deeply 3-lobed, lateral lobes oblong, somewhat obtuse, median lobe much larger, roundish-rhomboidal, broadly clawed, somewhat apiculate at the apex, margin slightly curled below, veins raised, branching, diverging, disk with two thick calli; column quite free. (Pl. 16, f. 1–4.)

In fl. Feb.; Holly Mount, Mt. Diabolo, *Harris*! Fl. Jam. 10,467.
Plant 7–9 in. high. *Pseudobulbs* 1·5–2 cm. l., ovoid, covered by scarious sheaths. *Leaves* linear-ligulate, obtuse, conduplicate at the base, with many nerves scarcely prominent, 12–22 cm. l., 7–10 mm. br. *Scape* laxly few-flowered, rugulose. *Bracts*, sterile sheathing, small, somewhat acute, 8 mm. l.; floral half-clasping, somewhat acute, 2–3 mm. l. *Ovary* and pedicel covered with minute warts, 1·5–1·7 cm. l. (apparently fertilised). *Pedicels* 1 cm. l. *Sepals* reticulate-6–7-nerved, lanceolate-oblong, somewhat acute, somewhat thick, dark green, 1 cm. l., 3 mm. br. *Petals* reticulate-3–4-nerved, oblanceolate-oval, obtuse, green, 9 mm. l., 3 mm. br. *Lip* brownish-yellow, 9 mm. l.; median lobe, 6 mm. l., 5–6 mm. br.; veins raised. *Column* 5 mm. l., stout, winged, at the apex auriculate.

26. **E. belvederense** *Fawc. & Rendle op. cit.* 123; pseudobulbs with two leaves and panicle much longer than leaf; ovary smooth; lip reticulate-nerved, the three median nerves thick, prominent, especially on the disk, 3-lobed, lateral lobes short, obtusely triangular, much smaller than the median, median deltoid or triangular-ovate, shortly apiculate; column quite free. (Pl. 16, f. 8–10.)

On trees; in fr. Dec.–Jan.; Belvedere, Hanover, 500 ft.; Lancaster, 2500 ft.; *Harris*! Fl. Jam. 7541, 7620; between Mandeville and Black River, J.P. 532, *Syme*!
Plant 1½ ft. high. *Pseudobulbs* about 4 cm. l., about 1 cm. br., lanceolate-ovoid, covered by scarious sheaths. *Leaves* linear-ligulate, obtuse, conduplicate at base, with many prominent nerves, 16–18 cm. l., 15–18 mm. br. *Panicle* laxly few-flowered. *Bracts*, sterile sheathing, small,

obtuse, 7–8 mm. l.; floral deltoid, half-clasping, shortly acute, 2–3 mm. l. *Pedicels* 1 cm. l. *Sepals* 5–7-nerved, oblong-lanceolate, subacute, 1·3 cm. l.; lateral slightly keeled below the apex, 3·5 mm. br.; median 3 mm. br. *Petals* reticulate-8-nerved, oblanceolate, 1·3 cm. l., 3 mm. br. *Lip* 1·2 cm. l., 9 mm. br.; lateral lobe 2 mm. l. *Column* stout, flattish anteriorly, 7·5 mm. l. (including anther); clinandrium 3-toothed, auriculate anteriorly. *Capsule* spindle-shaped, 3 cm. l., 1 cm. br.

27. **E. angustifolium** *Sw. Prodr.* 123 (1788); pseudobulbs with one grass-like leaf, and panicle much longer than leaf; lip 3-lobed, lobes subequal, lateral half-round, middle ovate; column connate with lip for one-third of its length.—*Sw. Fl. Ind. Occ.* 1512; *Griseb. op. cit.* 613.

On trees and rocks; Dolphin Head, *Purdie*! Dolphin Head, 1600 ft., *Harris*! Fl. Jam. 10,440.

Plant 8 in. to 1 ft. high. *Pseudobulbs* clustered, 1·5 cm. l., 6–7 mm. br., ovoid, acuminate. *Leaf* 12–22 cm. l., 2 mm. br., linear, exceeded by the scape. *Scape* about 3 dm. l., filiform, naked. *Bracts* 2·3–1 mm. l., minute, pointed. *Raceme* lax, branched at the base. *Pedicels* 4 mm. l., as long as the ovary. *Flowers*, perianth less than ¼ in. l. *Sepals* about 5 mm. l., about 1·5 mm. br., oblong-lanceolate, 3-nerved. *Petals* about as long as the sepals, linear-lanceolate, 1-nerved. *Lip* about as long as the sepals. *Column* nearly as long as the lip, auricled below the clinandrium.

28. **E. brachyglossum** *Cogn. in Fedde Repert. vii.* 121 (1909); stem without pseudobulb, with one leaf at apex and an erect slender peduncle twice as long as leaf with solitary flower; column shortly connate with lip above base.

On trees; in fl. Sept.; Green Hill Wood, 4000 ft., *Harris*! Fl. Jam. 7783.

Plant 8–12 in. high, with brown spots on sheaths, bracts and peduncle. *Stem* 4–7 cm. l., 1·5–2 mm. br., with sheaths below. *Leaf* sessile, somewhat leathery, linear-ligulate, tapering downwards shortly above the base, apex somewhat obtuse and obliquely emarginate, with numerous nerves prominent beneath, especially the middle nerve, 6–10 cm. l., 6–9 mm. br. *Sheaths* three, scarious, free portion short, triangular, obtuse, uppermost 2–3 cm. l., lowest 1–1·3 cm. l. *Bracts* 3–5, closely sheathing, free portion short, triangular, obtuse; floral bract shortly spathaceous, shorter than the ovary. *Pedicels* about 5 mm. l. *Flowers* orange-scarlet, ¾–1 in. l., solitary, erect, thinly membranous. *Ovary* cylindrical, smooth, with pedicel nearly 3 cm. l. *Sepals* 7-nerved; lateral oblong-elliptical, apex somewhat obtuse and conduplicate, margin slightly undulate, 1·8–2·3 cm. l., 7–8 mm. br.; median elliptical, apex rounded, 1·7–2 cm. l., 8–9 mm. br. *Petals* a little oblique, 5–6-nerved, elliptical, tapering to base, apex rounded, shorter than the sepals, 1·9 cm. l., 9 mm. br. *Lip* very short, not half as long as petals, 9 mm. l., 7 mm. br., embracing column, 3-lobed, lateral lobes rounded, median lobe somewhat smaller than lateral, indistinctly 3-lobed; disk with two broad longitudinal lamellæ and warty protuberances above them. *Column* as long as the lip, broader above than below, winged; clinandrium with two large auricles, apex toothed. *Capsule* (not ripe) narrowly spindle-shaped, 3-angled, 3-keeled, keels 2-edged, 2·4 cm. l.

29. **E. serrulatum** *Sw. Prodr.* 121 (1788); stem without pseudobulbs, with one leaf at apex and a spike shorter than leaf;

column shortly connate with lip above base.—*Sw. Ic. ined. t.* 43 ; *Griseb. op. cit.* 615. Cymbidium serrulatum *Sw. in Nov. Act. Upsal. vi.* 72 (1799) & *Fl. Ind. Occ.* 1465. (Pl. 15, f. 4.) Type in Herb. Mus. Brit.

On trees; in fl. and fr. throughout year; summit of Blue Mountains, *Swartz*! *Macfadyen*! Blue Mt. Peak, *Purdie*! Blue Mountains, 6000–7300 ft., J.P. 20, *Morris*! *Jenman*! near Cinchona, 5000 ft.; Morse's Gap, 5000 ft.; Sir John Peak, 6000 ft.; Blue Mt. Peak, 7420 ft.; *Harris*! Fl. Jam. 7860, 10,452; summit of Sir John's Peak, *Mrs. E. G. Britton*, 3878!

Plant 1–4 in. high. *Stem* ·5–3 cm. l., covered with sheaths. *Leaf* linear, tapering below, sessile, apex acute, serrulate, 2–7 cm. l., 3–5 mm. br., turning blackish in drying. *Sheaths* one longer than stem, somewhat spathaceous above, and one or two smaller and scarious below. *Bracts* ovate-triangular, acute, half-clasping, 3–4 mm. l. *Flowers* greenish, sometimes yellowish above, changing to reddish-yellow, sessile or subsessile, perianth $\frac{1}{8}$ in. l. *Ovary* thicker above. *Sepals* fleshy, lanceolate, obtuse, keeled, 5 mm. l. *Petals* 3-nerved, narrowly oblong, obtuse, 4 mm. l. *Lip* indistinctly 3-lobed, lateral lobes rounded, median lobe triangular, disk with three prominent nerves, 4 mm. l., 3 mm. br. *Column* very short. *Capsule* ellipsoidal, 10–11 mm. l.

30. E. **globosum** *Jacq. Enum. Pl. Carib.* 29 (1760); stem short, simple, without pseudobulbs, with a few linear leaves, and a few minute flowers (rarely one) with very short stalks clustered at apex; column free.—*Jacq. Sel. Stirp. Amer.* 222, *t.* 133, *f.* 1, *Ed. pict. t.* 203; *Sw. Prodr.* 121; *Griseb. op. cit.* 619; *Reichb. f. in Saund. Ref. Bot. ii. t.* 140; *Cogn. in Fl. Bras. iii. pt.* 5, 43, *t.* 9, *f.* 1 & *in Symb. Ant. iv.* 173. Cymbidium globosum *Sw. in Nov. Act. Upsal. vi.* 72 (1799) & *Fl. Ind. Occ.* 1467. Isochilus globosum *Lindl. Gen. & Sp. Orch.* 112 (1831).

On trees; in fl. and fr. throughout year; *Macfadyen*! *Imray*; *Purdie*! Moneague, *Prior*! New Carmel, *Wullschlaegel*, 1092! Wag Water River, J.P. 2338, *Syme*! Troy, 1600 ft., *Harris*! Fl. Jam. 10,445, 10,486.—Cuba, Porto Rico, Saba, St. Eustatius, St. Kitts, Guadeloupe, Dominica, Martinique, St. Vincent, Grenada, Trinidad, Nicaragua, B. Honduras, Mexico, B. Guiana, Brazil.

Plant 1–6 in. high. *Stems* 2·5–14 cm. l., numerous, clustered, slender, many-jointed, covered with sheaths. *Leaves* short, fleshy, rigid, linear, semiterete, flattened when dried, obtuse, jointed to the sheath, 1–2 cm. l., 1·5–2 mm. br. *Peduncles* one-flowered, 1–2 mm. l. *Flowers* yellow, red-variegated, perianth $\frac{1}{8}$ in. l. *Bracts* sheathing at the base, 1–1·5 mm. l. *Sepals* fleshy, elliptical-ovate, obtusely acute, 3 mm. l., median slightly broader, 2 mm. br., lateral very shortly connate at base, 1·5 mm. br. *Petals* yellow, fleshy, ovate-subrhomboid, obtusely angled, apex somewhat acute, a little shorter than the sepals. *Lip* fleshy, slightly shorter than the sepals, undivided, obovate-rhomboid, apex obtuse, slightly gibbous. *Column* short, thick, $1\frac{1}{2}$ mm. l. *Capsule* nodding, ellipsoidal-ovoid, apex shortly beaked, 4 mm. l., 3 mm. br.

31. E. **teretifolium** *Sw. Prodr.* 121 (1788); stem long, simple, without pseudobulbs, with a few linear fleshy leaves, and one flower at apex; column connate with lip about half its length.—*Sw. Ic. ined. t.* 44; *Griseb. op. cit.* 619. Cymbidium teretifolium

Sw. in Nov. Act. Upsal. vi. 72 (1799) & *Fl. Ind. Occ.* 1463. Isochilus teretifolium *Lindl. Gen. & Sp. Orch.* 112 (1831). Type in Herb. Mus. Brit.

On trees and rocks; in fl. Aug., in fr. Oct., Nov.; mountains, *Swartz*! J.P. 2054, *Morris*! near Chester Vale, *Watt*! Cinchona, 5000 ft.; Mt. Moses, 3000 ft.; Greenwich, 4500 ft.; *Harris*! Fl. Jam. 7627, 7778, 10,142.—Cuba.

Plant 6–11 in. high. *Stems* 15–28 cm. l., numerous, clustered, somewhat robust, many-jointed, covered with sheaths below, naked above and somewhat compressed. *Leaves* rigid, semiterete, obtuse, jointed to the sheath, 2–4·5 cm. l., on lower half or two-thirds of stem, one leaf with very much abbreviated sheath at apex. *Peduncle* solitary, terminal, short, one-flowered, 7 mm. l. *Bracts* at base of peduncle, four spathaceous bracts outside and two sheathing bracts inside, about 8 mm. l. *Flower* yellowish-green, perianth about ½ in. l. *Sepals* 12 mm. l., 1·6 mm. br., fleshy, linear, tapering towards the thickened obtusely acute apex, somewhat convex. *Petals* half as long as sepals, 6–7 mm. l., 1·5 mm. br., stalked, elliptical, thickened at the rounded apex, 3-nerved, with the median nerve forming a keel, green. *Lip* shorter than sepals, from junction with column 8 mm. l., about 2 mm. br., clawed, lower half of limb broadening upwards, concave, disk 3-nerved with a median keel at base, above 3-edged and thickened. *Column* shorter than petals, 5 mm. l., 2-winged at apex. *Capsule* ellipsoidal-spindle-shaped.

31. BROUGHTONIA R. Br.

Epiphytic herbs, with ovoid or globose pseudobulbs, which are covered with a few sheaths, and have one or two leaves at the apex. Leaves leathery or fleshy, oblong or elongate. Peduncles terminal, with adpressed sheaths. Raceme lax, simple or sparingly branched. Flowers medium-sized or rather large, shortly stalked. Bracts small, scale-like. Sepals equal, free, somewhat spreading. Petals larger than the sepals. Lip sessile at the base of the column, free from it, erect, broad, embracing the column. Spur adnate to ovary, long, slender. Column several times shorter than the sepals, erect, broadly 2-winged; clinandrium with short teeth. Anther terminal, opercular, incumbent, convex, 4-celled; pollinia 4, waxy, in two pairs, ovate, equal, compressed from the side, connected by a granular-viscous appendage. Capsule ovoid, beaked with the neck of the ovary.

Species 2, natives of the West Indies.

Flowers crimson, ½–nearly 1 inch l. .... 1. *B. sanguinea.*
Flowers lilac, 1–1¼ inch l. .... 2. *B. domingensis.*

1. **B. sanguinea** *R. Br. in Ait. Hort. Kew. ed.* 2, *v.* 217 (1813); flowers crimson, ½ to nearly 1 in. l.; lip cordate at base, glabrous on disk.—*Lindl. Gen. & Sp. Orch.* 118; *Hook. in Bot. Mag. t.* 3076. B. coccinea *Hook. in Bot. Mag. t.* 3536 (1836). Viscum radice bulbosa minus &c. *Sloane Cat.* 119 & *Hist. i.* 250, *t.* 121, *f.* 2. Epidendrum sanguineum *Sw. Prodr.* 124 (1788).

E. nodosum var. *β* *L. Sp. Pl.* 953 (1753). E. cochleatum *L. Sp. Pl. ed.* 2, 1351 (in part) (1763). Dendrobium sanguineum *Sw in Nov. Act. Upsal. vi.* 82 (1799) & *Fl. Ind. Occ.* 1529 (Pl. 14, f. 10–14.) Type in Herb. Mus. Brit.

On trees; in fl. Nov.-April; Spanish town and neighbourhood, *Sloane* Herb. iv. 118! *Swartz*! *Bancroft*! *Distin*! Westmoreland, *Purdie*! *Lane*! *Wilson*! *Prior*! Manchester, *Wullschlaegel*, 1059! lowlands up to 500 ft., *Morris*! *Syme*! Content Road, *Moore*! Manchester, *Bassett-Key*! Halberstadt, 2000 ft., *Miss B. Gossett*! near Tweedside, 2500 ft.; Great Goat Island, 100 ft.; *Harris*! Fl. Jam. 9020, 9209.—Cuba.

Plant 12–15 in. high. *Pseudobulbs* 2·5–5 cm. l., 1·5–2·5 cm. br., varying in shape from somewhat globular to nearly cylindrical, clustered, with two leaves and peduncle much longer than leaves. *Leaves* leathery, oblong, somewhat acute, with numerous nerves visible when dry, 8–20 cm. l., 1·5–4·5 cm. br. *Sheaths* 2 or 3, 3–6 cm. l. *Peduncle* long, 2–4 dm. l., sometimes branched with lax raceme (7–8 cm. l.) at apex. *Bracts* sterile, short, sheathing, adpressed, triangular, somewhat obtuse, few, distant, 5–6 cm. l.; floral minute. *Pedicels* 1–1·5 cm. l. *Flowers* 5–12. *Sepals* oblong-lanceolate, 1·5–2·4 cm. l., 4–5 mm. br., lateral broader and slightly falcate. *Petals* roundish-elliptical, tapering to base and to apex, which is either acute or obtuse, as long as sepals, 7 mm. br. *Lip* nearly as long as sepals and petals, and broader than long, roundish, truncate, margin eroded, yellow at base with purplish veins. *Column* about 9 mm. l., slightly curved, thicker and winged above.

2. **B. domingensis** *Rolfe in Gard. Chron. v.* 491 (1889); flowers rosy-lilac, 1–1¼ in. l.; lip hairy on the central veins. B. lilacina *Henfr. in Gard. Mag. Bot. iii.* 201 (1851). Cattleya domingensis *Lindl. Gen. & Sp. Orch.* 118 (1831). Læliopsis domingensis *Lindl. in Paxt. Fl. Gard. iii.* 155, *t.* 105 (1852, 53). (Pl. 14, f. 15–19.)

On trees; Westmoreland, *Purdie*! *Edwards*! *Tomlinson*!—Hispaniola.

Plant 1–2 ft. high. *Pseudobulb* 2–5 cm. l., 1·5–2·5 cm. br., ellipsoidal to globose, with two leaves (rarely one), and peduncle much longer than leaves. *Leaves* leathery, oblong-ligulate, apex oblique, mucronate, with numerous nerves apparent only when dry, 6–20 cm. l., 8–30 mm. br. *Sheaths* 2 or 3, 2·5–5 cm. l. *Peduncle* very long, 3–5 dm. l., rarely branching, with a lax raceme (5 cm. l.) at apex. *Bracts* sterile, short, sheathing, adpressed, triangular, somewhat obtuse, very few, distant, 6 mm. l.; floral minute, triangular. *Flowers* 8–14. *Sepals* linear-lanceolate, somewhat acute, 2 to nearly 3 cm. l., 4–6 mm. br. *Petals* elliptical tapering to base and to somewhat acute apex, about as long, or a little longer, and twice as broad as sepals. *Lip* a little longer than sepals, broadly emarginate, margin wavy, denticulate. *Column* about 7 mm. l. *Capsule* (unripe) ellipsoidal, beaked.

32. HORMIDIUM Lindl.

Small epiphytic herbs, with small pseudobulbs crowded along the creeping rhizome, with two or three membranous sheaths and two leaves at the apex. Leaves small, shortly sheathing at base, and jointed with sheath. Flowers small, stalked, in a terminal cluster, the cluster sometimes reduced to a single flower. Sepals, all of equal length, suberect or at length spreading, the median free, the lateral broader, at the base adnate to the columnar cup. Petals similar to the median sepal, but narrower. Lip connate at the base with the column into a cup, the blade erect-spreading, 3-lobed. Column short, with the dilated margins connate with the lip into a cup; clinandrium short. Anther terminal, opercular, incumbent, with four parallel cells; pollinia 4, waxy, ovoid, distinct, without an appendage, on the dehiscence of the anther connected at the apex by a sparse viscid substance. Capsule ovoid or subglobose, not beaked, the ribs winged.

Species 7, natives of tropical America from the West Indies and Mexico to Brazil.

Hormidium tripterum *Cogn. in Fl. Bras. iii. pt.* 5, 29, *t.* 8, *f.* 2 (1898); *Ames Orch. ii.* 7, *t.* 20. Cœlogyne triptera *Brongn. in Duperr. Voy. Coq. Bot.* 201, *t.* 42 *A* (1829).—Epidendrum pygmæum *Hook. in Bot. Mag. t.* 3233 (1833) & *Journ. Bot. i.* 49, *t.* 118 (1834); *Lindl. Fol. Orch. Epid.* 27; *Griseb. op. cit.* 615. Hormidium pygmæum *Hemsl. in Gard. Chron. xix.* 700 (1883). (Pl. 17, f. 1–5.)

On rocks covered with peat and moss in shade; in fl. and fr. Oct.-Feb.; Manchester, *Purdie*! Springfield, *Wullschlaegel* 1084! J.P. 2011, *Syme*! *Morris*! Clydesdale, *Cradwick*! Cascade, *Watt*! Bridge Hill, 1500 ft.; Chester Vale, 3000 ft. and 3500 ft.; *Harris*! Fl. Jam. 7103, 7535, 7823. Mabess River, *G. Nichols*!—Cuba, Porto Rico, Trinidad, Mexico, B. Guiana, Brazil, Peru.

Plant rising 3 or 4 in. from the long, creeping rhizome. *Pseudobulbs* 2–3 cm. l., ovoid-oblong or narrowly spindle-shaped. *Leaves* narrowly elliptical or oval, obtuse, minutely apiculate, shortly sheathing at the base and jointed with sheath, blade 3–5 cm. l., 1–1·5 cm. br. *Bracts* somewhat longer than the pedicels, 6 or 7 mm. l. *Flowers* yellowish-white, perianth ⅕ in. l. *Sepals* elliptical, acuminate, shallowly keeled, 5–6 mm. l., 2 mm. br.; lateral shortly connate at base. *Petals* linear, acuminate, 4–4·5 mm. l. *Lip* extending about half as long as sepals, 4·5–5 mm. br., median lobe triangular, acuminate, much smaller than the roundish lateral lobes. *Column* half as long as lip. *Capsule* about 1·3 cm. l.

33. LÆLIA Lindl.

Epiphytic herbs, with fleshy generally oblong pseudobulbs, not, however, present in the Jamaican species, which have a few sheaths, and at the apex one or two leaves. Leaves somewhat leathery, not plaited. Peduncles terminal, almost always only one-flowered, with a few sheaths at intervals. Bracts like the sheaths, the uppermost spathaceous. Sepals almost equal, free, spreading. Petals wider than the sepals. Lip erect at the base of the column; lateral lobes enclosing the column; the median lobe spreading. Column rather long, narrowly 2-winged; clinandrium crenulate. Anther attached to a posterior tooth, opercular, incumbent, convex, distinctly 2-celled, each cell imperfectly divided again into two cells by a longitudinal partition; pollinia 8, four in each cell, all compressed in the same direction, the ascending portion of the inferior pair connected with the descending portion of the upper pair in each cell by a linear granular appendage.

Species about 39, natives of the warmer parts of America from the West Indies and Mexico to Brazil.

L. monophylla *N. E. Brown in Gard. Chron. xviii.* 782 (1882); *Hook. f. in Bot. Mag. t.* 6683. Trigonidium monophyllum *Griseb. Fl. Br. W. Ind.* 629 (1864). Octadesmia monophylla *Benth. in Benth. & Hook. f. Gen. Pl. iii.* 526 (1883). (Pl. 17, f. 6–12.)

On trees; in fl. Aug.–Nov.; *Bancroft*! Mt. Moses, J.P. 2405, *Syme*! *Morris*! Greenwich Woods, *Fawcett*! Tweedside; Green Hill Woods, 4000 ft.; *Harris*! Fl. Jam. 7783, 7996.

Plant 4–8 (sometimes 12) in. high. *Stem* slender, 2–9 cm. l., covered with three speckled cylindrical sheaths springing from the nodes, and with one leaf at apex. *Leaf* linear-oblong, obtuse, 5–10 cm. l., 7–10 mm. br. *Peduncle* with 4–6 bracts like the sheaths, 5–22 cm. l. *Flowers* vivid orange-scarlet, except the anther-cap which is purple, about $\frac{3}{4}$ in. l. *Sepals*, lateral elliptical, obtuse, 17–22 mm. l., 5·5–7·5 mm. br.; median ovoid-elliptical, 16–19 mm. l., 6·5–7·5 mm. br. *Petals* obovate-elliptical, 14–17 mm. l., 6–9 mm. br. *Lip* obovoid in outline, 7–9 mm. l., continuous with wings of column below, 3-lobed near apex, lateral lobes rounded, median lobe semicircular, somewhat apiculate, disk papillose, with a sac-like outgrowth along the median line. *Column* 6·5–8·5 mm. l., winged; clinandrium broadly winged, crenulate on the dorsal margin.

34. SCHOMBURGKIA Lindl.

Epiphytic herbs; pseudobulbs with several scarious sheaths, 2-leaved at the apex. Leaves oblong, elongated, thick, leathery. Raceme simple on a terminal elongated sheathed peduncle, flowers stalked. Bracts persistent, rigidly membranous. Sepals almost equal, free, spreading, more or less undulate. Petals similar and equal to the sepals. Lip very shortly connate quite at the base with the column, erect; the sides at first loosely embracing the column; the upper margin undulate. Column straight or incurved, winged; clinandrium lobed. Anther attached to the apex of the median tooth, incumbent; imperfectly 8-celled; pollinia 8, waxy, superposed in pairs, broadly ovate, compressed, connected by a slender granular appendage. Capsule ovoid.

Species about 13, natives of tropical America, including the West Indies.

S. Lyonsii *Lindl. in Gard. Chron.* 1853, 615; *Hook. in Bot. Mag. t.* 5172. S. carinata *Griseb. Fl. Br. W. Ind.* 621 (1864). (Pl. 19, f. 1–6.)

On rocks and trees; in fl. Aug.–Nov., in fr. Dec.; St. Ann, *Purdie*! St. Ann, *Prior*! Castleton, J.P. 535, *Morris*! Radnor, *Cradwick*! Ramble,

Hanover, 600 ft., *Fawcett*! Content Road, 3000 ft., *Harris*! Savanna-la-Mar, *Lister Clark*. Fl. Jam. 10,391, 10,427.

Plant rising to 3 ft. high and more. *Pseudobulb* narrowly spindle-shaped, compressed, of several joints, 12–35 cm. l., 2·5–3 cm. br. *Sheaths* 8–13 cm. l., reduced to 2 cm. l. at base. *Leaves* 1–3 dm. l., 4–6 cm. br. *Peduncle* 6–9 dm., with sheaths which become spathaceous and crowded and spreading at base of raceme; sheaths 5–6·5 cm. l., acute, sometimes with purplish-brown spot. *Raceme* 7–12 cm. l. *Bracts* linear-lanceolate, acuminate, reflexed, 5·5–7·7 cm. l., to 6 mm. br. *Flowers* white spotted with purple, perianth about 1 in. l. *Pedicels* 5–5·5 cm. l., slender. *Ovary* a little over 1 cm. l. *Sepals* 7-nerved, ovate-oblong, obtuse, very shortly apiculate, with undulate, crispate margin, 2·2–2·3 cm. l., 9–10 mm. br., lateral somewhat falcate and broader than median. *Petals* 5-nerved, similar to sepals but longer and slightly broader, apex rounded, 2·6 cm. l., 11 mm. br. *Lip* 5-nerved, outer nerves branching, nerves prominent, sometimes lamellate; with yellowish edge, otherwise somewhat the same colouring as the sepals and petals, sometimes dull violet with crimson ribs, shorter than sepals, entire, ovate, very shortly clawed, with undulate crispate margin, apex bluntish, 1·6 cm. l., about 12 mm. br. (flattened). *Column* curved, about 1 cm. l.; lateral lobes of clinandrium somewhat larger than median. *Anther* with 2 horn-like processes on the top. *Capsule* 3–4 cm. l., and 1·7–1·8 cm. br.

35. BRASSAVOLA R. Br.

Epiphytic herbs. Stems scarcely thickened, sympodial, springing from the contracted base of the next older stem, with one or two sheaths covering the contracted base, with a few lax ash-coloured scarious sheaths above, and one leaf. Leaves thick and fleshy, conduplicate, almost half cylindrical, linear, tapering to a subulate apex. Flowers terminal, on an elongated peduncle, racemose, stalked. Bracts short. Ovary very slender. Sepals free, equal, spreading, long, linear, acuminate. Petals similar to the sepals, but slightly shorter and narrower, narrowly linear-lanceolate. Lip with a long involute serrate claw; blade roundish, acuminate, entire. Column very short, clinandrium 3-lobed, the median toothed at the apex. Anther attached to the median lobe of the clinandrium, opercular, incumbent, with two distinct cells, each cell divided into two smaller cells by a longitudinal partition; pollinia 8, four in each cell, broadly ovate and compressed in the same direction, connected by a granular, linear or blade-shaped appendage. Capsule large, ellipsoidal, 3-keeled, beaked.

Species about 22, natives of tropical America from the West Indies and Mexico to Argentine.

Flowers few, perianth 2–3 inches l. .... 1. *B. nodosa.*
Flowers many, perianth 1¼–1¾ inch long......... 2. *B. cordata.*

1. **B. nodosa** *Lindl. Gen. & Sp. Orch.* 114 (1831); flowers few; sepals longer than ovary with pedicel, 5–7·5 cm. l.; blade of lip roundish, 3–5 cm. l., claw 17–25 mm. l.—*Lindl. in Bot. Reg.*

t. 1465; *Griseb. Fl. Br. W. Ind.* 621; *Rolfe in Orch. Rev. x.* 65. B. grandiflora *Lindl. in Bot. Reg. xxv. Misc.* 16 (1839). B. venosa *Lindl. in Bot. Reg. xxvi. t.* 39 (1840); *Hook. in Bot. Mag. t.* 4021. Epidendron curassavicum &c. *Herm. Parad.* 207 (errore 187). Viscum arboreum &c. *Pluk. Alm.* 390, *t.* 117, *f.* 6. Epidendrum nodosum *L. Sp. Pl.* 953 (1753) (excl. reference to Sloane & var. *β.*); *Jacq. Sel. Stirp. Amer.* 226, *t.* 140, *Ed. pict.* 109, *t.* 213.

On rocks and trees; in fl. Aug., Sept.; *Wright*! *Dickson*! *March*! Cedar Valley, *Moore*!—Mexico, Honduras, Guatemala, Panama, Venezuela, Colombia, Peru.

Plant 8–13 in. high. *Stem* 6–7 cm. l., with two long joints, each covered with a sheath. *Sheaths* increasing in size from base (4 cm. l.), the topmost triangular, acute, 3·5–5·5 cm. l. *Leaf* 14–22 cm. l., 8–18 mm. br. *Peduncle* with three sterile joints below, followed by the few-flowered raceme, 10–27 cm. l. *Bracts*, sterile sheathing, 7–10 mm. l.; floral triangular, acute, 4–7 mm. l. *Ovary* with pedicel 3·5–4·5 cm. l. *Flowers* large, spreading, perianth 2–3 in. l. *Sepals* and *petals* pale green. *Lip* white. *Column* about 9 mm. l. *Capsule* about 5 cm. l.

2. **B. cordata** *Lindl. in Bot. Reg.* 1914, *t.* 1913 (1836); flowers many; sepals as long as ovary with pedicel, 3·5–4·5 cm. l.; blade of lip subcordate-roundish, 1·8–2 cm. l., claw 12–16 mm. l.—*Hook. in Bot. Mag. t.* 3782; *Griseb. op. cit.* 620. B. nodosa *Hook. in Bot. Mag. t.* 3229 (1833). B. subulifolia *Rolfe in Orch. Rev. x.* 66 (1902); *Griseb. loc. cit.* (in part). Viscum delphinii &c. *Sloane Cat.* 120 & *Hist. i.* 251, *t.* 121, *f.* 3. (Pl. 19, f. 7–11.)

On trees; in fl. June, Sept., Jan.; Spanish Town, *Sloane* Herb. iv. 120! *Wright*! *Bancroft*! *Macfadyen*! St. Thomas in Vale, *McNab*! St. John, *Purdie*! *Parke*! White River, *Cowan*! Fairfield, *Wullschlaegel*, 1065! lowlands, J.P. 2398, *Syme*! Mandeville, *Lady Blake*! Lacovia, *Tomlinson*! Belvedere, Hanover, 500 ft.; Halls Delight, 900 ft.; near Troy, 2000 ft., 1600 ft.; *Harris*! Walderston, *Mrs. Longstaff*! near Mandeville, *Harris & Britton*! Fl. Jam. 7661, 10,412, 10,448, 10,450, 10,451, 10,626.

B. subulifolia Lindl. Gen. & Sp. Orch. 114 was described from a plant from Nevis. Lindley says "species distinctissima." We have not seen a specimen.

Plant 8–21 in. high. *Stem* 12–18 cm. l., with three or four long joints, each covered with a sheath. *Sheaths* increasing in size from base (2 cm. l.), the topmost becoming spathaceous, free portion oblong, tapering to an acute apex, or oblong-lanceolate, 5·5–8 cm. l. *Leaf* 20–35 cm. l., 7–12 mm. br. *Peduncle* one or two sterile joints below, followed by the many-flowered raceme, somewhat crowded above, 8–20 cm. l. *Bracts* sterile, sheathing, free portion triangular, acute, 7–10 mm. l.; floral lanceolate, acuminate, 6–9 mm. l. *Ovary* with pedicel 3·5–4·5 cm. l. *Flowers* of medium size, spreading, perianth 1¼–1¾ in. l. *Sepals* and *petals* pale green or greenish-yellow. *Lip* white, with a long involute serrate claw. *Column* about 7 mm. l., thicker upwards. *Capsule* 3·2 cm. l., 1·6 cm. br.

36. TETRAMICRA Lindl.

Terrestrial herbs. Stems leafy, scarcely thickened, no pseudobulbs. Leaves from the rhizome within overlapping sheaths at the base of the scape, 1–3, linear, fleshy, semiterete, short and thick. Peduncle terminal, elongate, thin, rigid, with membranous adpressed sheaths at intervals. Raceme simple, lax, with small stalked flowers. Bracts minute. Sepals somewhat equal, free, spreading. Petals similar to the sepals but narrower. Lip attached at the base of the column, free, spreading from base; lateral lobes distinct on a short claw, ample, spreading; median broad, entire, contracted at the base. Column erect, broadly 2-winged above; lateral lobes or teeth of the clinandrium often membranous, median small. Anther terminal, opercular, incumbent, convex, with two distinct cells, which are imperfectly divided into two smaller cells; pollinia perfect, 4, waxy, segregated in pairs in the cells, compressed in the same direction, connected by appendages. Capsule oblong, not beaked.

Species 7, natives of tropical America from the West Indies to Brazil.

T. **parviflora** *Lindl. ex Reichb. f. in Walp. Ann. vi.* 439 (1861); *Griseb. Fl. Br. W. Ind.* 622. Bletia parviflora *Reichb. f. loc. cit.* (Pl. 18, f. 7–10.)

In dry rocky ground; in fl. Feb.-April; Stony Hill, 1000 ft., J.P. 2033, *Morris*! *Syme*! near constant Spring, 600-1000 ft., *Campbell*! *Le Ray*! *Harris*! *Fawcett*! Fl. Jam. 7655, 10,409.—Bahamas.

Plant 7-15 in. high. *Stem* very short, 2-4 mm. l., with two or three small leaves and a long peduncle. *Leaves* 1-4 cm. l., 3-6 mm. br. *Peduncle* 18-38 cm. l., very slender, with a lax raceme (4–10 cm. l.) at apex. *Bracts*, sterile sheathing, free part very short, acute, 3·5–5 mm. l.; floral half-clasping, ovate, very acute, 2–2·5 mm. l. *Flowers* small, different shades of pink, sometimes white with a tinge of pink at the top of the column, perianth less than ⅛ in. l. *Sepals* green, sometimes striped inside with fine lines of purplish-brown, 5-nerved, oblong, apex acute to obtuse, 4·5–5 mm. l., 1·6–2·1 mm. br. *Petals* 3-nerved, linear-lanceolate, somewhat obtuse, 4–5 mm. l., ·75–·9 mm. br. *Lip* 5–6 mm. l., 3-lobed, lateral lobes broadly elliptical-ovate; median obovate, 5-nerved, 3·5 mm. br. *Column* 2·5 mm. l.

37. HOMALOPETALUM Rolfe.

Epiphytic herb, with a creeping rhizome, giving off 1-leaved pseudobulbs at intervals. Leaves sessile, fleshy. Peduncle solitary from apex of pseudobulb, longer than leaf, with one flower. Bracts few, small, distant. Sepals equal, free, connivent. Petals similar to sepals. Lip nearly similar to petals, free, undivided. Column thicker upwards, semiterete, incurved, with a falcate-oblong auricle at each side of the base, without a foot; clinandrium margined; rostellum ovate-oblong, descending. Anther terminal, opercular, convex; cells two, distinct, imperfectly divided by a longitudinal division; pollinia, perfect 4, waxy, in pairs in the cells, with linear granular-viscous appendiculæ, at the apex of which are small imperfect pollinia.

Species 1, known only from Jamaica.

H. vomeriforme* comb. nov.—H. jamaicense *Rolfe in Hook. Ic. Pl. t.* 2461 (1896). Epidendrum vomiforme *Sw. Prodr.* (1788) & *Ic. ined. t.* 48. E. vomeriforme *Sw. Fl. Ind. Occ.* 1511 (1806); *Reichb. f. in Walp. Ann. vi.* 416. Brassavola vomeriformis *Reichb. f. ex Griseb. Fl. Br. W. Ind.* 621 (1864). (Pl. 18, f. 1–6.) Type in Herb. Mus. Brit.

On trees; in fl. Dec.–April; mountains, *Swartz*! Colthirst's Run, St. George, J.P. 238, *Morris*! Abbey Green, Blue Mts., 4000–5000 ft; near Cinchona, 5000 ft.; *Harris*! Fl. Jam. 6107, 7658, 7816, 9783, 10,494 (in part).

Plant 1½–4 in. high. *Pseudobulbs* cylindrical, ellipsoidal or somewhat globose, 4–6 mm. l. *Leaves* ovate-oblong to oblong, shrivelling on drying, keeled, somewhat obtuse, 1–2 cm. l., 4–5 mm. br. *Peduncle* filiform, 1–6·5 cm. l. *Bracts*, sterile sheathing, distant, 2–3 mm. l.; floral somewhat spathaceous, 3–3·5 mm. l. *Flowers* sessile, perianth ⅓–⅔ in. l. *Sepals* 1·4–2 cm. l., linear-lanceolate, acuminate. *Petals* similar, but shorter by about 2 mm. and narrower. *Lip* as long as the petals, elliptical (when flattened), shortly acuminate, disk smooth. *Column* 4 mm. l., slender below; clinandrium with one tooth behind and two in front, and a rounded lobe on each side. *Capsule* (unripe) ellipsoidal with a very short beak, 11 mm. l.

38. ELLEANTHUS Presl.

Terrestrial herbs, tall, simple or branched, leafy. Leaves sessile above a sheath, plaited, lanceolate, nerves somewhat prominent below, highest leaf bract-like. Flowers sometimes dense in the form of a head or in a dense two-ranked spike. Sepals almost equal, free, erect, somewhat rigid. Petals as long as the sepals, generally narrower. Lip attached at the base of the column, erect, as long as the sepals or longer, generally enclosing the column, concave at the very base, with two prominent calli in the hollow, generally slightly constricted above the hollow, above roundish or inconspicuously trilobed, margin entire or denticulate. Column erect, semiterete or winged, without a foot. Anther attached to the median tooth, opercular, somewhat convex, slightly incumbent, afterwards erect, distinctly 2-celled, the cells not divided; pollinia 4 in each cell, waxy, ovoid, hardly superposed in pairs, connected at the apex by a sparse viscid substance or by a small granular appendage. Capsule shortly cylindrical, erect or spreading.

Species 45, natives of tropical America from the West Indies and Central America to Brazil and Peru.

Flowers in a head .... 1. *E. capitatus.*
Flowers in a spike .... 2. *E. longibracteatus.*

1. **E. capitatus** *Reichb. f. in Walp. Ann. Bot. vi.* 475 (1861); flowers in a head.—*Cogn. in Fl. Bras. iii. pt.* 5. 325. Bletia

* The original spelling was *vomiforme*, an error in orthography, subsequently corrected by Swartz.

capitata *R. Br. in Ait. Hort. Kew. ed.* 2, *v.* 206 (1813); *Lindl. Gen. & Sp. Orch.* 123. Evelyna capitata *Poepp. & Endl. Nov. Gen. & Sp. i.* 32, *t.* 56 (1835); *Griseb. Fl. Br. W. Ind.* 622. (Pl. 20, f. 1–7.) Type in Herb. Mus. Brit.

On trees; in fl. July, Aug.; *Masson*! *Wiles*! *Macfadyen*! *March*! J.P. 232, *Morris*! near Claverty Cottage, Blue Mts., *Moore*! Vinegar Hill, St. George, *Harris*! Fl. Jam. 7717.—Guadeloupe, Martinique, Grenada, Trinidad, Guatemala, Mexico, Venezuela, Ecuador, Peru.

Plant rising to 4 ft. and over. *Stem* to 12 dm. l., 3–5 mm. br., simple, somewhat stout, erect, covered with leaf sheaths. *Leaves* papery, 7–9 nerved, 1–2 dm. l., 2–5 cm. br.; sheaths 5–6 cm. l., smaller near apex. *Head* dense, many-flowered, hemispherical, 2·5–5 cm. br. *Bracts*, outer without flowers, closely overlapping, ovate-triangular, acuminate, 2·5–4 cm. l., 1–1·5 cm. br.; inner bracts subtending the flowers half as short and two or three times narrower. *Flowers* rosy-purple, perianth ½ in. l., lip ¾ in. l. *Sepals* 3–5-nerved, ligulate, shortly tapering to an acute apex, 12–13 mm. l., 3–3·5 mm. br., lateral united and gibbous at base. *Petals* similar to sepals. *Lip* longer than the sepals, roundish with a narrower base, enclosing the column, saccate at base, apex broadly emarginate, margin denticulate-ciliate, with two large, subglobose calli at base. *Column* 9–10 mm. l., thicker above, with a projecting, obtuse process on the anterior side below the stigma.

2. **E. longibracteatus** *Fawc. Fl. Pl. Jam.* 38 (1893); flowers in a spike.—Evelyna longibracteata *Lindl. ex Griseb. loc. cit.* (1864). (Pl. 20, f. 8–12.) Type in Herb. Lindl. at Kew.

On peaty banks and rocks; in fl. March–May; *Wright*! *Macfadyen*! *March*! J.P. 528, *Syme*! Greenwich, *Moore*! Fairfield, *Oehler*! Lancaster, 2500 ft.; near Troy, 1400 ft; Cinchona, 10,414 ft.; *Harris*!—Dominica, Colombia, Ecuador.

Plant 2–3 ft. high. *Stem* 4·5–8 dm. l., 2–4 mm. br., simple, erect, nearly covered with leaf-sheaths. *Leaves* papery, 5-nerved, 8–17 cm. l., 1–2 cm. br.; sheaths 1·5–3·5 cm. l. *Spike* somewhat dense, 4–8 cm. l., several-flowered. *Bracts* ovate, acuminate, spathe-like, longer than the flower, 1·5–2·5 cm. l. *Flowers* cream-colour or pale yellow, perianth ⅓ in. l., lip longer. *Sepals* 9-nerved, lateral oblong-triangular, mucronate, saccate and gibbous at base, 7·5–8·5 mm. l., 4–5 mm. br. at base; median oblong, apex rounded, mucronate, 7 mm. l., 4 mm. br. *Petals* 6-nerved, ligulate, obtuse, nearly 7 mm. l., 2·3 mm. br. *Lip* not exserted beyond the sepals, saccate, enclosing the column, indistinctly 3-lobed, upper margin denticulate, somewhat eroded, curled, lobes rounded, nearly equal, with two large ovate calli at base, and two transverse linear calli at base of lobes, nearly as long as the lateral sepals. *Column* 6 mm. l., winged; clinandrium with a spreading lobe at each side. *Capsule* cylindrical, slightly puberulous, 8 mm. l.

[39. **PHAIUS** Lour.]

[Terrestrial herbs, rather tall, with stems crowded on a caudex. Leaves elongate, plaited, at the base narrowed or contracted into a long stalk, with sheaths generally striate. Scape springing from the caudex. Flowers in a raceme, several, stalked. Bracts sometimes small, linear, sometimes rather large, membranous or herbaceous. Sepals free, almost equal, spreading or somewhat erect. Petals similar to the sepals, but narrower. Lip erect, concave, produced at the base into a short spur, continuous with the base of the column; lateral lobes ample, erect, enclosing the column; median lobe expanded, generally short and broad, spreading or recurved. Column somewhat long, subterete, 2-angled or 2-winged, thicker upwards, without a foot; clinandrium short, oblique, sinuate on the margin. Anther attached near the margin of the clinandrium, opercular, incumbent, convex, distinctly 2-celled, cells divided imperfectly into four smaller cells; pollinia 8, obovate or oblong, compressed from the side, applied by the faces in pairs, the lower pair of each cell generally longer than the upper, connected at the apex by a granular appendage almost filling the cells in the younger anther. Capsule oblong, without a beak, ribs very prominent.

Species 50, natives of tropical Asia, but spreading to tropical Africa, the Mascarene Islands, Australia, the islands of the Pacific Ocean, China and Japan.

P. Tancarvilleæ Blume (*Tankervillii*) *Mus. Bot. Lugd.-Bat. ii.* 177 (1856). P. grandifolius *Lour. Fl. Cochinch.* 529 (1790); *Lindl. Gen. & Sp. Orch.* 126; *Griseb. Fl. Br. W. Ind.* 622; *Salisb. Ic. ined. vi. fol.* 573 Limodorum Tancarvilleæ *Banks ex L'Hérit. Sert. Angl.* 28 (1788). L. Tankervilliæ *Dryand. in Ait. Hort. Kew. iii.* 302, *t.* 12 (1789). Bletia Tankervilliæ *R. Br. in Ait. Hort. Kew. ed.* 2, *v.* 205 (1813); *Bot. Mag. t.* 1924. (Pl. 20, f. 13–17.)

In damp shady places; in fl. throughout year; Bethany, *Wullschlaegel*, 1061! *Prior*! Blue Mts., J.P. 7, *Morris*! Vinegar Hill, *Moore*! Clydesdale, near Cinchona, *Fawcett*! Farm Hill, *Miss B. Gosset*! near Catadupa, 1800 ft.; Wallenford, 3000 ft.; *Harris*! Fl. Jam. 7637, 9183.—China, Australia, N. Caledonia, Rarotonga.

Plant 1½–3 ft. high. *Stem* tuberous. *Leaves*, blade oblong-lanceolate to lanceolate, acute, 3–5 dm. l., 7–10 cm. br.; stalks 5–15 cm. l. *Scape* 5–9 dm. *Flowers* chocolate-brown and white, 1½ in. l. *Bracts* large, spathe-like, obtuse, about 3·5 cm. l. *Sepals* 7-nerved, oblong-lanceolate, 4–5·5 cm. l., 11–14 mm. br. *Lip* embracing the column, indistinctly 3-lobed above, median lobe slightly reflexed with margin curled; disk with two parallel keels, 3·5–4 cm. l. *Spur* slightly curved, 5 mm. l. *Column* 1·5–2 cm. l.

This species cannot be considered indigenous, but it is so completely naturalised and so common in some parts of Jamaica, that a full description has been included. It is also recorded from Cuba.]

40. CALANTHE R. Br.

Terrestrial herbs; stem very short, rhizomatous. Leaves few, large, broad, veins prominent, contracted into a sheathing petiole. Scape tall. Raceme simple, lax-flowered. Bracts very long, narrow. Sepals nearly equal, free, spreading. Petals somewhat similar, smaller. Lip clawed, claw connate with the column forming a tube, produced at the base into a short spur;

blade entire. Column short, erect, broadly winged at the apex above the tube. Anther subterminal, incumbent, 2-celled. Pollinia 8, waxy, narrowly pyriform. Capsule ellipsoidal, large.

Species about 138, widely distributed in tropics of the old world, especially Asia, one or two in tropical America.

C. mexicana *Reichb. f. in Linnæa xviii.* 406 (1844), *Ic. Fl. Germ. xiii. & xiv. t.* 355, *f.* 5, 6, 7, & *Xen. Orch. i.* 205, *t.* 79, *i.* 1–3. Ghiesbreghtia calanthoides *A. Rich. & Gal. in Ann. Sc. Nat. ser.* 3, *iii.* 28 (1845). (Pl. 20, f. 18–22.)

Damp shady places; in fl. June–Nov.; near Cinchona, 4800 ft., *Cradwick*! *Watt*! John Crow Peak, 5600 ft., *Fawcett*! Morse's Gap, 4800–5000 ft.; Mabess River, 3500 ft.; *Harris*! Fl. Jam. 7790, 10,437.—Cuba, Mexico.

Plant more than a foot high. *Stem* with abundant roots and two sheaths; upper sheath 9 or 10 cm. *Leaves* two, oblong to oval, acute, 25–40 cm. l., 7–11 cm. br. *Scape* 40–50 cm. *Raceme* 6–8 cm. *Flowers* white, ten or more, perianth ½ in. l., all parts downy. *Pedicels* 3–5 mm. *Bracts* 13–30 mm. l., linear, acuminate. *Sepals* 11–12 mm. l., 5–4 mm. br., oblong, shortly acute, 5-nerved, median rather shorter and broader than the lateral. *Petals* 8 mm. l., 3 mm. br., obovate-lanceolate, obtuse, 3-nerved. *Lip* (measuring from junction with tube) 6 mm. l., 4 mm. br., subrotund, shortly apiculate, entire (thus differing from the 3-lobed form generally characteristic of the genus). *Capsule* 2–2·5 cm. l.

41. BLETIA Ruiz & Pavon.

Terrestrial, erect herbs. Leaves springing from the top of a corm, few, very long, lanceolate, contracted into a sheathing petiole. Scape rising from the side of the corm, leafless with sheathing sterile bracts. Flowers numerous, stalked, somewhat large, in a terminal generally branched raceme. Bracts small. Sepals free, converging or somewhat spreading, subequal. Petals similar to the sepals, but broader. Lip attached to base of the column, free, erect, middle lobe spreading, broad, emarginate; lateral lobes somewhat broad, converging over the column; disc with five to seven crests. Column long, semiterete, 2-winged, incurved, auricled at the base, without a foot. Anther incumbent, two-celled. Pollinia 8, waxy, obovate. Capsule oblong, erect.

Species about 45, mostly in tropical America.

Flowers rosy-purple or purplish-crimson; lateral lobes of lip broader below than above, and with narrow apex .... 1. *B. purpurea.*
Flowers dark purple; lateral lobes of lip rounded above and narrowing towards the base .... 2. *B. florida.*

1. **B. purpurea** *DC. in Mém. Soc. Phys. & Hist. Nat. Genève, ix. pt.* 1, 100 (1841); flowers rosy-purple or purplish-crimson; lip with seven crests, only five of which pass into the middle

lobe, lateral lobes with a bluntly triangular apex above the sinus and increasing in breadth towards the base, middle lobe roundish.—*Griseb. Fl. Br. W. Ind.* 622. B. verecunda, *R. Br. in Ait. Hort. Kew. ed.* 2, *v.* 206 (1813); *Griseb. loc. cit.*; *Salisb. Ic. ined. vi. fol.* 572, 573. B. acutipetala *Hook. in Bot. Mag. t.* 3217 (1833). B. alta *Hitchcock in Rep. Miss. Bot. Gard. iv.* 132 (1893). Limodorum purpureum *Lam. Encycl. Méth. iii.* 515 (1789); *Redouté Liliacées i.* 83. L. tuberosum *Jacq. Coll. iv.* 108 (1790), non *L.** L. altum *Jacq. Ic. Rar. iii.* 17, *t.* 602 (1786–93), non *L.** L. verecundum *Salisb. Prodr.* 9 (1796). Cymbidium verecundum *Sw. in Nov. Act. Upsal. vi.* 75 (1799). (Pl. 21, f. 1–7.)

Dry stony banks, rocky land, on rocks in moist woods, 600-5000 ft.; in fl. and fr. Oct.-May; *Houstoun*! Fairfield, 1059, *Wullschlaegel*! J.P. 9, *Morris*! Cinchona, 5000 ft.! *Fawcett*! *G. Nichols*! Farm Hill, *C. Nicholls*! Hanover, *Tomlinson*! Silver Hill, 3500 ft., *Harris*!—Cuba, Bahamas, Florida, Central America.

Plant 1¼–2½ ft. high; leaves 1½ ft. long. *Corm* 2 cm., globose. *Leaves* 30–50 cm. l., 1–2 cm. br., narrowly lanceolate, long-tapering at base and apex, narrowing into a basal sheath. *Flowers*, perianth ½–¾ in. l. *Bracts*, sterile 1–1·3 cm. l., ovate, blunt, adpressed, sheathing at the base; floral 4–5 mm. l., broadly ovate, acuminate, somewhat clasping. *Pedicels* (including ovary) 1–1·5 cm., of fruit about half as long. *Sepals* 15–18 mm. l., 5–7 mm. br.; median ovate-lanceolate, acute; lateral ovate, acute, slightly gibbous. *Petals* broader and slightly shorter than the sepals. *Lip* shorter than the petals, veins numerous and branching, middle lobe emarginate, crenate, curled. *Column* winged. *Capsule* 2·7 cm. l., cylindrical, erect.

2. **B. florida** *R. Br. in Ait. Hort. Kew. ed.* 2, *v.* 206 (1813); flowers dark purple, sometimes rosy with white lip; lip with five crests, of which only three generally stretch far into the middle lobe, lateral lobes rounded above, narrowing gradually to the base, middle lobe generally cuneate in form.—*Griseb. op. cit.* 621. B. Shepherdii *Hook. in Bot. Mag.* 3319 (1834); *Griseb. loc. cit.* B. verecunda *DC. loc. cit.* non *R. Br.* Limodorum floridum *Salisb. Prodr.* 9 (1796) & *Ic. ined. vi. fol.* 572. L. altum *Sims in Bot. Mag. t.* 930 (1806). (Pl. 21, f. 8–11.)

Dry stony banks 1000–5000 ft.; in fl. and fr. Oct.–May; *Sloane* Herb. iv. 117*! St. Anns Bay, *Prior*! J.P. 2145, *Morris*! *Lane*! Bryans Hill, *Sullivan*! Silver Hill, 3500 ft.; Wallenford, 2500 ft.; Constant Spring; Green Valley; *Harris*! Fl. Jam. 7622, 7653, 7682.—Cuba.

Plant 2–3½ ft. with leaves 1½ to 2 ft. long. *Corm* 1·5–2·5 cm., ovate-globose. *Leaves* 20–50 cm. l., 2–9 cm. br., narrowly lanceolate to broadly lanceolate or oblanceolate, acuminate, narrowing into a basal sheath which persists after the fall of the leaf, many-nerved. *Flowers*, perianth less than an inch to 1½ in. long. *Bracts*, sterile 1–2·5 cm. l., ovate, blunt, adpressed, sheathing at the base; floral 3–10 mm. l., ovate,

---

* *Limodorum tuberosum* L. Sp. Pl. 950 is *Calopogon pulchellus* R. Br. *L. altum* L. Syst. ed. 12, ii. 594 is *Eulophia alta*.

acuminate, somewhat clasping. *Pedicels* (including ovary) about 2 cm., of fruit less than 1 cm. *Sepals* 2–2·5 cm. l., 7–9 mm. br.; median oblong, apiculate; lateral ovate-elliptical. *Petals* about the same length as the sepals but broader, oval. *Lip* shorter than the petals, middle lobe emarginate, crenate, curled. *Column* about ⅔ length of lip, winged. *Capsule* 4 cm. l. narrowly oblong, erect.

Limodorum floridum was founded by Salisbury on a specimen sent "from Jamaica by Wm. Shakespeare." The original specimen is not extant, but among Salisbury's drawings and Mss in the Department of Botany are comparative descriptions and sketches of parts of the flower of this species and of *L. verecundum*; the drawing of the lip of *L. floridum* is reproduced in our plate. Brown followed Salisbury in regarding the two species as distinct, but there are no specimens of *Bletia florida* authenticated by him. In our opinion *B. florida* is identical with the plant subsequently described by Hooker as *B. Shepherdii.*

42. EULOPHIA R. Br.

Terrestrial herbs, with rhizome thickened into a tuber-like pseudobulb, and stem with a few leaves. Leaves long, plaited, many-veined, contracted into the stalk. Scapes erect from the rhizome, tall, leafless, with several sheaths. Raceme simple; flowers medium-sized, shortly stalked. Bracts linear or lanceolate. Sepals free from each other, almost equal, the lateral broader at the base and more or less decurrent into the foot of the column. Petals somewhat similar to the median sepal, but a little broader and shorter. Lip attached at the apex of the foot of the column, forming a gibbous sac or spur between the lateral sepals, 3-lobed, lateral lobes embracing the column, the median rounded, spreading, with curled margin. Column erect, produced at the base into a foot, semiterete, with acute scarcely winged angles; clinandrium oblique, entire. Anther terminal, opercular, incumbent, obtusely conical, with a 2-lobed appendage, imperfectly 2-celled. Pollinia 2 or 4, more or less connate in pairs, broadly ovoid or globose, waxy, without appendage, attached to a gland of the rostellum, which is broad, generally membranous, or produced into a short thick stalk. Capsule ellipsoidal, pendulous, without a beak, ribs thick and prominent.

Species about 180, widely diffused through the Tropics, most numerous in Africa, and rare in Malaya, Polynesia, Australia and Tropical America.

E. alta *comb. nov.*—E. Woodfordii *Rolfe in Fl. Trop. Afr. vii.* 68 (1897); *Ames Orch. S. Fla.* 19. Limodorum altum *L. Syst. ed.* 12, *ii.* 594 (1767). Limodorum foliis nervosis lanceolatis &c. *Plum. Pl. Amer.* (*Burm.*) *t.* 189 & *Ic. ined. iii. t.* 192. Satyrium foliis liratis &c. *Browne Hist. Jam.* 325. Dendrobium longifolium *H. B. & K. Nov. Gen. & Sp. i.* 360 (1815). Cyrtopodium Woodfordii *Sims in Bot. Mag. t.* 1814 (1816); *Lindl. in Bot. Reg. t.* 1508. Cyrtopera Woodfordii *Lindl. Gen. & Sp. Orch.* 183

(1833); *Griseb. op. cit.* 630. C. longifolia *Cogn. in Fl. Bras. iii. pt.* 5, 354, *t.* 76. (Pl. 22, f. 4-8.) Type in Herb. Linn.

Ground Coco, Wild Coco.

In wet clayey soil in shady places; in fl. & fr. Nov.-Feb.; *P. Browne*! *Purdie;* Bethabara, *Wullschlaegel*, 1057 bis! St. Mary, J.P. 2397, *Syme*! Brokenhurst, *Wynne*! Clydesdale, *Fawcett*! Wallenford; near Castleton; *Harris*! Christiana, *Tomlinson*! Fl. Jam. 7638, 10,435.—Cuba, Florida, Hispaniola, Porto Rico, Guadeloupe, Dominica, Martinique, St. Vincent, Trinidad, Venezuela, Brazil, Colombia, Peru, West Africa.

Plant 2-4 ft. high. *Leaves* appearing after the flowers, three or four at a time, very long, lanceolate, acuminate, below tapering into a long sheath, 7-9-nerved, plaited, 3-5 dm. l., 3-10 cm. br. *Scape* ½-1 m. l. and sometimes more, much longer than the leaves. *Sheaths* long; free portion triangular, obtuse or somewhat acute. *Raceme* somewhat laxly many-flowered. *Bracts* linear-lanceolate or linear-subulate, acuminate, much shorter than the stalked ovary, 8-13 mm. l. *Pedicels* about 5 mm. l. *Ovary* about 1 cm. l. *Flowers* green with brownish-green lip; perianth ¾-1 in. l. *Sepals* 5-nerved, outer nerve branching, oblong to oblong-oblanceolate, apex bluntish; lateral oblique at apex, about 2 cm. l., 5·5 mm. br.; median about 1·6 cm. l., 4·5 mm. br. *Petals* with seven branching nerves, oblong, broadening slightly upwards, obtuse, about 1·4 cm. l., 6 mm. br. *Lip* brownish-green, shaded and veined inside with purple, about 1·8 cm. l., with several branching nerves; lateral lobes short, rounded; disk with warty calli on nerves. *Column* about 9 mm. l. to base, about 13 mm. l. from apex to end of foot. *Capsule* 4·5 cm. l., 1·8 cm. br., pendulous.

43. GOVENIA Lindl.

Terrestrial herbs, with a rhizome sometimes thickened into tuber-like pseudobulbs. Stems erect, with few leaves and some stalked sheaths at the base, of which one is often inflated and utricle-like. Leaves ample, plaited, with many veins, with a long sheathing base. Raceme terminal on a leafy stem, simple; peduncle with a few sheaths near the base. Flowers medium-sized, crowded, or at length distant, very shortly stalked. Bracts small, ovate or narrow, a little shorter than the ovary. Sepals almost equal, connivent; median incurved, erect; lateral falcate or decurved, at the base forming a short chin with the foot of the column. Petals somewhat similar to the sepals. Lip attached at the foot of the column, jointed, incumbent at the base, concave, spreading at the apex, undivided. Column incurved, broadly semiterete, winged, produced into a short foot at the base; clinandrium truncate. Anther terminal, opercular, incumbent, very convex, sometimes crested on the back, 1-celled; pollinia 4, waxy, broadly ovate, compressed, without an appendage, on the dehiscence of the anther attached by an oblong or broad stalk, gland small or dilated. Capsule oblong, without a beak.

Species about 17, natives of tropical America from the West Indies and Mexico to Brazil and Bolivia.

G. utriculata *Lindl. in Bot. Reg. xxv. Misc.* 47, *under no.* 66 (1839); *Hook. in Bot. Mag. t.* 4151; *Griseb. Fl. Br. W. Ind.* 628.

Limodorum utriculatum *Sw. Prodr.* 119 (1788). Cymbidium utriculatum *Sw. in Nov. Act. Upsal. vi.* 75 (1799) & *Fl. Ind. Occ.* 1477. (Pl. 22, f. 1–3.) Type in Herb. Mus. Brit.

In damp woods; in fl. Sept.–Dec.; *Swartz*! *Macfadyen*! Manchester, *Purdie*! Fairfield, *Wullschlaegel*, 1056! *Prior*! near Cinchona, J.P. 33; *Morris*! Red Hills, J.P. 2413, *Syme*! Trelawney, *Miss T. M. Barrett*! Hamilton Town, *Moore*! above Mt. Hybla; above Bridge Hill; *Fawcett*! Old England; near Ewarton, 1000 ft.; *Harris*! Fl. Jam. 6719, 10,433.—Cuba, Hispaniola, Porto Rico, Mexico, Venezuela.

Plant 1½–2 ft. high. *Stem* 3–4·5 dm. *Sheaths* inflated, one 10–20 cm. l., 2 or 3 cm. br., outer 4–8 cm. l.; cauline, free-portion oblong-ovate, obtuse, 2·5–5 cm. l. *Leaves* one or two, blade elliptical, tapering to base and apex, 15–25 cm. l., 5–9 cm. br.; the tubular sheathing base 13–18 cm. l. *Raceme* 6–14 cm. l., laxly few- to many-flowered. *Bracts* linear-lanceolate, lower floral, 1·5–2 cm. l., upper shorter. *Flowers* rather more than ½ in. l., cream-coloured. *Sepals* 5-nerved; lateral lanceolate, 12 mm. l., 3 mm. br.; median oblong, shortly acuminate, about 15 mm. l., 2·75 mm. br. *Petals* 6-nerved, outer nerves branching, broadly oblanceolate, acute, broader than sepals, 14 mm. l., 5 mm. br. *Lip* 5-nerved, nerves branching, ovate, acute, very shortly clawed, 9 mm. l., 4·3 mm. br. *Column* 6 mm. l. *Capsule* 2·5 cm. l.

44. LYCASTE Lindl.

Epiphytic herbs. Stem short, with few leaves, sheathed at the base, becoming thickened into fleshy pseudobulbs. Leaves ample, plaited, many-veined. Scapes erect at the sides of the pseudobulbs, closely sheathed, with one flower, rarely two or three. Flowers large, nodding. Sepals nearly equal, erect-spreading, the lateral broader, forming a chin. Petals similar to the sepals but smaller. Lip attached at the foot of the column, shorter than the sepals; lateral lobes erect; the median spreading. Column rather long, curved, semi-terete, produced at the base into a foot; clinandrium short. Anther terminal, opercular, incumbent, very convex, 1-celled; pollinia 4, oblong or ovoid, without an appendage, attached by a long linear stalk, gland small. Capsule oblong or spindle-shaped, erect.

Species about 40, natives of tropical America from the West Indies and Mexico to Peru.

L. Barringtoniæ *Lindl. in Bot. Reg. xxx. Misc.* 43 (1844); *Griseb. Fl. Br. W. Ind.* 627. Epidendrum Barringtoniæ *Smith Ic. pict. t.* 15 (1793); *Salisb. Ic. ined. fol.* 579. Dendrobium Barringtoniæ *Sw. Nov. Act. Upsal. vi.* 82 (1799); *Hook. Exot. Fl. t.* 119. Maxillaria ciliata *Lindl. in Bot. Reg. t.* 1206 (1829). M. Barringtoniæ *Lodd. Bot. Cab. t.* 1824 (1832); *Lindl. Gen. & Sp. Orch.* 147. Type in Herb. Smith. in Linn. Soc. (Pl. 23, f. 1–3.)

On rocks and trees; in fl. April–June; *Macfayden*! *Purdie*! Beaufort, *Wullschlaegel*, 1062! Mt. Moses, J.P. 2328, *Syme*! near Radnor, *Miss B. Gosset*! Content Road; Silver Hill Woodland; *Harris*! Fl. Jam. 10,392.—Cuba.

Plant 1–2 ft. high. *Pseudobulb* ellipsoidal with two or three leaves at apex, about 8 or 9 cm. l., 5 cm. br. *Leaves* long-elliptical, tapering to a short stalk and to apex, 3–5 dm. l., 5–12 cm. br. *Scapes* 8–12 cm. l. *Bracts* 4–6, sheathing, loose, ovate, obtuse, uppermost nearly as long as stalked ovary, 1·5–3 cm. l. *Flowers* drooping, nearly 2 in. l., olive-green. *Stalked ovary* about 3 mm. l. *Sepals* 4–4·5 cm. l., about 1·7 cm. br., oblong-ovate, lateral cohering and forming a bluntly conical chin, median narrower. *Petals* similar to sepals but smaller. *Lip* light buff in colour, about 3·5–4·5 cm. l., with a long claw broadening upwards, lateral lobes at apex of claw, narrowly falcate; median ovate-oblong, blunt, with fimbriate margin, and terminating abruptly by a broad deeply furrowed callus. *Column* bordered, thicker upwards, about 2 cm. l. (to end of foot over 3 cm. l.).

45. XYLOBIUM Lindl.

Epiphytic herbs. Stems short, with several sheaths, becoming thickened into fleshy pseudobulbs with one or two leaves. Leaves ample or elongate, plaited, many-veined, contracted into the stalk. Scapes at the base of pseudobulbs, erect, simple. Flowers in racemes, medium-sized, very shortly stalked. Bracts linear, generally rather long. Sepals nearly equal, erect, at length somewhat spreading; lateral broader than the median, adnate at the base to the foot of the column, forming a chin. Petals similar to the median sepal but smaller. Lip somewhat jointed with the foot of the column, sessile or contracted at the base and incumbent, then erect; lateral lobes or sides erect, embracing the column; the median lobe or upper portion short, broad, spreading; the face lamellate or callose. Column erect, semi-terete, anteriorly concave or narrowly 2-winged, produced at the base into a foot; clinandrium obliquely truncate. Anther terminal, opercular, incumbent, very convex, 1-celled; pollinia 4, ovoid, without an appendage, on the dehiscence of the anther attached to a stalk which is sometimes rather long; gland transverse, scale-like. Capsule oblong, erect.

Species about 12, natives of tropical America.

Pseudobulb with one leaf .... 1. *X. palmifolium*.
Pseudobulb with two leaves .... 2. *X. stachyobiorum*.

1. **X. palmifolium** *Fawc. Fl. Pl. Jam.* 39 (1893); flowers about $\frac{3}{4}$ in. l., lateral sepals only slightly keeled at back, lip entire.—Epidendrum palmifolium *Sw. Prodr.* 123 (1788). Dendrobium palmifolium *Sw. in Nov. Act. Upsal. vi.* 82 (1799) & *Fl. Ind. Occ.* 1527. Maxillaria decolor *Lindl. in Bot. Reg. t.* 1549 (1832); *Hook. in Bot. Mag. t.* 3981. M. palmifolia *Lindl. Gen. & Sp. Orch.* 148 (1832); *Griseb. op. cit.* 627. (Pl. 23, f. 4–7.) Type in Herb. Mus. Brit.

On trees and rocks; in fl. Sept.; *Swartz*! Mt. Moses, 2500 ft., J.P. 2315, *Syme*! Trelawney, *Miss T. M. Barrett*.—Cuba.

Plant over 1 ft. high. *Pseudobulb* cylindrical-ovate, 5·5–7·5 cm. l., with one leaf. *Leaves* (generally) broadly lanceolate, blade 26–42 cm. l.,

2·7–3 cm. br. *Scape* laxly few-flowered, about 10 cm. l. *Bracts*, sterile sheathing, ovate-oblong, acute or acuminate, 3–3·5 cm. l.; floral linear, acuminate, 2·5–3 cm. l. *Flowers* yellowish-white, sweet-scented. *Pedicels* about ·5 cm. l. *Ovary* about 1·3 cm. l. *Sepals* 7-nerved; median oblong, tapering towards an acute apex, 1·8 cm. l., 5 mm. br.; lateral very oblique and broad at base, falcately triangular-lanceolate, 2·2 cm. l., 9 mm. br. *Petals* 5-nerved, linear-oblong, tapering to apex, about 1·4 cm. l., about 3·5 mm. br. *Lip* white, about 15-nerved, outline obovate-oblong, about 1·6 cm. l., about 9 mm. br.; terminal portion fleshy, warty, retuse, 4 mm. l., 7 mm. br.; sides incurved above, margin crisped; disk with four or five median nerves thickened, extending beyond the middle of the disk. *Column* (from apex to end of foot) 1·4 cm. l.

2. **X. stachyobiorum** *Hemsl. in Biol. Centr. Amer. Bot. iii.* 252 (1883); flowers rather more than ½ in. l., lateral sepals with broad keel extending beyond apex, lip 3-lobed above.—Maxillaria stachyobiorum *Reichb. f. in Bot. Zeit. x.* 735 (1852).

On peaty banks; in fl. Jan.–March; near Castleton, *Fawcett*! Silver Hill, 3500 ft., *Harris*!—Nicaragua, Panama.

Plant over 1 ft. high. *Pseudobulb* large, ovate, compressed, 5·5–6 cm. l., with two leaves. *Leaves* lanceolate to oblanceolate, acute or acuminate, narrowed into the stalk, 30–45 cm. l., 5–6 cm. br. *Scape* laxly many-flowered, 14–17 cm. l. *Bracts*, sterile sheathing, ovate-oblong, acute or acuminate, about 3 cm. l.; floral subulate, about 1·5 cm. l. *Flowers* cream-coloured, irregularly speckled with brown. *Pedicels* 4–5 mm. l. *Ovary* about 1 cm. l. *Sepals* 5-nerved with less well-marked intermediate nerves, median oblong-ovate, shortly acuminate, 14 mm. l., about 4 mm. br.; lateral oblong-ovate, tapering and keeled towards apex, 15 mm. l., 5·5 mm. br. *Petals* 5-nerved, oblong-lanceolate, apex somewhat acute and fleshy, falcate, 11 mm. l., 3·5 mm. br. *Lip* with seven branching nerves, broadly oblong, 11 mm. l., 7 mm. br., shallowly 3-lobed above, lateral lobes rounded, terminal fleshy, broader than long, rounded, larger than lateral lobes; disk with the three or five median nerves elevated. *Column* (from apex to end of foot) nearly 1 cm. l. *Capsule* 2·5 cm. l.

46. ZYGOPETALUM Hook.

Epiphytic herbs, stemless, leafy. Leaves membranous, with numerous prominent veins, plaited. Scapes 1-flowered. Sepals nearly equal, spreading, free or scarcely very shortly connected at the base, lateral attached to the short foot of the column. Petals similar to the sepals. Lip attached to the foot of the column, forming with it a very short chin; lateral lobes erect, embracing the column; median lobe generally broad; disk with a transverse crest generally very prominent, fleshy. Column incurved, semiterete, produced at the base into a short foot; clinandrium oblique, entire, or membranous and hooded with fimbriate margin. Anther opercular, incumbent, 2-celled; pollinia 4, obovoid, segregated in pairs in the cells, without an appendage. Capsule ovoid or oblong, without a beak.

Species about 90, natives of warmer America from the West Indies and Central America to Brazil.

Z. flabelliforme *Reichb. f. in Walp. Ann. vi.* 652 (1861).—Z. cochleare *Lindl. in Bot. Reg. t.* 1857 (1836); *Hook. in Bot. Mag. t.* 3585; *Griseb. Fl. Br. W. Ind.* 629 (excl. distr.). Epidendrum flabelliforme *Sw. Prodr.* 123 (1788) & *Ic. ined. t.* 45. Cymbidium flabelliforme *Sw. in Nov. Act. Upsal. vi.* 73 (1799) & *Fl. Ind. Occ.* 1471. (Pl. 23, f. 17–19.) Type in Herb. Mus. Brit.

On trees; in fl. Oct.–March; mountains, *Swartz*! Manchester, *Purdie*! St. Andrew and St. Mary, J.P. 2099, *Morris*! Spanish River, *Watt*! St. Ann, *Miss T. M. Barrett*! Cedar Hurst, *Moore*! Clydesdale, 3500 ft.; Dollwood; Silver Hill Woodland; Catherine's Peak, 5000 ft.; Moody's Gap, 3000 ft.; above Hope Mines, 900 ft.; *Harris*! Fl. Jam. 7576, 7736, 7808.

Plant 8–12 in. high. *Leaves* oblanceolate to narrowly oblong, abruptly acute, tapering below into a sheathing stalk with which the blade is jointed, 12–30 cm. l., 2·5–4·2 cm. br. *Sheaths* below leaves, 3–6 cm. l. *Scape* from the axils of the outer leaves, erect, 3–10 cm. l. *Bracts* several at and just above base of scape, and two at base of flower, spathe-like, 1–2 cm. l. *Ovary* white, curved, 2–3 cm. l. (pedicel forming about one-third). *Flowers* white with blue lip, with fragrance like lily-of-the-valley, perianth 1–1¼ in. l. *Sepals* oblong or oblong-lanceolate, greenish-white, 3–3·5 cm. l., 9–10 mm. br. *Petals* similar and slightly smaller. *Lip* very large, somewhat square in outline when flattened, with a small short claw, at base broad, cordate, ventricose, margin involute and curled, apex emarginate; at the base within is a large lunate fleshy depressed crest, lobed and crenated, marked with purple lines continued into disk; 2·5–3 cm. l. and br. *Column* short, with two small wings above, white streaked with red in front, 1 cm. l. *Capsule* cylindrical, 4–5 cm. l., 10–13 mm. br.

47. BULBOPHYLLUM Thou.

Epiphytic herbs. Stem more or less covered with scarious sheaths. Pseudobulbs sessile at the axils of the sheaths, bearing one or two leaves at the apex. Flowering scapes at the sides of the pseudobulbs, leafless, simple, with numerous sheaths at intervals. Flowers minute, numerous on a terminal spike. Bracts small. Median sepal free, erect or spreading, about as long as the lateral; lateral at the base obliquely dilated, and adnate to the foot of the column, above spreading, free, or connate somewhat high up. Petals shorter than the sepals. Lip on the upper side of flower, contracted at the base, jointed with the foot of the column, or mobile, incumbent on the foot; entire or plumose-ciliate, recurved, often variously thickened. Column erect, generally short, produced at the base into a rather long foot, 2-winged above; clinandrium short, posteriorly often produced into a tooth which bears the anther, anteriorly 2-armed, 2-toothed or 2-angled, and sometimes the wings of the column below the arms appendaged on both sides with a somewhat small tooth. Anther terminal, opercular, incumbent, flattish-hemispherical or obtusely conical, generally 2-celled; pollinia waxy, normally four, segregated in pairs in the cells, sometimes one pair very small or wanting, without an appendage, or more rarely cohering by a small, subgranular viscid body. Capsule globose, ovoid or oblong, not beaked.

Species about 550, scattered for the most part through the tropical regions of Africa and Asia, very few in West Indies, South America or Australia, one in New Zealand.

Flower-stalk not swollen, petals ciliate........ 1. *B. jamaicense.*
Flower-stalk swollen, petals glabrous .... 2. *B. pachyrrhachis.*

1. **B. jamaicense** *Cogn. in Fedde Repert. vii.* 122 (1909); with flowering-stalk not swollen, petals ciliate, lip pilose below. (Pl. 23, f. 8–14.)

On trees; in fl. and fr. March; near Tweedside, south St. Andrew, 2500 ft., *Harris* ! Fl. Jam. 7998.

Plant 4 in. high. *Pseudobulbs* small, densely clustered, globose, 1-leaved, 6–8 mm. l. *Leaves* leathery, oblong-lanceolate, acute, 1–3$\frac{1}{2}$ cm. l., 4–8 mm. br. *Scape* 6–10 cm. l., much longer than the leaf, very slender, with a few short (5–6 mm. l.) obtuse sheaths at intervals; the upper portion (2–3 cm. l.) a spike with slender rachis. *Flowers* reddish, few (4–10), lax, small, $\frac{1}{8}$–$\frac{1}{5}$ in. l. *Bracts* ovate, acute, 2–3 mm. l., about as long as ovary. *Sepals* 3-nerved; median concave, ovate, acute, 4·5–5 mm. l., 2 mm. br.; lateral connate nearly to apex, elliptical, 5 mm. l., 3 mm. br. *Petals* 1-nerved, linear, tapering to an acute apex, ciliate, 3–4 mm. l., ·7–1 mm. br. *Lip* shortly clawed, somewhat thick, oblong-elliptical, with sides recurved, obtuse, papillose, pilose below, 3 mm. l. *Column* very short, ·5 mm. l., posterior tooth short, bristle-like, lateral subulate, erect, as long as column, branching at apex. *Capsule* obovoid, 5–6 mm. l.

2. **B. pachyrrhachis** *Griseb. Fl. Br. W. Ind.* 613 (1864); flowering-stalk swollen, petals and lip glabrous.—*Cogn. in Symb. Ant. iv.* 179. Pleurothallis? pachyrachis *A. Rich. Fl. Cub. ii.* 234 (1853), *Ic. t.* 74 (1863). Bolbophyllaria pachyrrhachis *Reichb. f. in Walp. Ann. vi.* 241 (1861). (Pl. 23, f. 15, 16.)

On trees; in fl. and fr. Jan.; St. Elizabeth, J.P. 525, *Syme* ! Belvedere, Hanover, *Harris* ! Fl. Jam. 7634.—Cuba, Porto Rico, Trinidad.

Plant 4–18 in. high. *Pseudobulbs* with two or more scarious sheaths, on a creeping rhizome, ovoid-conical, 4-winged, 2-leaved, 5–2 cm. l. *Leaves* somewhat leathery, ligulate, obtuse, 7–19 cm. l., 8–23 mm. br. *Scape* 10–45 cm. l., much longer than the leaves, with short (8–10 mm. l.) obtuse sheaths at intervals; the upper portion (5–30 cm. l.) a spike with more or less swollen and fleshy rachis. *Flowers* many, lax, small, $\frac{1}{8}$–$\frac{1}{5}$ in. l. *Bracts* very broad at the clasping base, triangular, obtuse, 3–4 mm. l., a little shorter than the flower; two, ovate, acute, at apex of ovary, 1 mm. l. *Sepals* 3-nerved, median concave, ovate, very acute, 4·5–5 mm. l.; lateral connate at base, ovate, acuminate, 4·3-5 mm. l., 1·5–2 mm. br. *Petals* 1-nerved, oblong-elliptical, obtuse, 1·75–2 mm. l., ·9–1·2 mm. br. *Lip* thick and fleshy, 3-angled in section, undivided, with a very thick, 3-angled claw, disk at right angles to claw, oblong-elliptical, obtuse, margin entire and glabrous, slightly keeled at back but not serrate, disk with a longitudinal furrow in the middle; claw about 1 mm. l., disk 2–2·4 mm. l. *Column* short, posterior tooth small, lateral teeth longer, subulate, slightly incurved, shorter than the petals, 1·5–1·75 mm. l. *Capsule* ellipsoidal, with six broad keels, 8 mm. l., 5 mm. br.

48. MAXILLARIA Ruiz & Pavon.

Epiphytic herbs. Stems sometimes very short on the rhizome becoming thickened into a fleshy pseudobulb with one or two leaves, sometimes elongated at the apex of the rhizome with the leaves close together in two ranks. Leaves leathery, thin or somewhat fleshy, veins thin except the central nerve. Scapes or peduncles at the base of the pseudobulbs or in axils of the leaves, solitary, with two to numerous sheaths, always 1-flowered. Flowers large or medium-sized. Sepals nearly equal, free amongst themselves, lateral adnate at the base to the foot of the column, generally forming a chin, not very prominent, spreading or more rarely somewhat erect. Petals similar to the sepals or sometimes smaller. Lip concave, erect at the apex of the foot of the column with a very short inflexed claw; lateral lobes erect, median ovate-oblong spreading, shorter than the sepals; disk with a tubercle or naked. Column with a short foot, erect, often slightly incurved, thick, semiterete, not winged, anteriorly concave; clinandrium somewhat thick, concave, entire. Anther terminal, opercular, incumbent, conical or semiglobose, 1-celled or imperfectly 2-celled, often pubescent; pollinia 4, ovate, compressed, without an appendage, on the dehiscence of the anther attached by a flat, short stalk, with a scale-like gland. Capsule ovoid or obovoid-oblong, erect, not beaked.

Species about 240, natives of tropical America from the West Indies and Mexico to Brazil.

Pseudobulbs conspicuous.
 Pseudobulbs clustered, ovoid-cylindrical, with membranous sheaths .... 1. *M. rufescens.*
 Pseudobulbs distant, ellipsoidal-compressed, with hard bases of foliaceous sheaths .... 2. *M. alba.*
Pseudobulbs poorly developed and inconspicuous, bases of leaves thickened .... 3. *M. sessilis.*

1. **M. rufescens** *Lindl. in Bot. Reg. under t.* 1802 (1835) & *t.* 1848; pseudobulbs clustered, ovoid-cylindrical, with membranous sheaths; lip yellow with numerous purple spots, distinctly 3-lobed, lateral lobes springing from a little below middle of lip, triangular-falcate, subacute, terminal lobe much larger, broadly oblong, emarginate, disk with an oblong callus half length of lip.—*Reichb. f. in Saund. Ref. Bot. ii. under t.* 79 & *t.* 133; *Griseb. Fl. Br. W. Ind.* 626; *Cogn. in Fl. Bras. iii. pt.* 6, 12. (Pl. 25, f. 13–16.)

On trees; in fl. May; cult. in Hope Gardens from plant collected at Lancaster, *Harris*, drawing by Miss H. Wood!—Trinidad, Costa Rica, Colombia, Venezuela, B. Guiana, Brazil.

Plant 6–8 in. high. *Stem*, primary almost obsolete. *Pseudobulbs*

3–3·5 cm. l., 1·5 cm. br., at first covered with three or more scarious membranous sheaths (to 3 cm. l.) with one leaf at apex. *Leaves* ligulate, shortly complicate at base, apex somewhat abruptly acute, 14–18 cm. l., 2 cm. br. *Peduncles* 2–3 cm. l., several, 1-flowered, with scarious sheaths at base; pedicel shorter than peduncle or ovary, with one sheathing bract (18 mm. l.). *Flowers* creamy-yellow, perianth nearly 1 in. l. *Ovary* and pedicel together 3 cm. l. *Sepals* oblong, about 2 cm. l., lateral spreading. *Petals* somewhat oblong, nearly as long as median sepal. *Lip* nearly as long as sepals.

Var. **minor** *Fawc. & Rendle in Journ. Bot. xlviii.* 108 (1910); flowers about half the size; disk of lip with a linear-tapering deep crimson callus ¾ length of lip, and two nerves on each side. (Pl. 25, f. 17–19.)

On trees and rocks; in fl. Aug.–Nov.; Mt. Moses, 2500 ft., J.P. 2009, *Syme*! Mabess River, 3000 ft., *Harris*! Fl. Jam. 7615.—Cuba, tropical America.

Pedicel with one sheathing bract (7–8 mm. l.). *Flowers*, perianth ⅝ in. l. *Sepals* 1–1·3 cm. l., 4·5 mm. br.

2. **M. alba** *Lindl. Gen. & Sp. Orch.* 143 (1832); pseudobulbs distant, ellipsoidal, compressed, with hard bases of foliaceous sheaths; lip somewhat fleshy, elliptical, indistinctly 3-lobed, terminal lobe oblong-ovate, slightly hooded and thickened at apex, callus extending from base to beyond the middle.—*Griseb. op. cit.* 626. Dendrobium album *Hook. Exot. Fl. t.* 142 (1823–1827). (Pl. 25, f. 1–6.)

On trees and banks; in fl. Aug.–March; *Sloane* Herb. iv. 116* & Ic! *Wiles*! *March*! near Cedarhurst; Mt. Moses; J.P. 465; *Morris*! Trelawney, *Miss T. M. Barrett*! Lancaster, 2500 ft., *Harris*! Fl. Jam. 7209.—Cuba, Trinidad, Guatemala, B. Guiana, Surinam, Brazil.

Stem stout, 7 cm. and more long. *Pseudobulbs* to 4 or 5 cm. l., covered at first with several sheaths, with one leaf at apex. *Sheaths* in two ranks, imbricate, the outer triangular, acute, short, the inner becoming gradually foliaceous. *Leaves* ligulate, obliquely and slightly emarginate, to 2·5 dm. and more, 1·5–2 cm. br. *Peduncles* several, solitary in the axil of a sheath, 1-flowered, bearing several two-ranked scarious narrow acuminate bracts. *Flowers*, perianth rather more than ⅞ in. l., white. *Ovary* with pedicel (if any) 2–3 cm. l. *Sepals* somewhat leathery, 9-nerved, oblong-ligulate, acuminate, about 2 cm. l., nearly 5 mm. br. *Petals* 9-nerved, oblong-elliptical, somewhat acute, 1·6 cm. l., 4 mm. br. *Lip* 1·1–1·2 cm. l., 4·5 mm. br. *Column* 8·5 mm. l. *Capsule* cylindrical-tapering, 3 cm. l., bearing the hard persistent column.

3. **M. sessilis** *comb. nov.*; pseudobulbs poorly developed and inconspicuous, bases of leaves thickened; lip somewhat fleshy, elliptical, with a narrow terminal portion, obtuse, and with a broad thickened margin, disk with a broad densely tomentose band along the central line to the apex.—M. crassifolia *Reichb. f. Bonpl. ii.* 16 (1854); *Griseb. op. cit.* 626. Epidendrum sessile *Sw. Prodr.* 122 (1788). Heterotaxis crassifolia *Lindl. in Bot. Reg.*

t. 1028 (1826); *Lodd. Bot. Cab. t.* 1651. Dicrypta Baueri *Lindl. Gen. & Sp. Orch.* 44 & 152 (1830 & 1832); *Bauer & Lindl. Ill. Orch., Gen. t.* 5. (Pl. 25, f. 7–12.) Type in Herb. Mus. Brit.

On rocks and trees; in fl. Aug.–March; *Swartz*! *McNab*! Petersfield, St. Mary, *Purdie*! Springfield, *Wullschlaegel*, 1087*a*! Wag Water R., J.P. 2 & 2399, *Morris*! & *Syme*! Trelawney, *Miss T. M. Barrett*! Belvedere, Hanover, 500 ft.; Newcastle Road, 1500 ft.; Tweedside, south St. Andrew, 2500 ft.; *Harris*! Fl. Jam. 7681, 7841, 9006.—Cuba, Costa Rica, Mexico, Venezuela, Brazil.

Plant 10–14 in. high. *Stem* very short, almost obsolete, buds springing from base of old plant. *Pseudobulbs* oblong, much compressed, with one leaf at apex, 2·5 cm. l., 1·5 cm. br. *Leaves* radical (except one at apex of pseudobulb), in two ranks, 3-edged, complicate below, sheathing and much thickened at base, jointed above the sheath, the outer lower leaves with very shortly developed blade; blade of inner leaves ligulate, oblique at apex and with a very short curved apiculus, keeled, 18–29 cm. l., 17–26 mm. br. *Peduncles* 8 or 9 mm. l., several, two or three in a cluster from the axil of each leaf and enclosed by its sheath, each cluster enclosed by three or four scarious bracts, about 1·5 cm. l.; each peduncle with three sheathing bracts from base; pedicel as long as peduncle, solitary, 1-flowered, with one sheathing bract. *Sepals* leathery, oblong, tapering to a thickened and slightly hooded apex; lateral 1·5 cm. l., 5 mm. br.; median concave, 1·4 cm. l., 5 mm. br. *Petals* somewhat fleshy, oblanceolate, somewhat acute, 1·25 cm. l., about 3 mm. br. *Lip* 1·2 cm. l., 4·5 mm. br. *Column* 8 mm. l. *Capsule* ellipsoidal with six broad keels and furrows, 2·3 cm. l.

49. **NEO-URBANIA** Fawc. & Rendle.

Herb, on trees or procumbent on shady stony banks, with long leafy branching stems, the lower internodes rooting. Leaves alternate, narrow, with persistent tightly adpressed sheathing stalks. Flowers stalked, rather small, few, clustered in the axils of the leaves. Sepals nearly equal in length and breadth, connivent, median free, lateral adnate to the foot of the column, forming a chin under the lip. Petals similar to the sepals. Lip 3-lobed, jointed to the foot of the column, and incumbent on it, the lateral lobes above the base curved in towards the column, the blade oval, recurved-spreading. Column very short, erect, not winged, produced at the base into a long foot; clinandrium slightly concave, entire, the median line raised, and ending in the slightly pointed rostellum; anther terminal, opercular, incumbent, hooded, 2-celled; pollinia 4, waxy, pyriform, equal, compressed laterally, collateral in pairs in the cells, not appendiculate, free. Capsule long, oblong, shortly beaked.

Differs from *Ponera* in its free simple pollinia and axillary flowers; from *Pleuranthium* in the lip, which is free from and attached to the foot of the column, and in the pollinia; from *Camaridium* in habit, in character of the pollinia (absence of gland and stipes), and in the connivent perianth.

Dedicated to Prof. Dr. Ignatius Urban in recognition especially of his work on the botany of the West Indies.

Species 1, a native of Jamaica and Cuba.

N. adendrobium *Fawc. & Rendle in Journ. Bot. xlvii.* 125 (1909). Ponera adendrobium *Reichb. f. in Flora xlviii.* 278 (1865); under Pleuranthium *Benth. & Hook. f. Gen. Pl. iii.* 526 (1883); Camaridium parviflorum *Fawc. in Symb. Ant. i.* 472 (1900). (Pl. 24.)

In fl. and fr. Sept.–March; near Cedar Hurst (high up on trees), J.P. 480, *Morris*! Lancaster, 2500 ft. (on stony banks under shade); *Harris*! Fl. Jam. 7650, 7696, 7874.—Cuba.

Plant several feet in length; internodes 2–4 cm. l. *Leaves* linear-lanceolate, subacuminate, 12–15 cm. l., 2 cm. br. *Peduncle* 7 mm. l. *Pedicel* 7·5 mm. l. *Bracts* scarious, the one subtending the cluster, 1·1 cm. l.; that at the base of the peduncle 1 cm. l.; that at the base of the pedicel 8 mm. l.; bracteole below the ovary 1·3 mm. l. *Ovary* 8 mm. l. *Flowers* white, perianth about ⅛ in. l. *Sepals* 5·5 mm. l., subacuminate, median oblong, lateral oblong-elliptical. *Petals* 5 mm. l., oblong, acuminate, 3-nerved. *Lip* middle lobe, 3 mm. l., nearly 3 mm. br., margin undulate, apex emarginate; disk at the base with two calli, 1-nerved, at the back 3-nerved, callose at the base, and with a thick apiculus below the apex. *Column* 2 mm. l.; foot 2 mm. l. *Capsule* (unripe) 1·6 cm. l., with slightly prominent ribs.

50. ORNITHIDIUM Salisb.

Epiphytic herbs. Stems leafy, covered with overlapping sheaths in two ranks, and emitting here and there lateral or sub-lateral 1–2-leaved pseudobulbs. Leaves thinly leathery, in *O. vestitum* only from pseudobulb, elongate, in *O. proliferum* short, those from stem, in two ranks, jointed with the sheath. Peduncles 1-flowered, clustered at the axils. Flowers small. Sepals about equal, free, erect or spreading, acute, the lateral scarcely forming at the base a very short chin. Petals similar to the sepals but smaller. Lip sometimes erect from the base of the column with which it is continuous or very shortly connate, sometimes attached at the apex of the very short foot, erect from a claw which is very shortly horizontal; lateral lobes erect, parallel, often embracing the column, the median spreading, undivided. Column short or a little elongated, thick, without a foot or with a very short foot, not winged; clinandrium obliquely truncate. Anther terminal, opercular, incumbent, helmet-like, imperfectly 2-celled; pollinia 4, waxy, ovoid, sometimes free, sometimes attached to a linear stalk, with a small gland. Capsule oblong-ellipsoidal.

Species about 44, natives of tropical America from the West Indies and Mexico to Brazil.

Leaves, one on each pseudobulb, none on stem .... 1. *O. vestitum.*
Leaves, two on each pseudobulb, in two ranks on stem 2. *O. proliferum.*

1. **O. vestitum** *Reichb. f. in Walp. Ann. Bot. vi.* 491 (1863); *Griseb. Fl. Br. W. Ind.* 625; *Cogn. in Fl. Bras. iii. pt.* 6, 95 (in part).

Epidendrum vestitum *Sw. Prodr.* 124 (1788). Cymbidium vestitum *Sw. in Nov. Act. Ups. vi.* 70 (1779) & *Fl. Ind. Occ.* 1448; *Lindl. Gen. & Sp. Orch.* 168. (Pl. 26, f. 1–4.)

On trees, and rocks among peat; in fl. and fr. Aug.–March; "Jamaicæ temperatioris," *Swartz*! Radnor, *Purdie*! Springfield, *Wullschlaegel*, 1088! J.P. 24, *Morris*! Trelawney, *Miss T. M. Barrett*! Moneague, *Lady Blake*! Ellen Aire, Moy Hall, *Sullivan*! Cold Spring, *Moore*! Shooters Hill, Manchester, *Scharschmidt*! Cinchona, 5000 ft.; Lancaster, 2500 ft.; Mabess River, 3900 ft.; Vinegar Hill, 3900 ft.; *Harris*! Fl. Jam. 7210, 7567, 7654, 7839, 9008, 9552, 10,100.—Cuba, Venezuela, Surinam, Brazil, Peru.

Stem to 3 dm. l., branching, decumbent and rooting, covered with scales, which are triangular-ovate, acute, sheathing at base, imbricating, dry and becoming scarious, 1–1·5 cm. l. *Pseudobulbs* oblong-ellipsoidal, compressed, 1-leaved, 2 to nearly 3 cm. l. *Leaves* lanceolate-ligulate, apex acute or obtuse, shortly attenuated at the base, middle nerve prominent below, 1–2 dm. l., 1·5–2 cm. br. *Peduncles* clustered in the sheathing scales of the stem just below each pseudobulb, many-jointed, with a sheathing, spathaceous bract at each node. *Sepals*, lateral 5-nerved, broadly elliptical, apex acute, thickened, 5 mm. l., 3 mm. br.; median 7-nerved, elliptical, apex with a thickened apiculus, 5·5 mm. l., 2·4 mm. br. *Petals* 5-nerved, broadly lanceolate, acute, 4·5 mm. l., 1·8 mm. br. *Lip* broadly 3-lobed, clawed, lateral lobes rounded, median larger than lateral, shortly ovate, thickened at the apex with a dorsal apiculus, 4 mm. l. *Column* 1·75 mm. l. *Capsule* smooth, 1 cm. l.

2. **O. proliferum** *comb. nov.* Epidendrum proliferum *Sw. Prodr.* 124 (1788) & *Ic. ined. t.* 47. Cymbidium proliferum *Sw. in Nov. Act. Upsal. vi.* 71 (1799) & *Fl. Ind. Occ.* 1450. Isochilus prolifer *R. Br. in Ait. Hort. Kew. v.* 209 (1813), non. *Lindl. in Bot. Reg. t.* 825 (Scaphyglottis prolifera *Cogn.*). (Pl. 26, f. 5–7.) Type in Herb. Mus. Brit.

On rocks and trees, amongst moss; in fl. and fr. Dec.; "mont. Liguan. nov.," *Swartz*! near Cedar Hurst, J.P. 2121, *Syme*! *Harris*! Lancaster, *Harris*! Fl. Jam. 7566.

Stem ·5–2 dm. l., decumbent and rooting below. *Pseudobulbs* obovate-ellipsoidal, compressed, 2-leaved, 1·5–2 cm. l. *Leaves* lanceolate-oblong or oblong, apex obliquely emarginate, apiculate, middle nerve prominent below, those on pseudobulbs narrowed at base, those on stems jointed to strongly compressed sheath, 1·5–2·5 cm. l., 4–7 mm. br. *Sheaths* slightly keeled. *Peduncles* 7 mm. l., 1-flowered, with three sterile bracts at base; pedicel very short, with one floral bract. *Flowers* purple, perianth ⅓ in. l. *Ovary* cylindrical, 6-furrowed, very long. *Sepals* with three strong nerves and four faintly indicated, slightly keeled, obtuse, 8 mm. l., 3 mm. br.; lateral ovate-oblong; median elliptical. *Petals* 3-nerved, elliptical-oblong, obtuse, 7 mm. l., 2 mm. br. *Lip* similar to petals, erect, purple. *Column* with a short foot, 5 mm. l. *Capsule* elongate, half-concealed by the sheaths, narrowly spindle-shaped, 6-furrowed, 1·5 cm. l.

O. coccineum Salisb. is quoted by Grisebach as coming from Jamaica on the authority of Swartz, but we have seen no specimen from Swartz nor any citation of it by him.

51. MACRADENIA R. Br.

Small herbs. Pseudobulbs with one leaf at apex and scarious sheaths at base. Leaves flat, fleshy or leathery. Peduncles short, from base of pseudobulb. Racemes nodding or reflexed, lax.

Flowers medium-sized. Bracts small. Sepals equal, free, somewhat spreading. Petals a little smaller. Lip continuous with the base of the column, erect; lateral lobes broad, erect, embracing the column, the median spreading. Column erect, not winged, without a foot, furrowed in front, terminated by a thin erect or slightly inclined rostellum; clinandrium behind the rostellum. Anther erect in the bottom of the clinandrium, oblong, produced into a long membranous appendage embracing the rostellum; pollinia 2, waxy, ovoid, distinct, without an appendage, attached to the dilated apex of a long, very narrow stipes, gland small, ovate. Capsule ellipsoidal, tapering to base, shortly beaked.

Species 13, natives of the West Indies, Florida, Guatemala and Surinam.

M. lutescens *R. Br. in Bot. Reg. t.* 612 (1822); *Lindl. Gen. & Sp. Orch.* 195; *Reichb. Fl. Exot. t.* 117; *Griseb. Fl. Br. W. Ind.* 635; *Rolfe in Orch. Rev. iv.* 357 (excl. syn.); *Cogn. in Fl. Bras. iii. pt.* 6, 111; *Ames Orch. S. Fla.* 20, *t.* 11. Rhyncadenia cubensis *A. Rich. Fl. Cub. t.* 85 (1853). (Pl. 27, f. 7–10.)

On trees; in fl. Nov.–Jan., in fr. Jan.; near Castleton, J.P. 1, 2388, *Morris*! *Syme*! *Campbell*! Blue Mts. *Mrs. Maclaverty*! Belvedere, Hanover, *Harris*! Fl. Jam. 7679.—Cuba, Florida, Trinidad, Colombia, Venezuela.

Plant 5–8 in. high. *Pseudobulbs* slender, narrowly ovoid-cylindrical, 2–5 cm. l., 5–10 mm. br. *Leaf* oblong-lanceolate, acute, slightly tapering at base, 10–15 cm. l., 1·5–3 cm. br. *Peduncle* shorter than the leaf, sometimes drooping, with a few sheathing acuminate bracts below (1 cm. l.). *Pedicels* with ovary 1·5–2 cm. l., of fruit ·8–1 cm. l. *Bracts* triangular, acute, much shorter than ovary, 3–5 mm. l. *Flowers* 4–9, dingy yellow with brown or purplish blotches on the inside, shortly stalked, perianth nearly ⅓ in. l., sometimes longer. *Sepals* membranous, lateral 5–6-nerved, elliptical-lanceolate, acute, somewhat falcate, 1 cm. l., 3 mm. br.; median 7-nerved, very concave, elliptical, acute, 1·1 cm. l. *Petals* similar to sepals, but 5–7-nerved, elliptical-oblong, acute, somewhat falcate, 9 mm. l., nearly 3 mm. br. *Lip* roundish-obovate in outline below, 5 mm. l., 6 mm. br., with a narrow reflexed, terminal lobe; disk with three keels along centre as far as base of terminal lobe, and four nerves on each side; terminal lobe linear-lanceolate, 3-nerved, narrower at base than above, margin revolute, 4 mm. l., 1·5 mm. br. *Column* thicker above, 4·5 mm. l., 2 mm. br. above; rostellum linear, about 1·2 mm. l.; clinandrium membranous, cup-shaped, with the margin irregularly toothed. *Anther* shortly and imperfectly 2-celled.

52. IONOPSIS H. B. & K.

Epiphytic herbs. Stems leafy, very short, without pseudobulbs. Leaves few, narrow, leathery; sheaths persistent, overlapping, in two ranks. Peduncles 1–3, lateral or subterminal, elongated, thin, rigid, with few sheaths. Racemes slender, lax, simple or branching. Flowers medium-sized, with somewhat long stalks. Bracts very small. Sepals of nearly equal length, erect or spreading at the apex; median free, the lateral quite at the base under the lip connate into a short sac. Petals similar to the median sepal but wider. Lip attached to the base of the column, large, expanded, clawed, the claw bearing two pairs of flat calli, broadening above. Column short, erect, thick, anteriorly concave, not winged, without a foot; clinandrium short, truncate. Anther terminal, opercular, incumbent, semi-globose, 1-celled or imperfectly 2-celled, anteriorly shortly produced; pollinia 2, waxy, obovoid-subglobose, without an appendage, attached to a linear stalk, gland scale-like. Capsule oblong or ovoid, without a beak or with only a short one.

Species 9, natives of tropical America from the West Indies and Mexico to Brazil and Peru.

Lip 2-lobed, much longer than the sepals; leaves flat	1. *I. utricularioides.*
Lip entire, scarcely longer than sepals; leaves terete	2. *I. satyrioides.*

1. **I. utricularioides** *Lindl. Coll. Bot. t.* 39 *A.* (1825); leaves flat, linear-ligulate; lip twice as long as sepals.—*Lindl. Fol. Orch. Ionops.* 2; *Griseb. Fl. Br. W. Ind.* 636; *Cogn. in Symb. Ant. iv.* 180. I. tenera *Lindl. in Bot. Reg. t.* 1904 (1836), *in Paxt. Fl. Gard. ii.* 13, *f.* 14, & *Fol. Orch. Ionops.* 2. Epidendrum utricularioides *Sw. Prodr.* 122 (1788). Dendrobium utricularioides *Sw. in Nov. Act. Upsal. vi.* 83 (1799) & *Fl. Ind. Occ.* 1531. (Pl. 27, f. 4–6.) Type in Herb. Mus. Brit.

On trees; in fl. Jan.–June; *Swartz*! *Wright*! *Shakespear*! *Masson*! Wag Water River, St. Mary, J.P. 2284, *Morris*! *Syme*! Manchester; Mona, St. Andrew, 700 ft.; Belvedere, Hanover, 500 ft.; *Harris*! Fl. Jam. 7635.—Cuba, Hispaniola, Porto Rico, Martinique, St. Vincent, Trinidad, Guatemala, Mexico, Brazil, Peru.

Plant 7 in. to nearly 2 ft. high, occasionally with minute pseudobulbs. *Leaves* leathery, jointed to the sheathing base, 5–12 cm. l., 5–12 mm. br., with a few scarious sheaths on the rhizome below the leaves. *Panicle* 20–50 cm. l., much longer than the leaf, with a few distant adpressed sheaths (5–7 mm. l.). *Bracts* minute, triangular, acute, 1·5–2 mm. l. *Flowers* white or pale rose-colour, variegated or with purple veins; perianth about ¼ in. l., lip over ½ in. l. *Sepals* 3-nerved; lateral broadly lanceolate, acute, united at base into a very short simple obtuse sac, 6 mm. l., 2 mm. br.; median oblong, obtuse, 4·5 mm. l., 2 mm. br. *Petals* 3-nerved, oblong, obtuse, very shortly apiculate, 4·5 mm. l., 2 mm. br. *Lip* 2-lobed above, tapering below to base, with 2 pairs of wing-like calli on the narrow base opposite the column, the lower pair larger and less fleshy than the upper, 10–13 mm. l., and as broad or broader. *Column* very short, 2 mm. l. *Capsule* ellipsoidal, beaked, 2 cm. l., 7 mm. br.

2. **I. satyrioides** *Reichb. f. in Walp. Ann. vi.* 683 (1861); leaves terete, subulate; lip a little longer than sepals.—I. testiculata *Lindl. Gen. & Sp. Orch.* 193 (1833) & *Fol. Orch.*

Ionops. 1; *Griseb. loc. cit.* I. teres *Lindl. in Bot. Reg. xxiv. Misc.* 181 (1838), & *Fol. Orch. Ionops.* 2. Epidendrum satyrioides *Sw. Prodr.* 123 (1788). Dendrobium testiculatum *Sw. in Nov. Act. Upsal. vi.* 83 (1799) & *Fl. Ind. Occ.* 1533. (Pl. 27, f. 1–3.) Type in Herb. Mus. Brit.

On trees; in fl. April–Oct., in fr. April–Jan.; *Swartz*! Chepstow, J.P. 462, *Morris*! Spanish Town, *Hay*! near Hope Bay, *Moore*! Swift River, 500 ft.; Belvedere, Hanover, 500 ft.; Troy, 1500 ft.; *Harris*! Fl. Jam. 7621, 8666, 10,456, 10,463.—Hispaniola, Trinidad.

Plant 2 in. to a foot high, varying much in size. *Leaves* 3–14 cm. l., 2–3 mm. br. *Raceme* or panicle, 3–15 cm. l., generally longer than the leaf, with a few distant lanceolate acute or acuminate adpressed sheaths (2–8 mm. l.), and sometimes with a few small leaves. *Bracts* small, triangular, acute, half-clasping, concave, 1·5–3·5 mm. l. *Flowers* creamy-white, perianth ¼ in. l. *Sepals*, lateral 3-nerved, oblong-lanceolate, acute, united (for about 2 mm.) at the base and forming a 2-lobed sac, 7 mm. l., 2 mm. br.; median, 5-nerved, ovate-lanceolate, acute, a little shorter than the lateral, 2·3 mm. br. *Petals* 5-nerved, ovate-oblong, margin slightly undulate and curled, apex obtuse, 7 mm. l., nearly 3 mm. br. *Lip* 8·5 mm. l., obovate, long-tapering to base, with 5 nerves branching above, margin slightly undulate, apex obtuse, undulate and curled and so it may appear to be either 2-lobed or 3-lobed, with two minute calli at the outer edge of the base, and on the disk two lamellate calli which extend to about 2 mm. from base. *Column* short, 2 mm. l., thicker above. *Capsule* ellipsoidal, tapering to base, 13–16 mm. l., 5 mm. br.

53. COMPARETTIA Pœpp. & Endl.

Epiphytic herbs. Stem very short with two or three sheaths, becoming thickened into a small fleshy pseudobulb with one leaf. Leaf leathery. Scapes at the base of the pseudobulbs, erect, elongate, simple. Flowers laxly racemose, distinctly stalked. Bracts small. Sepals, all of equal length, erect-spreading, median free, lateral connate and produced at the base into a long spur free from the petals. Petals as long as the sepals but broader. Lip continuous with the base of the column, produced at the base into two long linear spurs, included within the spur of the sepals; spreading, much longer than the sepals, very broad, emarginate, clawed, the claw broadening below into a rounded base. Column erect, somewhat thick, semiterete, furrowed anteriorly, not winged, without a foot; clinandrium short, oblique, posteriorly somewhat acute. Anther terminal, opercular, incumbent, very convex, 1-celled; pollinia 2, waxy, broadly ovoid, furrowed, without an appendage, attached to a linear wedge-shaped stalk, gland ovate. Capsule ovoid or oblong, acute-angled, beaked.

Species 6, natives of the West Indies and the Andes, from Central America to Peru.

C. falcata *Pœpp. & Endl. Nov. Gen. & Spec. i.* 42, *t.* 73 (1836); *Hook. in Bot. Mag. t.* 4980; *Griseb. Fl. Br. W. Ind.* 636;

Lindenia iv. 41, *t.* 163; *Williams Orch. Alb. viii. t.* 359; *Cogn. in Symb. Ant. iv.* 181 & *in Fl. Bras. iii. pt.* 6, 181, *t.* 42, *f.* 2. (Pl. 28, f. 10–15.)

On trees; in fl. Nov.–March; Manchester, *Purdie*! St. Ann, *Prior*! J.P. 25, 2015 (partly), *Morris*! *Syme*! Greenhill Wood, *Moore*! Greenwich, *Watt*! Mt. Diabolo, 2300 ft.: Lancaster, 2500 ft.; *Harris*! Fl. Jam. 9009.—Cuba, Porto Rico, Costa Rica, Guatemala, Colombia, Venezuela, Ecuador, Bolivia, Peru.

Plant 6–15 in. high. *Pseudobulbs* small, slender, 1·5–2·5 cm. l., 4–5 mm. br.; sheaths triangular-lanceolate, 2·5–4·5 cm. l. *Leaf* somewhat leathery, oblong-elliptical to oblong-ligulate, obtuse, apiculate, 5–18 cm. l., 1·2–3·8 cm. br. *Scape* 1–3 dm. l., much longer than the leaf, simple or occasionally branching, with a few distant adpressed triangular acuminate or obtuse sheaths (4–10 mm. l.). *Pedicels* with ovary 1–1·5 cm. l. *Bracts* triangular, acute, clasping, 3–4 mm. l. *Flowers* bright pink or light crimson, lip ½ in. l. *Sepals* concave, 10 mm. l., lateral connate, oblong, tapering to an obtuse apex, 5-nerved; median oblong, acuminate, 5-nerved; spur 13 mm. l. *Petals* 7-nerved, obovate-elliptical, abruptly acute, 8–10 mm. l., 5 mm. br. *Lip* half as long again as the sepals, 15 mm. l., and nearly as broad, mid-vein forming a wing-like callus in the lower half of the claw. *Column* 5 mm. l.

54. TRICHOPILIA Lindl.

Epiphytic herbs, pseudobulbs with one leaf. Leaf fleshy, erect, at the base folded upon itself. Scape short, leafless, with few sheaths. Sepals subequal, free, narrow, somewhat erect. Petals somewhat similar to sepals. Lip adnate to base of column, spreading. Clinandrium hooded, large, membranous. Anther within the clinandrium, opercular, incumbent, 1-celled; pollinia 2, obovoid-oblong, without appendage, but affixed by a strap-shaped stipes with a small gland. Capsule ovoid-ellipsoidal.

Species about 27, natives of warmer parts of America, chiefly in Colombia, Central America and Mexico, with two species in the West Indies.

T. jamaicensis *Fawc. & Rendle in Journ. Bot. xlviii.* 107 (1910). (Pl. 29, f. 1–3.) Type in Herb. Mus. Brit. & Jam. Herb.

On trees; in fl. Sept.; Mansfield, near Bath, *Harris*! Fl. Jam. 7697.

Plant 6 or 7 in. high. *Pseudobulb* narrowly cylindrical, scarcely swollen, 1–2 cm. l., about 2 mm. br. *Leaf* linear, subterete and tapering above, 11–17 cm. l., about 4 mm. br. *Scape* about 2·5 cm. l.; sheath nearly 2 cm. l. *Bracts* elliptical-oblong, acuminate, 10–14 mm. l., about 4 mm. br. *Pedicel* and *ovary* 1 cm. l.; pedicel of fr. ·5 cm. l. *Flowers*, perianth nearly ¾ in. l. *Sepals* 3-nerved, with a few undulating pinnate nerves; lateral, linear-lanceolate, slightly folded on itself, forming a keel which extends separately as an acute projection beyond the apex, 19 mm. l., barely 2·5 mm. br.; median 4-nerved, slightly reticulate-nerved, ligulate, acuminate, 17 mm. l., 2·5 mm. br. *Petals* with nerves as in lateral sepals, similar to median sepal, but a little shorter and broader, 16 mm. l., 3 mm. br. *Lip*, general outline broadly oval-elliptical, somewhat spoon-shaped, obscurely 3-lobed, apex abruptly acute, saccate below, margin denticulate-eroded, 15–17-nerved, nerves branching, at base on inside 3 nerves prominent and callose, forming the inner walls of two shallow pockets, 14 mm. l., about 10 mm. br. *Column* 5 mm. l., thicker upwards, with a small tooth on each side at the base of the clinandrium; clinandrium 2·5 mm. l., with upper margin truncate and eroded; stigma deeply sunk, margin rounded. *Capsule* (unripe), 1·5 cm. l., 5–6 mm. br.

55. BRASSIA R. Br.

Epiphytic herbs. Stems short, terminated by a pseudobulb with one or two leaves. Leaves under the pseudobulb few or reduced to sheaths, leathery. Peduncles from under the pseudobulb simple. Racemes lax, generally few-flowered. Flowers rather large. Bracts short or spathe-like. Sepals free, spreading, narrow, acuminate or tailed, the lateral longer. Petals similar to the median sepal or smaller. Lip sessile at the base of the column, spreading, flat, undivided, shorter than the sepals, at the base 2-lamellate. Column short, erect, not winged, without a foot; clinandrium hardly prominent, truncate. Anther terminal, opercular, incumbent, very convex or semiglobose, 1-celled; pollinia 2, waxy, obovoid, without an appendage, attached to a flat oblong or linear stalk.

Species 40, natives of tropical America, from the West Indies and Mexico to Brazil and Peru.

Lip about 1½ in. l.; lateral sepals twice as long ...... 1. *B. maculata.*
Lip about 1 in. l.; lateral sepals 4 or 5 times as long 2. *B. caudata.*

1. **B. maculata** *R. Br. in Ait. Hort. Kew. ed.* 2, *v.* 215 (1813); lateral sepals twice as long as the lip; lip broadly ovate-roundish, acute, contracted into an oblong base; lamellæ tooth-like at apex, hairy at base.—*Bot. Mag. t.* 1691; *Lindl. Fol. Orch. Brass.* 1; *Griseb. Fl. Br. W. Ind.* 633; *Salisb. Ic. ined. vi. fol.* 580. B. guttata *Lindl. in Pl. Hartw.* 94 (1842). B. Wrayæ *Hook. in Bot. Mag. t.* 4003 (1843). (Pl. 28, f. 1–6.)

On rocks and trees; in fl. April–June, Oct.; Old England, St. Andrew, *Purdie*! Fairfield, *Wullschlaegel*, 1884 bis! *Oehler*! near Gordon Town, J.P. 6, *Morris*! *Sullivan*! *Harris*! Greenwich, *Moore*! near Radnor, *Miss B. Gosset*! Dolphin Head, *Harris*! Fl. Jam. 10,442.—Cuba, Guatemala.

Plant 1½–2 ft. high. *Pseudobulbs* large, oblong, sometimes contracted below, compressed, 2-edged, 6–10 cm. l., 2·5–3·5 cm. br. *Leaves* leathery, broadly ligulate, acute, tapering to base, 10–40 cm. l., 3–5 cm. br. *Peduncle* stout, longer than the leaves, with a few distant sheathing bracts, 3·5–6 dm. l., with 7–14 flowers. *Bracts*, sterile sheathing, obtuse, distant, 12–20 mm. l.; floral triangular, acute, much shorter than ovary, 6–10 mm. l. *Flowers* large, yellowish-green with brown or purple spots; lip, yellowish-white, 1¼–1½ in. l., sepals 2–3 in. l. *Sepals* narrowly linear-lanceolate, 4·5–7·5 cm. l., 4–6 mm. br. *Petals* similar to sepals 3–4 cm. l., 4–5 mm. br. *Lip* 3·5–4 cm. l., 2·5–3 cm. br.; lamellæ yellow, 5–6 mm. l. *Column* 5–6 mm. l., green.

2. **B. caudata** *Lindl. in Bot. Reg. t.* 832 (1824); lateral sepals very long, very acuminate, generally four or five times as long as lip; lip obovate-oblong, abruptly acuminate, lamellæ thickish, pubescent, with free tooth-like points and two teeth above distinct from the lamellæ.—*Lindl. Fol. Orch. Brass.* 3; *Hook. Exot. Fl. t.* 179 & *in Bot. Mag. t.* 3451; *Reichb. Fl. Exot. t.* 267; *Griseb. loc. cit.*; *Cogn. in Fl. Bras. iii. pt.* 6, 257, *t.* 58. Helleborine ramosissima &c. *Plum. Nov. Pl. Amer. Gen. Cat.* 9 (1703) & *Pl. Amer.* (*Burm.*) *t.* 177. Epidendrum caudatum *L. Sp. Pl. ed.* 2. 1349 (1763). (Pl. 28, f. 7–9.)

On trees; in fl. May–Aug.; *Macfadyen*! Iron River, Golden Spring, J.P. 2281, *Syme*! St. George, *Watt*! Clarendon, *Tomlinson*! Fl. Jam. 10,419.—Cuba, Guatemala, Mexico, Surinam.

Plant 10–15 in. high. *Pseudobulbs* large, oblong, compressed, 2-edged, 9–12 cm. l., 2·5–3 cm. br.; at apex 2-leaved. *Leaves* leathery, broadly ligulate, abruptly acute, tapering towards base, 18–30 cm. l., 4–5·5 cm. br. *Peduncle* stout, a little longer than the leaves, with a few distant sheathing bracts below, 2·5–3·5 dm. l., with 9–15 flowers. *Bracts* oblong-triangular, obtuse, shorter than the ovary, 8–10 mm. l. *Flowers* large, shortly stalked, greenish-yellow, with brown or purplish spots, lip about 1 in. l., lateral sepals 2 to 4 or 5 in. l. *Sepals* linear-lanceolate, very long, very acuminate, 5–6 mm. br., lateral 5–11 cm. l., median 4–5 cm. l. *Petals* lanceolate, long-acuminate, 2–2·5 cm. l., 4–6 mm. br. *Lip* 2·5–3 cm. l., 10–12 mm. br. *Column* 4 mm. l.

The floral parts vary much in size. In a living plant brought from Jamaica by Mrs. Longstaff, which flowered at Highlands, Putney, the measurements were as follows: sepals 6 mm. br., lateral 13·3 cm. l., median 5·2 cm. l.; petals 2·3 cm. l., 4·5 mm. br.; lip 2·8 cm. l., 1·2 cm. br.; column, including anther, 5 mm. l.

56. ONCIDIUM Sw.

Epiphytic herbs. Stems leafy, generally short or very short, with or without a pseudobulb. Leaves 3-edged or flat. Peduncles lateral, generally elongated and branching. Racemes lax. Bracts small. Sepals generally almost equal, spreading or reflexed, free or lateral connate nearly to apex. Petals similar to the median sepal or larger. Lip attached to the base of the column, continuous, median lobe spreading, very broad, emarginate or 2-fid; disk at the apex of the claw remarkably crested or tuberculate. Column short, thick, furnished with petaloid wings, without a foot; clinandrium sometimes very short, truncate, sometimes ovate, obliquely erect, entire or anteriorly shortly 2-dentate. Anther terminal, opercular, incumbent, very convex, semiglobose or hooded, 1-celled or imperfectly 2-celled; pollinia 2, waxy, obovoid, without an appendage, attached to a flat stalk, narrow and elongated or broad and very short. Capsule almost always ovoid, oblong or spindle-shaped, more or less beaked.

Species over 400, natives of tropical America, from the West Indies, Florida and Mexico to Brazil and Bolivia.

Leaves 3-edged, keeled, linear-ligulate; no pseudobulb; lateral sepals connate nearly to apex.
  Lip, middle lobe clawed .... 1. *O. guttatum.*
  Lip, middle lobe not clawed.
    Middle lobe entire .... 2. *O. triquetrum.*
    Middle lobe 2-lobed .... 3. *O. pulchellum.*
Leaves flat; lateral sepals free.
  Pseudobulbs wanting, or, if present, small; leaves very broad, oblong .... 4. *O. luridum.*
  Pseudobulbs present, leaves ligulate .... 5. *O. leucochilum.*

1. **O. guttatum** *comb. nov.*; leaves 3-edged, keeled, linear-ligulate; no pseudobulbs; lateral sepals connate nearly to apex; lip 3-lobed, basilar lobes small, obovate-oblong, terminal lobe many-nerved, kidney-shaped-2-lobed, broadly emarginate, apiculate, clawed, crest with three erect lobes, two behind parallel, the third concave in front.—O. quadripetalum *Sw. Vet. Handl. Stockh. xxi.* 240 (1800). O. tetrapetalum *Willd. Sp. Pl. iv.* 112 (1805); *Lindl. Gen. & Sp. Orch.* 198 & *Fol. Orch. Oncid.* 12; *Griseb. Fl. Br. W. Ind.* 631. O. tricolor *Hook. in Bot. Mag. t.* 4130 (1844). Ophrys bulbis filiformibus &c. *Plum. Amer.* (*Burm.*) 177, *t.* 182, *f.* 2. Viscum, delphinii flore albo guttato, minus, radice fibrosa *Sloane Cat.* 120 & *Hist. i.* 251, *t.* 148, *f.* 2. Epidendrum guttatum *L. Sp. Pl.* 953 (1753). E. tetrapetalum *Jacq. Sel. Stirp. Amer.* 230, *t.* 142 (1763), *Ed. pict. t.* 216; *Sw. Fl. Ind. Occ.* 1484. Cymbidium tetrapetalum *Sw. in Nov. Act. Upsal. vi.* 74 (1799). C. guttatum *Willd. Sp. Pl. iv.* 102 (1805). (Pl. 29, f. 18–21.)

On trees; in fl. Sept., Oct., March, April, in fr. March, April; Spanish Town, *Sloane*, Herb. iv. 122, 123! Macfadyen! *Purdie*! Luidas Vale, St. Thomas in Vale, *Prior*! Mile Gully, J.P. 246, *Morris*! Chapelton, *Fawcett*! Mona, 1000 ft.; Great Goat Is., 150 ft.; *Harris*! Fl. Jam. 7873, 9017, 10,438. —Cuba, Hispaniola, Porto Rico, St. Thomas, St. Martin, Saba, St. Eustatius, Mexico, Colombia.

Plant 9 in. to 2 ft. high. *Leaves*, margin entire to somewhat serrulate, 5–20 cm. l., 3–6 mm. br. *Peduncle* erect, simple or slightly branched above, much longer than the leaves, with a few distant short acute sheaths below (5–9 mm. l.), many-flowered above, 3–6 dm. l. *Pedicels* with ovary 1·3–2 cm. l. *Bracts* triangular, acute, much shorter than ovary, 3–5 mm. l. *Flowers* variable in colour, lip white tinted with pink, sepals and petals brownish with purplish markings, or flowers dingy-brown with rose-coloured markings, or brilliantly coloured white, green and crimson; lip over ½ in. l., ¾ in. br. *Sepals* spathulate, keeled, apex acute, 7·5–8 mm. l.; lateral 5-nerved; median 3-nerved, 3 mm. br. *Petals* 3-nerved, in general outline broadly elliptical, broadly clawed, more or less indistinctly 3-lobed, lateral lobes rounded, median lobe broadly triangular-ovate, apex shortly acute, 7·5–8 mm. l., 4·5–5 mm. br. *Lip* varying in size, 10–13 mm. l., 8·5–16 mm. br., lobes 2·5 mm. l. *Column* short, 3·5 mm. l., with a broad square wing in front of lower half; upper wings falcately ovate, acute, 3 mm. l.

O. variegatum Sw., from Hispaniola, differs in its obtuse sepals and emarginate petals. Although this species and *O. sylvestre* Lindl. are quoted by Grisebach from Jamaica we have not seen any authoritative specimens.

2. **O. triquetrum** *R. Br. in Ait. Hort. Kew. ed.* 2, *v.* 216 (1813); leaves 3-edged, keeled, linear-ligulate; no pseudobulb; lateral sepals connate nearly to apex; lip 3-lobed, lateral lobes rounded, median lobe much larger, somewhat deltoid, with an apiculate apex, callus at base in front smooth, rounded, behind united with column.—*Lindl. Gen. & Sp. Orch.* 205 & *Fol. Orch. Oncid.* 11; *Hook. in Bot. Mag. t.* 3393; *Griseb. op. cit.* 631. Epidendrum triquetrum *Sw. Prodr.* 122 (1788). Cymbidium triquetrum *Sw. in Nov. Act. Upsal. vi.* 74 (1799) & *Fl. Ind. Occ.* 1475. (Pl. 29, f. 12–17.) Type in Herb. Mus. Brit.

On trees; in fl. Jan., Feb., June, in fr. Jan.; *Swartz*! *Bligh; Macfadyen*! Iron River, Golden Spring, J.P. 2286, *Syme*! near Castleton, *Campbell*! Spaldings, *Sullivan*! near Browns Town, 1000 ft.; Belvedere, Hanover, 500 ft.; *Harris*! Fl. Jam. 7647.

Plant 4–8 in. high. *Leaves* 3–19 cm. l., 5–10 mm. br. *Peduncle* erect, simple or slightly branched above, shorter or longer than the leaves, with a few distant triangular acute keeled sheaths below (4·5–6 mm. l.), few-flowered above, 7–18 cm. l. *Pedicel* with ovary 11–15 mm. l., of fruit 8 mm. l. *Bracts* triangular, acuminate, much shorter than ovary, 2–4 mm. l., sterile longer. *Flowers* variable in colour, white and densely purple-spotted, or rose bordered with white and deeply stained and spotted with crimson, lip nearly ½ in. l. *Sepals* slightly concave, oblanceolate, apex acute, 9–10 mm. l., 4 mm. br. *Petals* triangular-ovate, clawed, apiculate, as long as sepals, 5·5–6 mm. br. *Lip* a little over 10 mm. l. and nearly as broad. *Column* very short, 3–4 mm. l., wings somewhat triangular with an acute apex.

3. **O. pulchellum** *Hook. in Bot. Mag. t.* 2773 (1827); leaves 3-edged, keeled, linear-ligulate; no pseudobulbs; lateral sepals connate nearly to apex; lip, median lobe 2-lobed, broader than long, lateral lobes large, rounded, varying from considerably smaller to almost the size of half the median lobe, crest at base of disk consisting of three narrow calli free at apex.—*Lindl. Gen. & Sp. Orch.* 206, *Fol. Orch. Oncid.* 12 & *in Bot. Reg. t.* 1787; *Griseb. op. cit.* 631; *Cogn. in Fl. Bras. iii. pt.* 6, 434. (Pl. 29, f. 4–8.)

On trees; in fl. April–July; *Bancroft*! *Macfadyen;* Blue Mts., 2000–3000 ft., J.P. 2385, *Morris*! Mt. Moses, 3500–4000 ft., J.P. 2385, *Syme*! Browns Town, *Levy*! Trelawney, *Woollett*! Moy Hall, *Clark*! near Troy, 1400 ft.; Mabess River, 3000 ft.; *Harris*! Fl. Jam. 7362, 8656.—Cuba, B. Guiana.

Plant 1–2 ft. high. *Leaves* 7–20 cm. l., 7–15 mm. br. *Peduncle* erect, simple or occasionally branched, much longer than the leaves, with a very few distant short acute sheaths below (7–12 mm. l.), many-flowered above, 3–6 dm. l. *Pedicels* with ovary 1·5–2 cm. l. *Bracts* triangular, acute, much shorter than ovary, 3–5 mm. l. *Flowers* white with rosy tinge, not large, but varying very much in size; lip ½–1 in. l. *Sepals* boat-shaped, lateral, 2-nerved, oblong or oblong-elliptical, tapering to base, 6–10 mm. l., 3·5–4 mm. br.; median 3-nerved, concave, obovate-cuneate, acute, 5·5–9 mm. l., 2·5–3 mm. br. *Petals* 6–10 mm. l., 4·5–7 mm. br., roundish-elliptical or obovate, tapering to base, 3-nerved, apex rounded or apiculate, margin undulate. *Lip* 1·3–2·5 cm. l., median lobe 1·2–3 cm. br. *Column* very short, 2 mm. l., wings somewhat oblong-triangular with rounded corners, about 3 mm. l., 4 mm. br.

4. **O. luridum** *Lindl. in Bot. Reg. t.* 727 (1823); leaves large, leathery, broadly ligulate-oblong; pseudobulbs wanting or occasionally occurring and quite small; lateral sepals free; lip much narrower at base than above, at base cordate, lateral lobes small, semi-elliptical, margin revolute; median lobe large, shortly clawed, kidney-shaped, at apex emarginate, sinus between the lobes roundish, rather broad, disk at the base with a thick fleshy crest composed in the lower half of two swollen calli uniting below where there is sometimes a small tubercle on each side, a third thin callus rises between these calli and is continued above with a 2-lobed tubercle on each side.—*Lindl. Gen. & Sp. Orch.* 201 (excl. syn. L. & Willd.), *Fol. Orch. Oncid.* 41 (excl. syn. L. & Willd.); *Hook. in Bot. Mag. t.* 3603; *Griseb. op. cit.* 632. O. luridum var. guttatum *Lindl. in Bot. Reg. xxv. t.* 16 (1839); *Ames Orch. S. Fla.* 21. O. carthaginense *Sw. in Vet. Handl. Stockh. xxi.* 240 (1800) & *Fl. Ind. Occ.* 1479 non *Auct.* O. guttatum *Reichb. f. in Walp. Ann. Bot. vi.* 782 (1863); *Cogn. in Fl. Bras. iii. pt.* 6, 411. Viscum radice bulbosa majus &c. *Sloane Cat.* 119 & *Hist. i.* 250, *t.* 148, *f.* 1. Epidendrum undulatum *Sw. Prodr.* 122 (1788), non *Sims in Bot. Mag. t.* 777. (Pl. 29, f. 9–11.)

On trees and rocks; in fl. April–Sept.; near Spanish town, *Sloane Herb.* iv. 177! *Wright*! *Poore*! *Masson*! *Bancroft*! *Macfadyen*! Wag Water River, *Purdie*! Fairfield, *Wullschlaegel*, 1095*a*! *March*! common, 200–3500 ft., J.P. 2396 *Morris*! Bog Walk, 300 ft., *Lehmann*! Cedar Hurst, *Watt*! *Moore*! Trelawney (yellow without spots), *Miss Jeffrey Smith*! Mansfield, Bath, 1000 ft.; Tyre, near Troy, 1500 ft.; *Harris*! Fl. Jam. 8646.—Cuba, Florida, Dominica, Martinique, St. Lucia, St. Vincent, Grenada, Trinidad, Margarita, Venezuela, Colombia, B. Guiana, Peru, Mexico.

Roots numerous, densely-clustered, long. *Sheaths* triangular-ovate, acute, 1–12 cm. l. *Leaves* 3–5 dm. l., 6–9 cm. br. *Panicle* 9–18 dm. l., much longer than the leaves, with several very short adpressed distant sheaths (to 1·5 cm. l.). *Bracts* very short, 3–5 mm. l., triangular, acute. *Flowers* rather large, long-stalked, segments spreading, margin curled-wavy, yellow colour spotted more or less with cinnamon-brown, rarely yellow without any spots, lip about ¾ in. l., and nearly an inch br. *Sepals* with 3–5 branching nerves of about equal length, rounded at the apex, median obovate-roundish or obovate-oblong, at the base abruptly long and narrowly clawed, 1·4–1·7 cm. l. (claw 4–5 mm. l.), 8·5–10 mm. br.; lateral narrower, broadly elliptical, at the base long tapering into a claw, 1·4–1·7 cm. l., 7–8 mm. br. *Petals* with four or five branching nerves, obovate-roundish, or obovate-oblong at the base abruptly, rather broadly, shortly clawed, about as long as sepals or slightly shorter, 1·5–1·6 cm. l., 9–10 mm. br. *Lip* 1·5–2 cm. l., 1·6–2·3 cm. br., at base about 8 mm. br. *Column* short, 4–5 mm. l., wings 2-lobed, anterior lobe hatchet-shaped, posterior small rounded. *Capsule* very large, obovoid-ellipsoidal, tapering to base, 5 cm. l., 2 cm. br.

5. **O. leucochilum** *Bateman ex Lindl. in Bot. Reg. under t.* 1920 (1837); leaves large, leathery, ligulate, acute; pseudo-

bulbs ovoid-oblong, furrowed, compressed, with two leaves; lateral sepals free; lip 3-lobed, lateral lobes small, oblong-obtuse; middle lobe broadly kidney-shaped, emarginate or retuse; crest with two long lobes on each side, and an erect recurved one in the middle.—*Bateman Orch. Mex. & Guat. t. I.* (1837); *Lindl. Fol. Orch. Oncid.* 48 & *in Paxt. Mag. Bot. vii.* 241 with plate.

On trees in the Blue Mts.; Monklands, *Moore*! near Newcastle, *Walker*!—Mexico, Guatemala.

Roots numerous, long, branching. *Pseudobulbs* 5–10 cm. l. *Leaves* 2–4 dm. l., 2·5–4 cm. br. *Panicle* 10–20 dm. l., much longer than the leaves, laxly-flowered, with several adpressed distant sheaths (about 1 cm. l.). *Bracts* triangular, acute, 5–7 mm. l. *Flowers* rather large, long-stalked; sepals and petals nearly equal, about ¾ in. l., greenish-brown barred with light yellow-green; lip white, about same length. *Sepals* and *petals* similar, elliptical-oblong, acute, spreading, about 2 cm. l. *Column* short, about 5 mm. l.; wings somewhat hatchet-shaped, acute or obtuse above, rounded below.

O. altissimum *Sw.* is credited to Jamaica by Swartz (Prodr. 123 & Fl. Ind. Occ. 1481). We have seen no specimen nor any other record of the species for Jamaica. Swartz's citation may have been due to the fact that Jacquin, in his first description of the species (*Epidendrum altissimum* Pl. Ins. Carib. 30, 1760), cited Sloane's figure (Hist. Jam. i. t. 148, f. 1), which, as we have seen, represents *O. luridum*. The species is known from Guadeloupe, Dominica, Martinique, St. Vincent, Mexico, Panama, Colombia, British Guiana, and Surinam.

A very closely allied species, *O. sphacelatum* Lindl., from Florida and Central America, has been grown under the name of *O. altissimum* in the Jamaica Botanic Gardens, and a flower is figured on Pl. 29, f. 22–25, but there is no satisfactory evidence that it is native in Jamaica.

57. LEOCHILUS Knowles & Westc.*

Small epiphytic herbs. Stems very short, terminated by a pseudobulb with one leaf, covered by membranous scarious sheaths or a few leaves under the pseudobulb. Leaves oblong or elliptical, flat, contracted into the stalk. Scapes under the pseudobulb, thin, simple or more rarely somewhat branched. Racemes lax, with small tender flowers, distinctly stalked. Bracts about as long as the stalks or shorter, narrow, acute. Sepals about equal, spreading, free or the lateral shortly connate at the base. Petals similar to the sepals or a little broader. Lip attached at the base of the column, continuous, spreading,

* We have retained the original spelling for this genus. Bentham and Hooker f., Gen. Pl. iii. 564 (1883), on etymological grounds, wrote *Leiochilus*, overlooking the fact that a genus of *Rubiaceæ* established by Hooker in an earlier volume of the same work, ii. 116 (1873), bears the same name.

oblong, undivided, sometimes longer than the sepals, the disk fleshy or callose. Column short, erect, without a foot, not winged but 2-auricled below the stigma; clinandrium truncate; rostellum elongated. Anther terminal, opercular, incumbent, 1-celled, anteriorly produced beyond the cell into a membranous hooded appendage; pollinia 2, waxy, globose, without an appendage, attached to a narrow elongated stalk, gland ovate. Capsule ellipsoidal, beaked.

Species 7, natives of the West Indies, Mexico, and Central America.

L. labiatus *Kuntze Rev. Gen.* 656 (1891).—*Cogn. in Symb. Ant. iv.* 182. L. cochlearis *Lindl. in Bot. Reg. xxviii. Misc.* 23 (1842); *Griseb. Fl. Br. W. Ind.* 634. Epidendrum labiatum *Sw. Prodr.* 124 (1788), *Fl. Ind. Occ.* 1493 & *Adnot. Bot. t.* 2, f. 6. Oncidium labiatum *Reichb. f. in Walp. Ann. vi.* 741 (1861). (Pl. 30, f. 1–5.)

On trees; in fl. March-June; *McNab;* St. Mary; St. Ann; *Purdie!* Beaufort, *Wullschlaegel,* 1091! Castleton, on river banks, J.P. 2324, *Syme!* near Gordon Town; near Hope Bay; *Moore!* near Browns Town, *Miss T. M. Barrett!* Christiana, *Tomlinson!* Castleton; Troy, 1500 ft.; *Harris!* Fl. Jam. 10,423, 10,426, 10,462.—Cuba, Hispaniola, Guadeloupe, Dominica, St. Vincent, Trinidad.

Plant 2–9 in. high. *Roots* numerous, very long. *Pseudobulb* somewhat globose, 7–15 mm. l. *Leaves,* from apex of the pseudobulb and from its base, somewhat leathery, 2–7 cm. l., ·5–1·5 cm. br. *Scape* with a few distant lanceolate keeled acute sheaths below (5–9 mm. l.), and a few flowers above. *Pedicels* about equalling ovary, together 8 mm. l. *Bracts* triangular, acute, 2–3 mm. l. *Sepals* green, spotted with dark reddish-brown, 3-nerved; lateral oblong, obtuse, united at a distance of 1·5–3 mm. from base, 4–4·5 mm. l., 1·3–1·5 mm. br. each at point of union; median elliptical, very shortly acuminate, 3·5 to nearly 4 mm. l., 1·5 to nearly 2 mm. br. *Petals* green, spotted with dark reddish-brown, elliptical-oblong, apex conduplicate, shortly acute, 3·25–3·5 mm. l., 1·5–1·75 mm. br. *Lip* yellow, streaked with dark reddish-brown at base, elliptical-oblong, contracted above the base, where there is a fleshy callus, and very concave at the base, apex generally emarginate, about 5 mm. l., 2·5 mm. br. *Column* 1·5 mm. l. *Capsule* 14–15 mm. l., 7 mm. br.

58. CRYPTARRHENA R. Br.

Small epiphytic herbs. Stems leafy, very short, without pseudobulbs. Leaves several, crowded, membranous-leathery, with leathery sheaths overlapping in two ranks. Peduncles axillary, erect, longer than the leaves, simple. Flowers small, laxly racemose. Bracts narrow, about equalling the pedicels. Sepals almost equal, free, spreading, acute. Petals somewhat like the sepals. Lip continuous with the base of the column, spreading, claw long, keeled-crested above; lateral lobes at the apex of the claw narrow, spreading or recurved; median shorter, broader, divided. Column short, thickened above, not winged, without a foot; clinandrium elevated, hooded, with small teeth on the margin. Anther within the clinandrium, opercular, incumbent, convex, shortly acuminate, imperfectly 2-celled; pollinia 4, waxy, incumbent in pairs without appendage, affixed by the dilated apex of a linear stipes, with a small gland. Capsule ovoid or oblong.

Species 2, one extending through the West Indies and Central America to Peru and Guiana, the other growing in Surinam.

C. lunata *R. Br. in Bot. Reg. t.* 153 (1816). C. pallidiflora *Reichb. f. in Bot. Zeit. x.* 766 (1852). (Pl. 30, f. 6, 7.)

On trees; in fl. Aug.–Sept., in fr. Oct.–Feb.; J.P. 2406, *Morris*! St. Ann, *Miss T. M. Barrett*! Cedar Hurst; Spanish River; *Moore*! Holly Mount, Mt. Diabolo, *Harris*! Fl. Jam. 8912, 10,390, 10,397, 10,430, 10,431. —Costa Rica, Mexico, Trinidad, B. Guiana, Peru.

Plant 4–10 in. high. *Leaves* linear-oblanceolate, acute, jointed to the sheathing base, 6–14 cm. l., 7–13 mm. br. *Peduncle* 10–18 cm. l. *Pedicel* and ovary 4–6 mm. l. *Bracts* lanceolate, acuminate, 2·5–5 mm. l. *Sepals* elliptical, tapering to an apiculate apex; lateral 3·5 mm. l., nearly 2 mm. br.; median 3·25 mm. l., 1·5 mm. br. *Petals* obovate-oblong, apex rounded, upper margin minutely eroded, about 3 mm. l., 1·5 mm. br. *Lip* about 4 mm. l., lateral lobes linear-subulate, curved backwards, more than 2 mm. l., terminal lobe 2-fid, upper margin slightly irregularly eroded, 1 mm. l., 2 mm. br. *Column* about 2 mm. l. *Capsule* cylindrical, tapering abruptly at base into the stalk, 6–8 mm. l., 3 mm. br.

59. DICHÆA Lindl.

Epiphytic herbs. Stems leafy, elongated, often pendulous, covered with the leaf-sheaths and leaves. Leaves in two ranks, somewhat membranous, spreading, short. Peduncles axillary, solitary, each with one small flower, with two or three scarious ample bracts, one above the other, near the base, the upper one acute, the lower obtuse, and bracteoles generally two at apex below the flower, the lower broadly roundish, apiculate, the upper linear, sometimes wanting. Sepals subequal, free, spreading, lateral, slightly oblique, forming at the base with the foot of the column a scarcely prominent chin. Petals similar to the sepals, but smaller. Lip fleshy, fixed to the base of the column, sometimes with a short claw; blade sometimes broadening upwards with angles produced sometimes into lobes or teeth, concave, undivided. Column erect, semiterete, very stout, at the base produced into a short or very short foot; clinandrium oblique, short, entire, rather broad. Anther terminal, opercular, incumbent, very convex, imperfectly 2-celled; pollinia 4, waxy, ovoid, closely incumbent in pairs on themselves, without appendage, affixed by a flat stipes, with a small gland. Capsule ovoid or oblong, muricate or smooth, ribs only slightly prominent.

Species about 35, natives of tropical America, chiefly the West Indies, Central America and Mexico.

Leaves persistent, not jointed below; capsule densely covered with long bristles.

Leaves oblong, tapering to an acute apex with a long tip, inclined slightly towards the end of the shoot .... 1. *D. echinocarpa.*

Leaves elliptical-oblong with a rounded apex with a short tip, inclined slightly away from end of shoot .... 2. *D. muricata.*

Leaves linear, inclined slightly away from end of shoot .... 3. *D. trichocarpa.*

Leaves jointed above the sheath, limb falling; capsule smooth.

Leaves oblong-linear, with a long tip at apex, spreading .... 4. *D. graminoides.*

Leaves oblong, with a very short tip, inclined towards end of shoot .... 5. *D. glauca.*

Leaves jointed above the sheath, limb falling; capsule densely covered with long bristles.

Leaves oblong or oblong-ligulate, with a short or long tip, ciliate at apex, spreading .... 6. *D. Morrisii.*

1. **D. echinocarpa** *Lindl. Gen. & Sp. Orch.* 208 (1833) (in part); leaves persistent, not jointed below, oblong, tapering to an acute apex with a long tip, inclined slightly towards the end of the shoot; lip white, variegated with blue, very shortly clawed, somewhat quadrangular-oblong, cuneate towards base, angles not produced into lobes, apex apiculate; capsule densely covered with long bristles which are swollen at the base.—*Griseb. Fl. Br. W. Ind.* 624. Epidendrum echinocarpon *Sw. Prodr.* 124 (1788). Cymbidium echinocarpon *Sw. in Nov. Act. Upsal. vi.* 71 (1799) & *Fl. Ind. Occ.* 1452. (Pl. 30, f. 26–30.) Type in Herb. Mus. Brit.

Amongst mosses on trees and rocks; in fl. Aug.–Nov., in fr. Oct.–April; mountains, *Swartz*! *Wiles*! *Macfayden*! *March*! J.P. 2036, *Morris*! Mt. Moses, 3000 ft.; Dollwood; Cinchona, 5000 ft.; Clydesdale, 4500 ft.; *Harris*! Fl. Jam. 7527, 7994, 9023.—Cuba, Porto Rico.

Plant to 1½ ft. l. *Stem* to 5 dm. l., simple or sometimes branched, decumbent, pendulous, emitting roots, somewhat slender, compressed, 1·5 mm. br. *Leaves*, limb 12-25 mm. l., 5-8 mm. br., the long apical tip readily rubbing off, median nerve very prominent below, forming a slight keel, 5–7 nerves on each side; sheaths 12–15 mm. l., compressed, keeled. *Peduncles* about 15 mm. l.; basal bracts, lower 2–4 mm. l., upper 5 mm. l.; bracteoles about 3 mm. l. *Flowers* pale yellow. *Sepals* 5-nerved, outside muriculate; lateral oblong, elliptical, tapering to a shortly acuminate apex, 9·2 mm. l., 4 mm. br.; median obovate-oblong, tapering to a shortly acuminate apex, 8–9 mm. l., 3·25 mm. br. *Petals* outside muriculate, oblanceolate-oblong, somewhat falcate, tapering to a shortly acuminate apex, nearly 8 mm. l., 2·5 mm. br. *Lip* 6 mm. l., about 4 mm. br. *Column* barely 2·5 mm. l., short and thick, hooded, stigmatic lip absent. *Capsule* ellipsoidal, somewhat tapering to base, 12–15 mm. l., 6–8 mm. br.

Lindley included two species under his *D. echinocarpa*, viz. *Limodorum pendulum* Aubl. Guian. t. 322, and *Epidendrum echinocarpon* Sw. Prodr. 124; in this he is followed by Cogniaux in Symb. Ant. iv. 182 (1903) and in Fl. Bras. iii. pt. 6, 487, under the name *D. pendula*. Swartz described the lip as entire, and the Jamaican specimens agree with this. Aublet's figure shows a trilobed lip, which accords better with Lindley's description of the lip as sagittate. Aublet's species should therefore bear the name *D. pendula*, while *D. echinocarpa* should be reserved for the Jamaican plant.

2. **D. muricata** *Lindl. Gen. & Sp. Orch.* 209 (1833); leaves persistent, not jointed below, elliptical-oblong, with a rounded apex with a short tip, inclined slightly away from end of shoot; lip blue, 7-nerved, shortly clawed, obovate-rhomboid, the outer angles with a recurved, long, acuminate lobe, apex tapering, long-mucronate; capsule densely covered with long bristles which are only slightly swollen just at the base.—*Griseb. op. cit.* 624; *Cogn. in Symb. Ant. iv.* 183 (in part) & *in Fl. Bras. iii. pt.* 6, 487 (in part). Cymbidium muricatum *Sw. in Nov. Act. Upsal. vi.* 71 (1799) & *Fl. Ind. Occ.* 1454. (Pl. 30, f. 22–25.)

On trees; in fl. Dec.–Jan., in fr. Jan.; Morse's Gap, *Purdie*! *Wilson*! Blue Mts., *Prior*! Bethabara, *Wullschlaegel*, 1212! Cedar Hurst; near Mabess River, 4000 ft.; *Harris*! Fl. Jam. 7649.—Cuba, Dominica, Martinique, St. Vincent, Grenada, Guatemala.

Plant to 1½ ft. l. *Stems* to 5 dm. l., simple, somewhat erect, emitting roots, slender, compressed, 1 mm. br. *Leaves*, limb 10–20 mm. l., 5–7 mm. br., median nerve prominent below or sometimes forming a slight keel, 5–8 nerves on each side; sheaths 7–10 mm. l., compressed, keeled. *Peduncles* about 2 cm. l.; bracts, lower 1·5–3·5 mm. l., upper 4·5 mm. l.; bracteoles 2·5–3 mm. l. *Sepals* incompletely 5-nerved, elliptical, acuminate, 7·5 mm. l., 2·5 mm. br. *Petals* incompletely 5-nerved, lanceolate-elliptical, shortly acuminate, spotted with blue, 7 mm. l., 2·1 mm. br. *Lip* 7-nerved, 6 mm. l., 3·25 mm. br. (just above the lobes). *Column* 2·5–3 mm. l., stigmatic lip depressed; clinandrium hooded. *Capsule* obovate, tapering to base, 14–17 mm l., 5 mm. br.

Cogniaux has included this species in Fl. Bras. as a native of tropical S. America, but as there seems to have been some confusion between the W. Indian and S. American species of *Dichæa*, we have cited only those localities from which we have seen specimens.

3. **D. trichocarpa** *Lindl. Gen. & Sp. Orch.* 209 (1833); leaves persistent, not jointed below, linear, acute, inclined slightly away from end of shoot; lip sessile, broadly oblong, with a triangular lobe on each side above, just above the base cuneate, apex truncate with a blunt mucro projecting at the back, with callus-like thickening at the apex on each side of the central line; capsule densely covered with long bristles, which are swollen nearly throughout their length.—*Griseb. op. cit.* 624. Epidendrum trichocarpon *Sw. Prodr.* 124 (1788). Cymbidium trichocarpon *Sw. in Nov. Act. Upsal. vi.* 71 (1799) & *Fl. Ind. Occ.* 1455. (Pl. 30, f. 17–21.) Type in Herb. Mus. Brit.

On trees; in fl. Aug., in fr. Dec.–April; mountains, *Swartz*! *R. Thomson*! Abbey Green; Mt. Moses; J.P. 2049, *Morris*! *Syme*! Mabess

River, *G. Nichols*! near Woodcutter's Gap, 4000 ft.; Clydesdale, 4500 ft.; near Cinchona, 5000 ft.; *Harris*! Fl. Jam. 7578, 7995, 9019.—Cuba.

Plant to 2 ft. l. *Stems* to 6 dm. l., generally branching, pendulous, emitting roots, slender, terete, a little over 1 mm. br. *Leaves*, limb 9–15 mm. l., 1–2·5 mm. br., slightly fleshy, median nerve of the younger leaves in drying more or less prominent below, two or three nerves on each side sometimes evident; sheaths, 5–7 mm. l., slightly compressed, sometimes slightly keeled. *Peduncles* about 8 mm. l.; bracts, lower 1·5–2·5 mm. l., upper about 5·5 mm. l., attached about 1 mm. above base; bracteoles 2·5–3 mm. l. *Sepals* 3-nerved; lateral oblong-elliptical, tapering to an acute point, nearly 7 mm. l., 2·5 mm. br.; median ovate-elliptical, tapering to a somewhat obtuse point, nearly 6 mm. l., 2·5 mm. br. *Petals* 3-nerved, oblanceolate-oblong, apex somewhat obtuse, slightly apiculate, 5 mm. l., 1·75 mm. br. *Lip* 4·5 mm. l., 4 mm. br. (incl. lobes). *Column* 2·5 mm. l., 2 mm. br., short, thick, slightly hooded above stigma, stigmatic lip obscurely 2-lobulate. *Capsule* narrowly obovate, ·8 cm. l., 4 mm. br.

4. D. **graminoides** *Lindl. Gen. & Sp. Orch.* 209 (1833); leaves deciduous, jointed above the sheath, oblong-linear, with a long tip at apex, spreading; lip 5-nerved above, upper portion deltoid, auriculate, shortly acuminate, auricles rounded, lower portion roundish, with a longitudinal callus along central nerve at the base; capsule smooth.—D. graminea *Griseb. Fl. Br. W. Ind.* 625 (1864). Epidendrum graminoides *Sw. Prodr.* 125 (1788). Cymbidium graminoides *Sw. in Nov. Act. Upsal. vi.* 71 (1799), *Fl. Ind. Occ.* 1459 & *Adnot. Bot. t.* 1, *f.* 1. (Pl. 30, f. 8–12.) Type in Herb. Mus. Brit.

On trees, decaying logs and rocks; in fl. Aug.–Dec.; mountains, *Swartz*! Morse's Gap, *Purdie*! Bethabara, *Wullschlaegel*, 1212! Mt. Moses; Cinchona; J.P. 2070, *Morris*! Moody's Gap, *Syme*. Mabess River, *G. Nichols*! Cinchona, 5000 ft.; below Vinegar Hill, 3500 ft.; *Harris*! Fl. Jam. 7464, 7524, 7766, 9899.

Plants 4–12 in. high. *Stems* 1–3 dm. l., simple or branching a little from base, somewhat erect, emitting roots, somewhat slender, compressed, 2 mm. br. *Leaves*, limb 3–4 cm. l., 3–5 mm. br., six to eight nerves on each side of the median, slightly prominent; sheaths 10–16 mm. l., compressed, slightly keeled. *Peduncles* about 2 cm. l.; bracts, lower 2·5 mm. l., upper about 6 mm. l., attached about 3 mm. above base; bracteoles about 2·5 mm. l. *Flowers* whitish. *Sepals* incompletely 5-nerved, elliptical, tapering to an acute apex, lateral 6 mm. l., 2·3 mm. br.; median slightly shorter. *Petals* incompletely 5-nerved, ovate, apex acute, nearly 5 mm. l., 2·2 mm. br. *Lip* 4·5 mm. l., 3·2 mm. br. (at lobes), 2 mm. br. at lower portion. *Column* about 2 mm. l., 1·2 mm. br., stigmatic lip forming a triangular obtuse lobe; anther broad, emarginate, apiculate, with acutely-pointed sides.

We have not seen any specimens elsewhere than from Jamaica that appear to agree with the description of the species.

5. D. **glauca** *Lindl. Gen. & Sp. Orch.* 209 (1833); leaves deciduous, jointed above the sheath, oblong, with a very short tip, inclined towards the end of the shoot; lip 7-nerved, with a thick short claw, broadly cuneate, with each lateral angle produced into a short ligulate lobe, apex retuse, at base a

roundish callus ; capsule smooth.—*Griseb. op. cit.* 625. Epidendrum glaucum *Sw. Prodr.* 124 (1788). Cymbidium glaucum *Sw. in Nov. Act. Upsal. vi.* 71 (1799) & *Fl. Ind. Occ.* 1457. (Pl. 30, f. 13–16.) Type in Herb. Mus. Brit.

On trees; in fl. July, Aug.; Blue Mts., *Swartz*! *Wiles*! *Macfadyen*! *R. Thomson*! J.P. 31, *Morris*! John Crow Peak, *G. Nichols*! near Portland Gap, *Capper*! near Cinchona, *Miss R. Capper*! Pleasant Hill, *Harris*! Fl. Jam. 7361.—Cuba, Mexico.

Plant to 1½ ft. *Stems* to 4·5 dm., simple, clustered, erect, somewhat stout, compressed, about 3 mm. br. *Leaves* of a glaucous colour, limb 3–5·5 cm. l., 9–13 mm. br., median nerve prominent or slightly keeled below, lateral nerves 13–19 on each side, very thin ; sheaths 1·3–1·8 cm. l., compressed, keeled. *Peduncle* about 2 cm. l.; bracts 5·5–6 mm. l., upper about 4 mm. from base; bracteole about 1·5 mm. l., broadly roundish, apiculate. *Flowers* whitish. *Sepals* 3-nerved; lateral oblong-ovate, tapering to an acuminate apex, about 8 mm. l., 3·5 mm. br.; median oval, tapering to an acuminate apex, 7·5 mm. l., 3·25 mm. br. *Petals* 3-nerved, lanceolate-elliptical, acute, 7 mm. l., 2·75 mm. br. *Lip* nearly 6 mm. l., about 5 mm. br. (exclusive of lobes). *Column* about 2·5 mm. l., 1·5 mm. br., short, thick, slightly hooded above stigma, apiculate, stigmatic lip triangular with a ligulate lobule at apex. *Capsule* obovate-ellipsoidal, tapering to base, 2 cm. l., 6 mm. br.

6. **D. Morrisii** *Fawc. & Rendle in Journ. Bot. xlviii.* 107 (1910); leaves jointed above the sheath, deciduous, oblong or oblong-ligulate, at apex rounded with a short or long tip, margin ciliate above, spreading; lip below linear-oblong, above dilated, anchor-shaped, margin minutely ciliate, apex apiculate, lateral lobes long, curved towards the base, subulate; capsule densely covered with long bristles, which are swollen at the base. (Pl. 30, f. 31–36.) Types in Herb. Mus. Brit., Bot. Gard. N. York & Yale Univ.

On trees; in fl. July; Mt. Moses, 3500 ft., J.P. 2269, *Morris*! Hardware Gap, *G. Nichols*!

Plant to 1½ ft. l. *Stems* to 4·5 dm. l., few, simple, stout, compressed, 4 mm. br. *Leaves*, limb 3·5–5 cm. l., 10–13 mm. br., median nerve prominent or slightly keeled below, lateral nerves 20–30 on each side, very thin; sheaths about 25 mm. l., compressed, keeled. *Peduncle* about 8 mm. l., bracts at the base about 1 cm. l. *Sepals*, margin ciliate; lateral 7-nerved, ovate, shortly acuminate, 11–13 mm. l., 5·5–7 mm. br.; median 9-nerved, oblong, tapering to an acute apex, as long as but narrower than lateral. *Petals* 7-nerved, like the median sepal, 10 mm. l., 3·5 mm. br. *Lip* white barred with brownish-red colour below, fleshy, about 9 mm. l. *Column* very short, thick, winged, 4 mm. l. *Capsule* about 15 mm. l., 10 mm. br.

This species is named after Sir D. Morris, who discovered it. Since printing our description in the Journal of Botany we have seen better material kindly lent by Mr. G. E. Nichols, and have made a few emendations.

60. CAMPYLOCENTRUM Benth.

Epiphytic herbs; with leafy stems, or stemless and leafless; with clustered roots. Leaves oblong, in two ranks. Flowers minute, in two ranks, crowded, in spikes, radical or lateral. Bracts small, persistent. Sepals free, subequal, erect, connivent. Petals as long as the sepals, but narrower. Lip as long as the sepals, sessile at base of column, produced at the base into a more or less clavate spur, often recurved; the blade somewhat quadrate below, with convolute margins, produced upwards into a triangular lobe. Column very short, without a foot; anther terminal, opercular, incumbent, convex, 2-celled; pollinia 2, globose, waxy, with filiform stalks. Capsule small, narrowly oblong, dehiscing in the middle by three valves.

Species 30 in number, natives of tropical America from Brazil to the West Indies and Mexico.

Leafy plants.
  Leaves deeply 2-lobed at apex. Capsule smooth.
    Lip 4·5 mm. l., terminal lobe about half length of lip........................ 1. *C. Barrettiæ.*
    Lip about 3 mm. l., terminal lobe less than half length of lip .... 2. *C. jamaicense.*
  Leaves entire or only slightly lobed at apex. Capsule ribbed .... 3. *C. minus.*
Leafless and stemless plants.
    Bracts large, hooded, serrate .... 4. *C. pachyrrhizum.*
    Bracts small, entire .... 5. *C. Sullivanii.*

1. **C. Barrettiæ** *Fawc. & Rendle in Journ. Bot. xlvii.* 127 (1909); leaves deeply 2-lobed at apex; spikes opposite to, and much shorter than, the leaves, solitary or sometimes geminate, densely flowered, 1–2·7 cm. l.; lip 4·5 mm. l., 5-nerved above, 7-nerved at base, obsoletely 3-lobed, as long as the sepals, 2 mm. br., middle lobe linear-lanceolate, acute; lateral lobes bluntly rounded, not developed anteriorly, middle lobe about half length of the lip, 2 mm. l., nearly 1 mm. br. at base; spur linear, narrowly clavate, slightly constricted at the origin, 2 mm. l., nearly 1 mm. br., much shorter than the sepals, straight or slightly curved in flower, spreading in fruit; capsule smooth. (Pl. 31, f. 9–15.)

On trees; in fl. Oct.–Jan.; 27, *Purdie*! New Part, St. Mary, *Wilson*! Iron River, Water Valley, St. Mary, J.P. 2325, *Syme*! near Browns Town, *Miss T. M. Barrett*! August Town, *Moore*! Charlemont, near Ewarton, 1000 ft.; near Gordon Town; Green Vale, 2000 ft.; near Browns Town, 1200 ft.; *Harris*! Albany, St. George, *Watt*! Fl. Jam. 6580, 7172, 7801, 10,405, 10,406.

Plant 2–9 in. high. *Roots* somewhat slender, long, flexuose, canescent, simple, 1–3 dm. l., 1–2 mm. br. *Stem* somewhat stout, slightly compressed, emitting roots, simple, covered by the leaf-sheaths, becoming bare below, 5–26 cm. l., 2–3 mm. br. *Leaves*, blade rather large, somewhat leathery, oblong to ligulate, tapering to the base, at the apex very obliquely 2-lobed, lobes rounded at the apex, in two ranks, numerous, caducous at the base of the stem, 3–7 cm. l., 8–15 mm. br., middle nerve prominent below, canaliculate above, lateral nerves numerous; sheaths closely adpressed, with numerous ribs, 6–15 mm. l. *Bracts* triangular, acute, 1 mm. l. *Flowers* sessile. *Sepals* ligulate, apex subacute, slightly apiculate, concave, 3-nerved, 4–5 mm. l., a little over 1 mm. br.; lateral, at the base slightly oblique and not constricted. *Petals* linear-lanceolate, acute, 3-nerved, with involute margins, 4 mm. l., 1 mm. br. *Capsule* (not ripe) elongated, narrowly oblong, straight, 1 cm. l.

2. **C. jamaicense** *Fawc. Fl. Pl. Jam.* 40 (1893); leaves deeply 2-lobed; spikes opposite to, and very much shorter than, the leaves, solitary, about 1 cm. l.; bracts triangular, acute, barely 1 mm. l.; lip 5-nerved above, 7-nerved at base, subquadrate with an oblong shortly apiculate lobe, about 3 mm. l., 1·75 mm. br.; apical portion less than half the length of the lip, a little over 1 mm. l., ·6 mm. br.; spur cylindrical, less than 2 mm. l., shorter than the sepals; capsule smooth.—Aeranthus jamaicensis *Reichb. f. & Wullschl. in Walp. Ann. vi.* 901 (1861). Aeranthes micrantha *Griseb. Fl. Br. W. Ind.* 625 (1864) (in part), non *Reichb. f.*

Fairfield, 1053, *Wullschlaegel!*

Plant 4 in. high. *Leaves* elliptical, at apex very obliquely 2-lobed, lobes rounded at the apex, in two ranks, numerous, caducous at the base of the stem, 2·5–3 cm. l., 1–1·3 cm. br.; middle nerve prominent below, canaliculate above, lateral nerves numerous; sheaths tubular, closely adpressed, with numerous ribs, 5–10 mm. l. *Sepals* ligulate, apex obtuse, shortly apiculate, concave, 3-nerved, about 3 mm. l., less than 1 mm. br. *Petals* similar to sepals. *Capsule* (not ripe) spindle-shaped, 8 mm. l.

3. **C. minus** *Fawc. & Rendle loc. cit.*; leaves, apex obtuse, occasionally slightly 2-lobed, obliquely or not; spikes opposite to the leaves and about as long or longer, 2–4-clustered, densely flowered, to 4 cm. l.; lip without nerves, obsoletely 3-lobed, middle lobe triangular, acute, with involute margins; lateral lobes rounded, projecting very slightly anteriorly; 2·5 mm. l., at lobes barely 2 mm. br.; middle lobe about 1 mm. l., less than half the length of the lip; spur ellipsoidal, much constricted at origin, 1·5 mm. l., ·7 mm. br., much shorter than the sepals, spreading; capsule 6-ribbed. (Pl. 31, f. 5–8.)

On trees; in fl. Dec., in fr. Jan.; Cedar Hurst, *Harris!*

Plant to 5 or 6 in. high. *Roots* slender, long, flexuose, canescent, simple, to 2 dm. l., 1–1·5 mm. br. *Stem* slightly compressed, emitting roots, simple, covered by the leaf-sheaths, to 16 cm. l., about 2 mm. br. *Leaves*, blade somewhat membranous, elliptical, in two ranks, numerous, caducous near the base of the stem, 3–4 cm. l., 7–12 mm. br., middle nerve prominent below, canaliculate above, lateral nerves numerous; sheaths closely adpressed, with numerous ribs, 6–8 mm. l. *Bracts* triangular, acute, apiculate, keeled, 2 mm. l., more than 2 mm. br *Flowers* sessile. *Sepals* without nerves, median linear-lanceolate, acute, shorter than the lateral, 2·8 mm. l., barely 1 mm. br. at base, ·7 mm. higher up; lateral linear-lanceolate, acute, 2·6 mm. l., ·7 mm. br. *Petals* linear-lanceolate, acute, without nerves, 2 mm. l., ·6 mm. br. *Capsule* (not ripe) oblanceolate, 6 mm. l.

4. **C. pachyrrhizum** *Rolfe in Orch. Rev. xi.* 246 (1903); leafless; bracts spathaceous, roundish, acuminate, cordate, margin minutely eroded, reddish-brown, 3–4 mm. l.—Aeranthus pachyr-

rhizus *Reichb. f. in Flora xlviii.* 279 (1865); A. spathaceus *Griseb. Cat. Pl. Cub.* 264 (1866).

On trees; Castleton; Cedar Hurst; J.P. 2326, *Syme*! Belvedere, Hanover, 500 ft., *Harris*! Fl. Jam. 7570.—Cuba, Trinidad, Guiana.

Plant less than 1 in. to 1½ in. high. *Aërial roots* hanging free or sometimes attaching themselves to the bark of trees, broad and flat, flexuose, simple, to 3 dm. l., to 4 mm. br. *Stem* reduced to a knob from which the aërial roots and the racemes spring. *Leaves* wanting. *Spikes* radical, several, 3-many-clustered, with numerous flowers; peduncle puberulous. *Flowers* sessile. *Sepals*, median narrowly triangular, 3-nerved, abruptly acute, 4·6 mm. l., about 1 mm. br.; lateral lanceolate, slightly oblique, 1-nerved, 5·2 mm. l., 1·2 mm. br. *Petals* subulate, 3-nerved, 4·5 mm. l., 1 mm. br. *Lip*, lower half suborbicular, upper triangular, acuminate; 4 mm. l., lower part about 2 mm. br., upper 1 mm. br.; 9-nerved below, 5-nerved above. *Spur* narrowly ellipsoidal, slightly constricted at the origin, about 2·5 mm. l., ·7 mm. br., spreading. *Capsule* ellipsoidal, 6-ribbed, about 8 mm. l.

5. **C. Sullivanii** *Fawc. & Rendle op. cit.* 128; leafless; bracts ovate, acute, clasping, 1·7 mm. l., 1·3 mm. br. Aeranthes filiformis *Griseb. Fl. Br. W. Ind.* 625 (1864) (in part, excl. syn.). (Pl. 31, f. 1–14.)

On trees; in fl. Jan.–March; Negril; Woodside, St. Mary; *Purdie*! Manchester Mts., *Wilson*! J.P. 34, *Morris*! Spaldings, *Sullivan*! Monklands, *Roberts*! Alexandria, Ramble, *Mrs. Rankine*! Belvedere, Hanover, 500 ft., *Harris*! near Hope Bay, *Moore*! Fl. Jam. 7523, 10,464.

Plant, raceme 1–2 in. high. *Aërial roots* hanging free or sometimes attaching themselves to the bark of trees, somewhat slender, long, flexuose, canescent, simple, to 3 dm. l., 1·5–2 mm. br. *Stem* reduced to a knob, from which the aërial roots and the racemes spring. *Leaves* wanting. *Spikes* radical, several, 4–7-clustered, with numerous flowers in two ranks, 3–6 cm. l., rachis minutely puberulous. *Pedicels* short, about ·7 mm. l. *Ovary* obovoid, 1·5–2 mm. l., nearly 1 mm. br. *Flowers* brownish-white. *Sepals* 1-nerved; median ovate, acute, apiculate, 1·4–1·7 mm. l., ·8 mm. br.; lateral ovate-lanceolate, apiculate, 1·7–1·8 mm. l., ·7–·8 mm. br. *Petals* 1-nerved, oval or ovate-elliptical, subacute, 1·25 mm. l., ·6–·7 mm. br. *Lip* 3-lobed, 7-nerved, 1·3 mm. l., 1·3 mm. br., middle lobe triangular, acute, apiculate, ·25 mm. l.; lateral rounded; spur linear, slightly clavate, 1·1–1·2 mm. l. *Column* green. *Capsule* broadly ellipsoidal, 9-ribbed, 4 mm. l., 2–3 mm. br.

C. filiforme *Rolfe* (*Epidendrum filiforme* Sw.; type in Herb. Mus. Brit. from S. Domingo) differs from this species in the lip, the capitate spur, and the obtuse sepals and petals.

61. HARRISELLA Fawc. & Rendle.

Epiphytic herbs, stemless and leafless; roots clustered. Flowers minute, in two ranks, in lax few-flowered radical racemes, jointed on a short pedicel which is swollen above. Bracts small, persistent. Sepals free, subequal, erect, slightly spreading above. Petals similar to, and as long as the sepals, but narrower. Lip slightly longer than the sepals, sessile at base of column, produced at the base into a spur, globose from a very constricted base; the blade somewhat globose, embracing the column, and hooded over it. Column very short, without a foot; anther terminal, opercular, incumbent, rather flat, indistinctly 2-celled; pollinia 2, globose, waxy, with short filiform stalks. Capsule small, shortly ellipsoidal, dehiscing from the apex by six valves which are alternately broad and narrow.

Species, only one known, native of Jamaica and Cuba.

The genus is very near to *Campylocentrum*, in which it has been included, but differs in the inflorescence, the lip, the form of the anther, and the form and dehiscence of the capsule.

It is dedicated to Mr. William Harris, F.L.S., Superintendent of Public Gardens, Jamaica, and an indefatigable collector, who has added much to our knowledge of the Jamaican Flora.

H. porrecta *Fawc. & Rendle in Journ. Bot. xlvii.* 266 (1909).—Aeranthus porrectus *Reichb. f. in Flora xlviii.* 279 (1865); *Griseb. Cat. Pl. Cub.* 265. Campylocentrum porrectum *Rolfe in Orch. Rev. xi.* 247 (1903); *Ames Orch. i.* 15, *t.* 4. (Pl. 31, f. 16–21.)

On trees; in fl. and fr. in Nov.; Manchester, *Purdie*! above Hope Mines, *Harris*! Fl. Jam. 7762.—Cuba, Porto Rico, Florida, Yucatan.

Plant 1½–2 in. high. *Aërial roots* hanging free or sometimes attaching themselves to the bark of trees, very slender, flexuose, white, canescent, simple, to 1 dm. l., ·5 mm. br. *Racemes* several, clustered, zig-zag at each node, 4–5 cm. l. *Bracts* tubular below, obtuse, 1·3 mm. l. *Pedicels* slightly exceeding the bracts, thicker upwards than the base of ovary. *Flowers* about 1/12 in. long, yellow-green. *Ovary* attenuated below, glandular along six vertical lines, 2·7 mm. l. *Sepals* reticulate-nervulose, slightly keeled, a little over 2 mm. l., median ·8 mm. br.; lateral 1 mm. br.; dorsal oval-elliptical, obtuse; lateral oval, oblique, acute. *Petals* a little over 2 mm. l., ·7 mm. br., linear, obtuse, apiculate, 1-nerved. *Lip* 2·5 mm. l., obtusely apiculate. *Spur* ·7 mm. *Column* terete; clinandrium 2-lobed anteriorly; anther-case, somewhat deltoid. *Capsule* 6 mm. l., 5 mm. br.

62. DENDROPHYLAX Reichb. f.

Epiphytic leafless herbs, with the roots densely clustered, a very short stem and no pseudobulb. Scapes thin, simple, with small distant bracts. Flowers solitary on the scape or a few in a raceme, very large, medium-sized or small. Sepals about equal, free, spreading. Petals somewhat similar to the sepals. Lip sessile at the base of the column, produced at the base into a long spur much dilated towards the mouth, continuous with the column, erect; limb entire or 2-lobed. Column very short, broad, without a foot. Anther terminal, opercular, incumbent, indistinctly 2-celled; pollinia 2, somewhat globose, waxy, each with a stalk, which is simple or flat and 2-partite. Capsule oblong or elongated.

Species about 5, natives of West Indies.

Sepals about ¾ inch l. .... 1. *D. funalis*.
Sepals about ¼ inch l. .... 2. *D. Barrettiæ*.

1. **D. funalis** *Fawc. Fl. Pl. Jam.* 40 (1893); lip very large, broad, with two rounded lobes, united by a short claw to the wide mouth of the spur, reticulate-9-veined, to apex of lateral lobe 1·7 cm. l., 2·2 cm. br.; spur filiform. tapering towards apex, nearly three times as long as the sepals, about 5·5 cm. l. —Epidendrum funale *Sw. Prodr.* 126 (1788). Limodorum funale *Sw. in Nov. Act. Upsal. vi.* 79 (1799) & *Fl. Occ. Ind. iii.* 1521. Œceoclades funalis *Lindl. Gen. & Sp. Orch.* 237 (1833). Angræcum funale *Lindl. in Bot. Reg. xxvi. under t.* 68 (1840); *Hook. in Bot. Mag. t.* 4295. Aeranthus funalis *Reichb. f. in Walp. Ann. vi.* 902 (1861). Aeranthes funalis *Griseb. Fl. Br. W. Ind.* 625 (1864). Polyrrhiza funalis *Pfitz. in Engl. & Prantl Nat. Pflanzenf ii. pt.* 6, 209 & 215 (1889). (Pl. 32.) Type in Herb. Mus. Brit.

On trees; in fl. Nov.-March; *Swartz*! Wag Water River, J.P. 2330, *Syme*! near Browns Town, *Miss T. M. Barrett*! *H. Levy*! near Christiana, *Mrs. Sewell*! Blue Mts., *Moore*! near Gordon Town; Hope Mines; *Harris*! Fl. Jam. 5655, 7763, 9005.

Plant 5 or 6 in. high. *Roots* to 3 dm. or more l., slender, compressed, flexuose, canescent, simple, occasionally branched, repeatedly forming new growths from the tips of the old roots. *Stem* very much reduced, extending occasionally to scarcely 2 cm., covered with brown sheathing scales, leafless, bearing long slender woody stolons. *Scape* simple, few-flowered, 5–10 cm. l. *Bracts* sheathing, wide at mouth, acute or acuminate, brown, 4–5 mm. l. *Pedicel* (of fruit) 1 to nearly 2 cm. l. *Sepals* elliptical, obtuse, reticulate-7-veined; median spreading, about 2 cm. l., 7 mm. br.; lateral deflexed, about 2·2 cm. l., 6 mm. br. *Petals* oblong-elliptical, subacute, reticulate-7-veined, spreading, about 2 cm. l., 5–6 mm. br. *Column* short, with two broad wings produced anteriorly. *Anther* subhemispherical, cells not divided, but indicated by an almost obsolete wall and by a longitudinal depression outside; pollinia on a ribbon-like process of the flat stalk. *Capsule* elongated, cylindrical, 6-angled, 7 cm. l., about ·5 cm. br.

2. **D. Barrettiæ** *Fawc. & Rendle in Journ. Bot. xlvii.* 266 (1909); lip undivided, saccate, above broadly ovate, shortly acuminate, extending about as far as lateral sepals; spur spreading, 2·3 cm. l., increasing in thickness from the middle towards the apex. Types in Herb. Kew. and in Herb. Mus. Brit.

On trees; in fl. in Aug.; near Browns Town, *Miss T. M. Barrett*! Holly Mount, Mt. Diabolo, 2600 ft., *Harris*!

A small stemless, leafless plant, about an inch high. *Roots* very slender, flexuose, simple, to 12 cm. l. *Scapes* several, clustered, simple, to 2·5 cm. l. *Bracts* few (three), tubular below, free part ovate, apiculate, 1·5 mm. l. *Flowers* small, one flower appearing at a time, greenish-white, the spur rather lighter in colour. *Sepals* spreading, 3-nerved, linear-lanceolate, subacute to acute, 5–6 mm. l., 1·3–1·5 mm. br. *Petals* erect, 3-nerved, similar to sepals, linear-ligulate, acute, nearly 5 mm. l., 1 mm. br. *Column* very short, auriculate at apex as in *D. funalis*, auricles toothed on upper margin; anther-cells somewhat ovate-rectangular in outline; pollinia with a simple stalk.

INDEX OF GENERA AND SPECIES

(*Synonyms in italics*)

L

PLATES

EXPLANATION OF PLATE 1

Habenaria macroceratitis

1. Flower, natural size.

H. distans var. *jamaicensis* (*H. jamaicensis*)

2. Flower × about 2.
3. Petal × 5.

Habenaria alata

4. Flower × nearly 2.
5. Column and lip with part of petals × 4.
6. Petal × 5.
7. Lip, under side × 7.

Habenaria monorrhiza

8. Flower × between 2 and 3.
9. Smaller flower do.
10. Petal × 5.

Pogonia macrophylla

11. Upper part of stem × $\frac{2}{3}$.

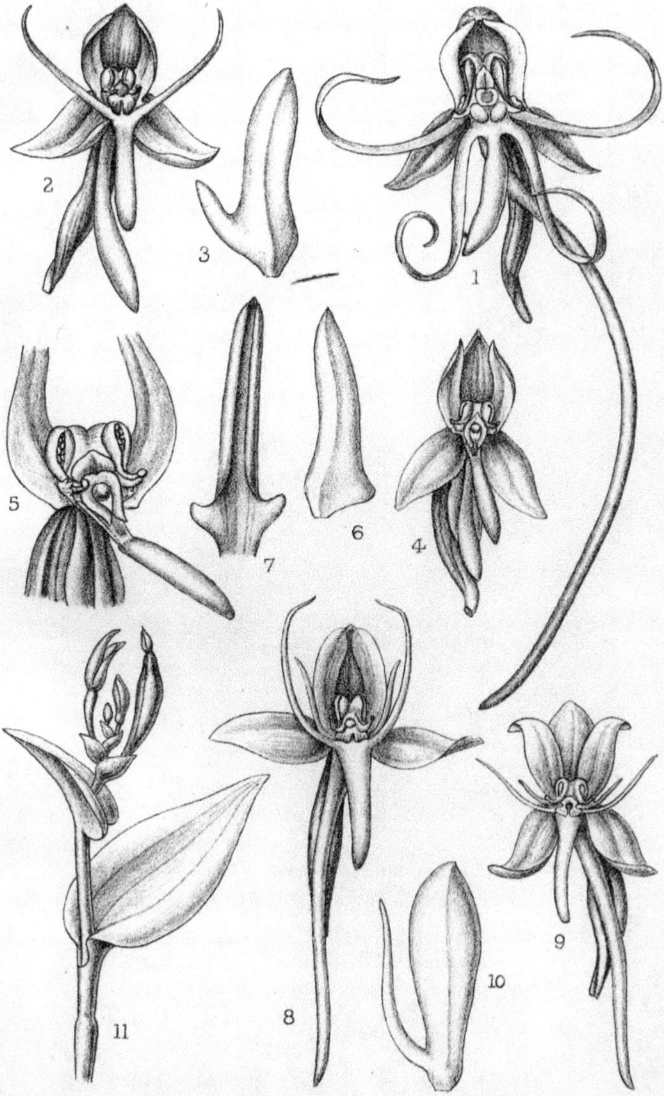

H.A. WOOD DEL. HIGHLEY LITH.

1 Habenaria macroceratitis; 2, 3 H. distans var.; 4-7 H. alata; 8-10 H. monorrhiza; 11 Pogonia macrophylla.

EXPLANATION OF PLATE 2

Habenaria Purdiei

1. Flower × 2.
2. Lateral sepal × 3.
3. Petal × 3.

Habenaria socialis

4. Flower × 2.
5. Lateral sepal × 4.
6. Petals × 4.
7. Lip × 4.

Habenaria troyana

8. Flower × 2.
9. Lateral sepal × about 5.
10. Petal × 4.
11. Another form of petal × 4.
12. Lip × 4.

Pogonia gentianoides

13. Plant, natural size (from Swartz's type).
14. Flower × 2.
15. Lip × 4.

Vanilla inodora

16. Lip and column × ½.

Vanilla Wrightii

17. Flower, with one sepal and one petal removed, × ½.

Vanilla claviculata (by error *parasitica*)

18. Lip and column × ½.
19. Column showing attachment of portion of lip × ½

Vanilla phæantha

20. Lip (cut open) and column × ½.

Pelexia setacea

21. Flower, nat. size.
22. Flower, with two sepals and one petal removed, nat. size.

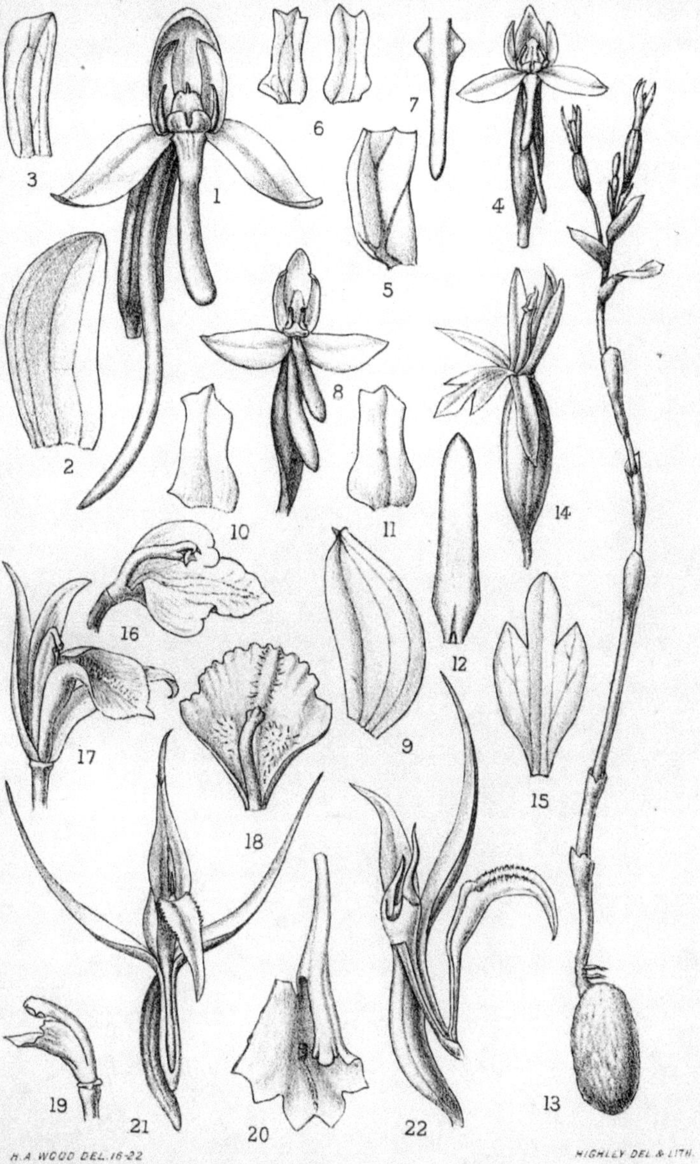

H. A. WOOD DEL. 16-22 HIGHLEY DEL & LITH.

1-3 Habenaria Purdiei; 4-7 H. socialis; 8-12 H. troyana; 13-15 Pogonia gentianoides; 16 Vanilla inodora; 17 V. Wrightii; 18-19 V. parasitica; 20 V. phæantha; 21, 22 Pelexia setacea.

EXPLANATION OF PLATE 3

Spiranthes tortilis

1. Flower × 4.
2. Lip × 4.
3. Column × about 5, from side.
4. Do. from back.
5. Do. from front.

Spiranthes Fawcettii

6. Flower × about 2½.
7. Lip × 3.
8. Column × 3, from front.
9. Do. from back.

Spiranthes elata

10. Lateral sepal × 4.
11. Median sepal × 4.
12. Petal × 4.
13. Another form of do.
14. Lip × 4.

Erythrodes plantaginea

15. Flower, with part of the sepals and one petal removed, × 3.

Erythrodes hirtella

16. Flower × 3.
17. Flower, with lateral sepals and one petal removed, × 3.
18. Column and spur × about 4.
19. Do. from side.

Erythrodes jamaicensis

20. Lip and spur × 3.

Stenorrhyncos squamulosus

21. Flower × 1½.

Stenorrhyncos lanceolatus

22. Flower × 1½.

Wullschlægelia aphylla

23. Plant, nat. size (from Swartz's type).

24. Flower, with lateral sepal pulled down and a petal removed, × 6.

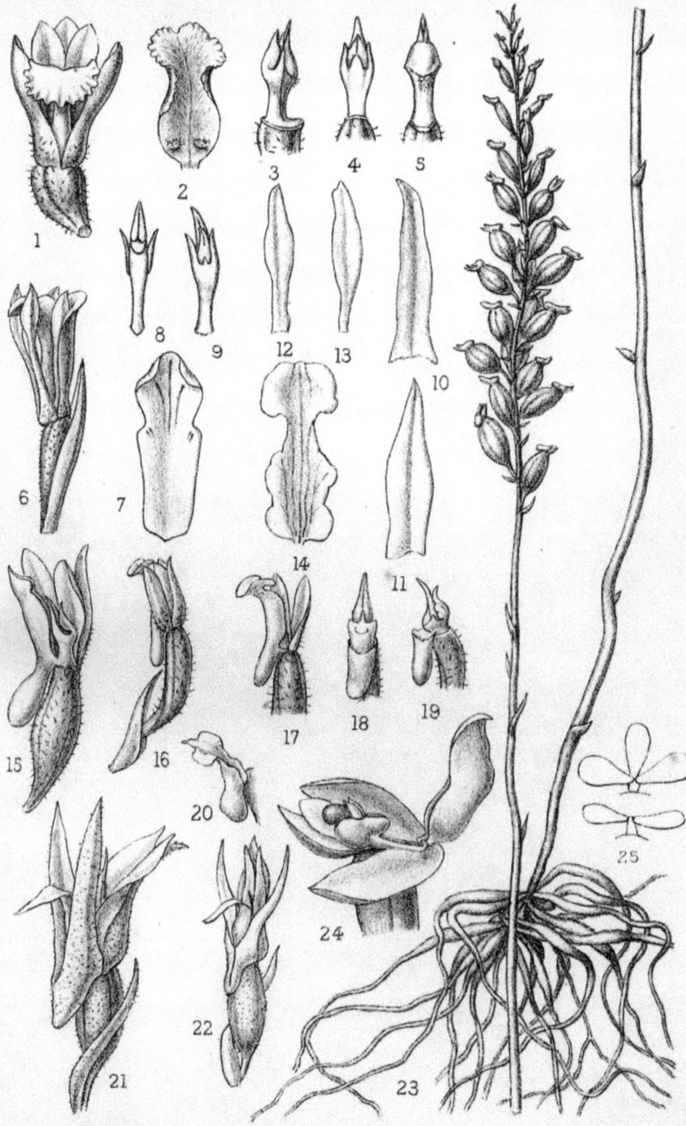

H. A. WOOD DEL. 1-9, 15-19, 21, 22 HIGHLEY DEL. & LITH.

1-5 Spiranthes tortilis; 6-9 S. Fawcettii; 10-14 S. elata; 15 Erythrodes plantaginea; 16-19 E. hirtella; 20 E. jamaicensis; 21 Stenorrhyncos squamulosus; 22 S. lanceolatus; 23-25 Wullschlægelia aphylla.

EXPLANATION OF PLATE 4

Pseudocentrum minus

1. Plant, half natural size.
2. Flower, about nat. size, from side
3. Do. from back.
4. Do. from front.
5. Flower × about 3.
6. Section of flower × between 2 and 3.
7. Column enlarged.
8. Column from side enlarged.
9. Anther enlarged.
10. Anther, with pollinia, enlarged.
11. Pollinia enlarged.

H. A. WOOD DEL. HIGHLEY LITH.

Pseudocentrum minus.

EXPLANATION OF PLATE 5

Cranichis pilosa

1. Raceme, natural size.
2. Flower × about 4½.
3. Column, with lip thrown back, × about 10.

Prescottia oligantha

4. Flower, with one sepal and one half and one petal removed, × 16.

Stenoptera ananassocomos

5. Plant, nat. size.
6. Flower × 8.

Ponthieva Harrisii

7. Raceme, nat. size.
8. Flower, with sepal thrown back, × about 3.
9. Flower, with one sepal and one half and one petal removed, × about 6.
10. Anther enlarged, from different points of view.
11. Pollinia enlarged.

Ponthieva pauciflora

12. Flower, with one sepal and one petal removed, × about 6.
13. Lip × about 6.

Corymborchis flava

14. Flower, with lip and lateral sepals spread out, nat. size.

Tropidia polystachya

15. Flower × 2.

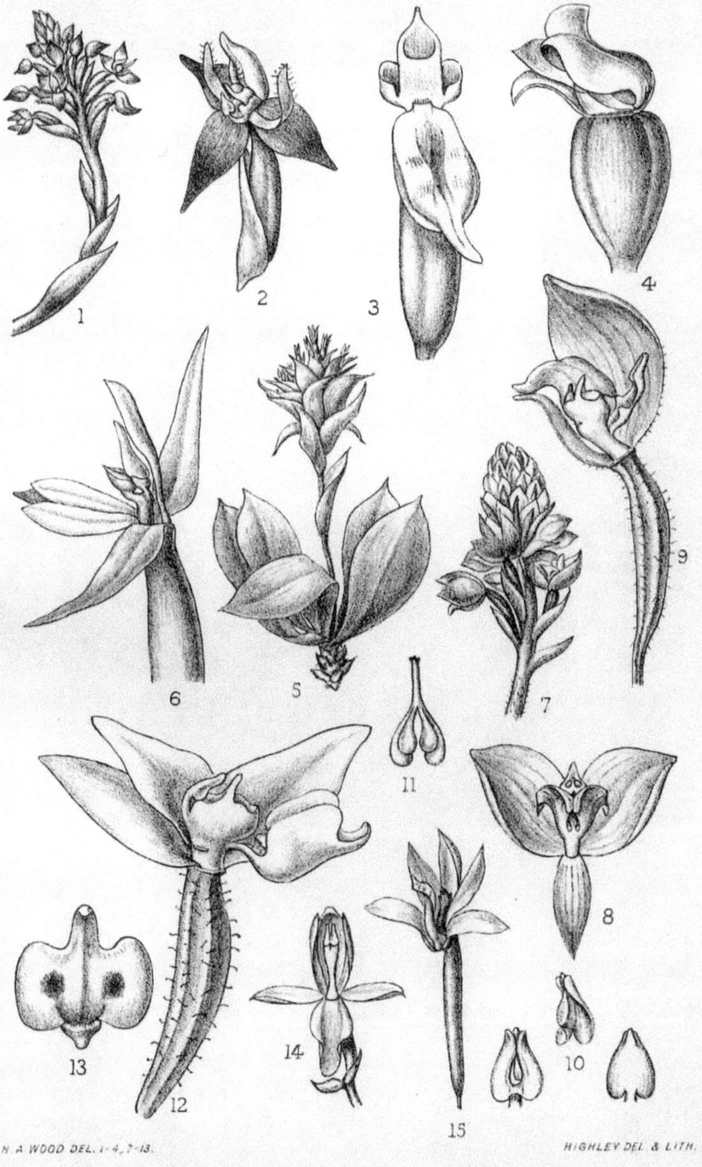

1-3 Cranichis pilosa; 4 Prescottia oligantha; 5, 6 Stenoptera ananassocomos; 7-10 Ponthieva Harrisii; 12, 13 P. pauciflora; 14 Corymborchis flava; 15 Tropidia polystachya.

EXPLANATION OF PLATE 6

Liparis Harrisii

1. Top of flower-spike, natural size.
2. Sepal pressed flat × 2½.
3. Lip pressed flat × 3.
4. Column × 5, from the front.
5. Do. from the side.

Liparis Saundersiana

6. Top of flower-spike, nat. size.
7. Column, from the front, enlarged.
8. Column, from the side, enlarged.
9. Anther case enlarged.
10. Pollinia enlarged.

Malaxis integra

11. Plant, lower portion, nat. size.
12. Top of flower-spike, nat. size.
13. Flower × about 6, from front.
14. Do. from back.
15. Lip × about 8.
16. Column enlarged.
17. Pollinia enlarged.

Malaxis Grisebachiana

18. Plant, lower portion, nat. size.
19. Top of flower-spike, nat size.
20. Flower × about 6, from front.
21. Do. from back.
22. Column enlarged.
23. Pollinia enlarged.

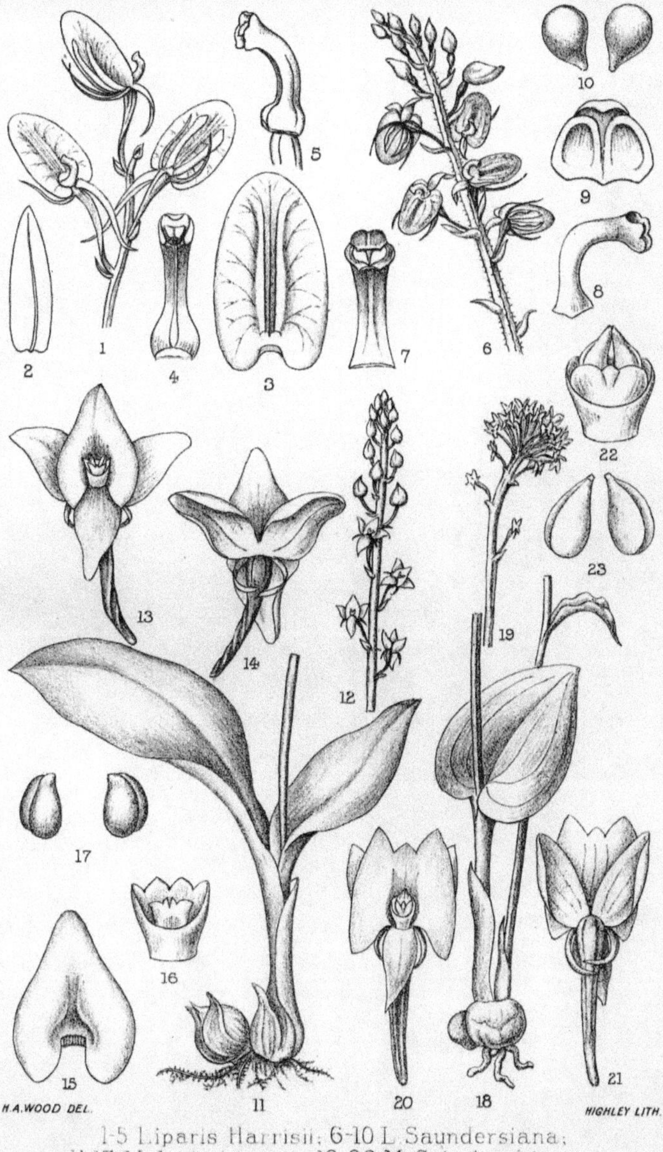

H.A. WOOD DEL. HIGHLEY LITH.

1-5 Liparis Harrisii; 6-10 L. Saundersiana; 11-17 Malaxis integra; 18-23 M. Grisebachiana.

EXPLANATION OF PLATE 7

Galeandra Beyrichii

1. Part of peduncle with one flower, about natural size.
2. Lip torn away at base and flattened, with column, about nat. size.
3. Anther enlarged.

Polystachya minuta

4. Flower, with one sepal and one half removed, and lip turned back, × about 4.
5. Lip spread out × about 5.

Polystachya minor

6. Flower, with one sepal and one half and one petal removed, and lip turned back, × about 4.
7. Lip spread out × about 5.

Cryptophoranthus atropurpureus

8. Flower, nat. size.
9. Flower, with sepals removed and one petal turned back, × 3.
10. Lip × 3.

Stelis ophioglossoides

11. Raceme and part of leaf, nat. size.
12. Flower × 5.
13. Petals, lip and column × 16.
14. Column viewed more from the anterior × 16.

Stelis micrantha

15. Raceme and part of leaf, nat. size.
16. Flower (Fl. Jam. 10,087) × 5.
17. Petals, lip and column of do. × 16.
18. Column of do., viewed more from behind, × 16.
19. Flower (Fl. Jam. 7835) × 5.
20. Petals, lip and column of do. × 16.
21. Petal of do. lying flatter × 16.
22. Lip of do. lying flatter × 16.
23. Bud of do. × 16.
24. Column of do. × 16.
25. Pollinia × 16.
26. Flower of Sowerby's specimen × 5.
27. Flower of Swartz's specimen × 5.
28. Flower (J.P. 23) × 5.

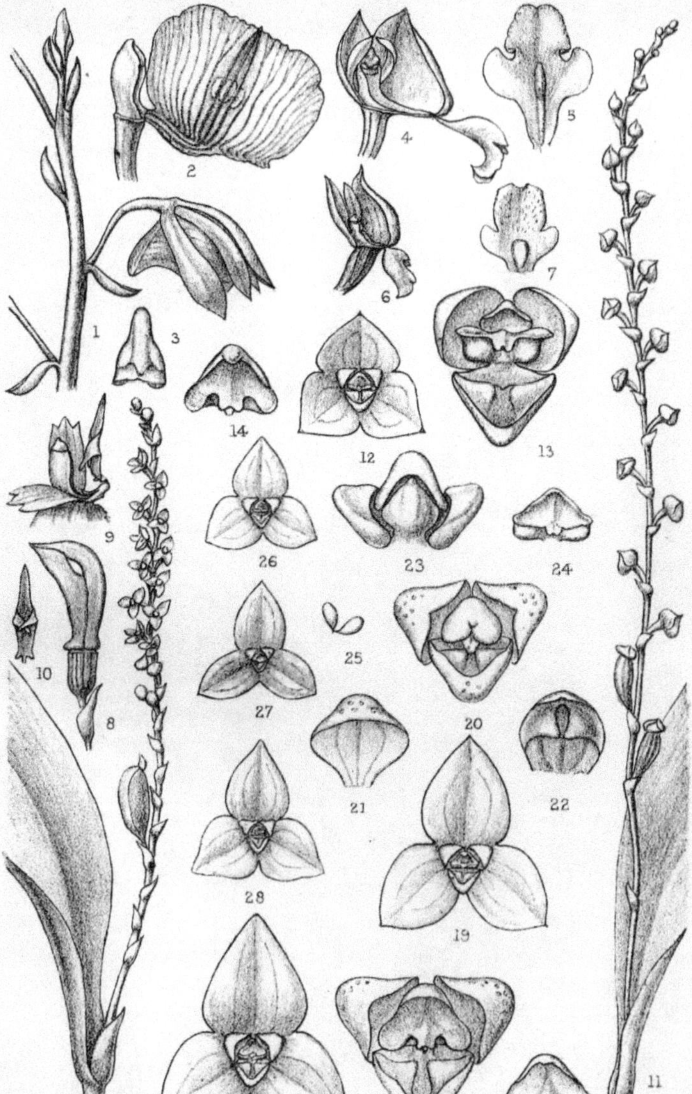

H. A. WOOD DEL. 11. 15. HIGHLEY DEL. & LITH.

1-3 Galeandra Beyrichii; 4,5 Polystachya minuta; 6,7 P. minor; 8-10 Cryptophoranthus atropurpureus; 11-14 Stelis ophioglossoides; 15-28 S. micrantha.

EXPLANATION OF PLATE 8

Pleurothallis trilobata

1. Plant, natural size.
2. Flower × 5.
3. Petal × 9.
4. Lip × 10.
5. Column × 10.
6. Pollinia enlarged.

Pleurothallis brachypetala

7. Plant, nat. size.
8. Flower with part of the sepals removed × 7.
9. Lip × 10.
10. Lip spread out × 10.

Pleurothallis rotundifolia

11. Plant, nat. size.
12. Flower × 4.
13. Petal × 7.
14. Lip × 6.
15. Lip spread out × 6.
16. Column × 7.

Pleurothallis velaticaulis

17. Portion of stem with leaf and raceme, nat. size.
18. Flower × 5.
19. Dorsal sepal × 5.
20. Lateral sepal × 5.
21. Petal × 8.
22. Lip × 7.

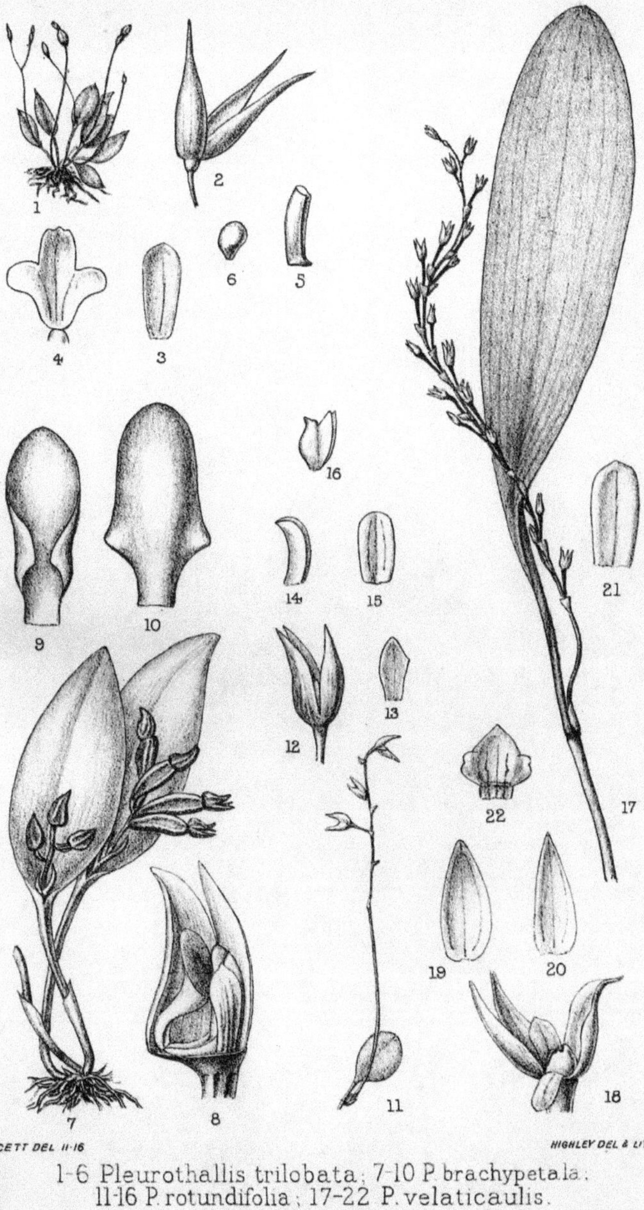

FAWCETT DEL 11·16 HIGHLEY DEL & LITH.

1-6 Pleurothallis trilobata; 7-10 P. brachypetala; 11-16 P. rotundifolia; 17-22 P. velaticaulis.

EXPLANATION OF PLATE 9

Pleurothallis Morrisii

1. Plant, natural size.
2. Petal × 15.
3. Lip × 15.
4. Column × 15.

Pleurothallis confusa

5. Portion of plant, nat. size.
6. Petal × 15.
7. Lip × 9.

Pleurothallis Wilsonii

8. Lip × 9.

Pleurothallis Helenæ

9. Plant, nat. size.
10. Flower spread out × 6.
11. Lip × 19.
12. Ovary and column × 15.
13. Anther case enlarged.
14. Pollinia enlarged.

Pleurothallis hirsutula

15. Portion of plant, nat. size.
16. Flower × 4.
17. Petal × 5.
18. Lip × 6.
19. Column × 7.

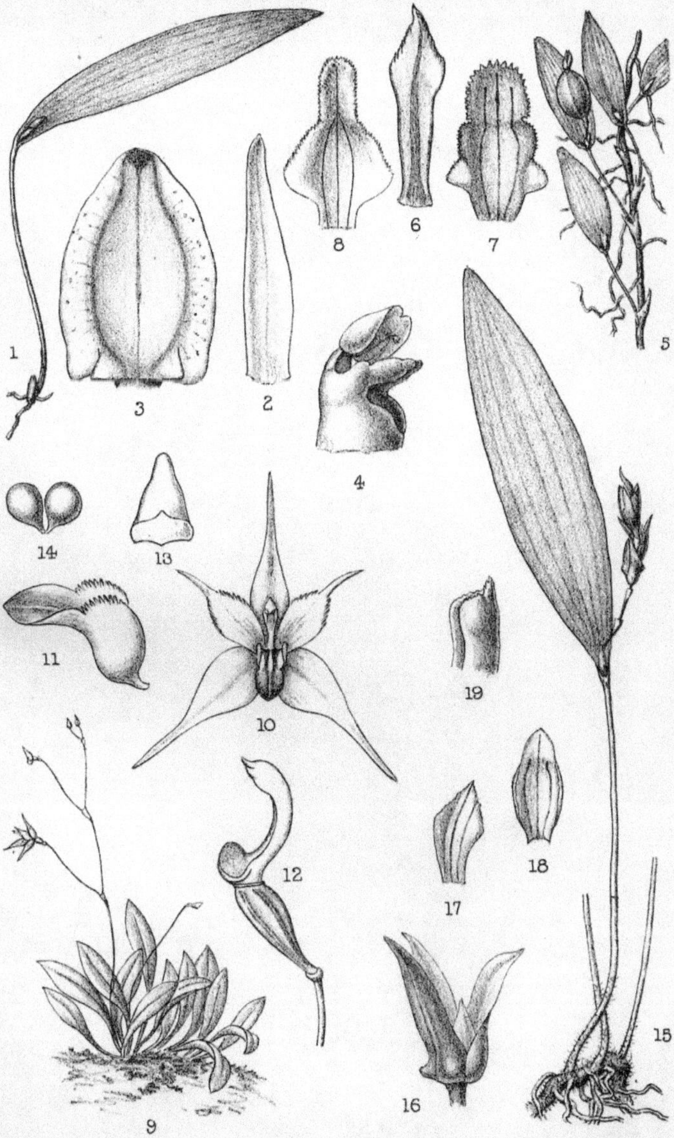

H. A. WOOD DEL. 9·14 HIGHLEY DEL & LITH.

1-4 Pleurothallis Morrisii; 5-7 P. confusa; 8 P. Wilsoni; 9-14 P. Helenæ; 15-19 P. hirsutula.

EXPLANATION OF PLATE 10

Pleurothallis uncinata

1. Plant, half natural size.
2. Capsule and flower, nat. size.
3. Petal $\times$ about 4 times.
4. Lip $\times$ about 4 times.
5. Column enlarged.

HIGHLEY DEL. & LITH.

Pleurothallis uncinata.

EXPLANATION OF PLATE 11

Lepanthes arcuata

1. Plant, natural size.
2. Flower with dorsal sepal in front × 11.
3. Do. with lateral sepal in front × 11.
4. Flower flattened out × 13.
5. Petal flattened out × 14.
6. Anther case enlarged.
7. Pollinia enlarged.

Lepanthes Harrisii

8. Plant, nat. size.
9. Flower flattened out × 9.
10. Anther case enlarged.
11. Pollinia enlarged.

Lepanthes Woodiana

12 & 13. Plant nat. size.
14. Flower flattened out × 12.
15. Capsule enlarged.
16. Anther case enlarged.
17. Pollinia enlarged.

Lepanthes elliptica

18. Plant, nat. size.
19. Flower flattened out × 5.
20. Petals and column enlarged.
21. Lip enlarged.

Lepanthes brevipetala

22. Plant, nat. size.
23. Flower flattened out × 12.

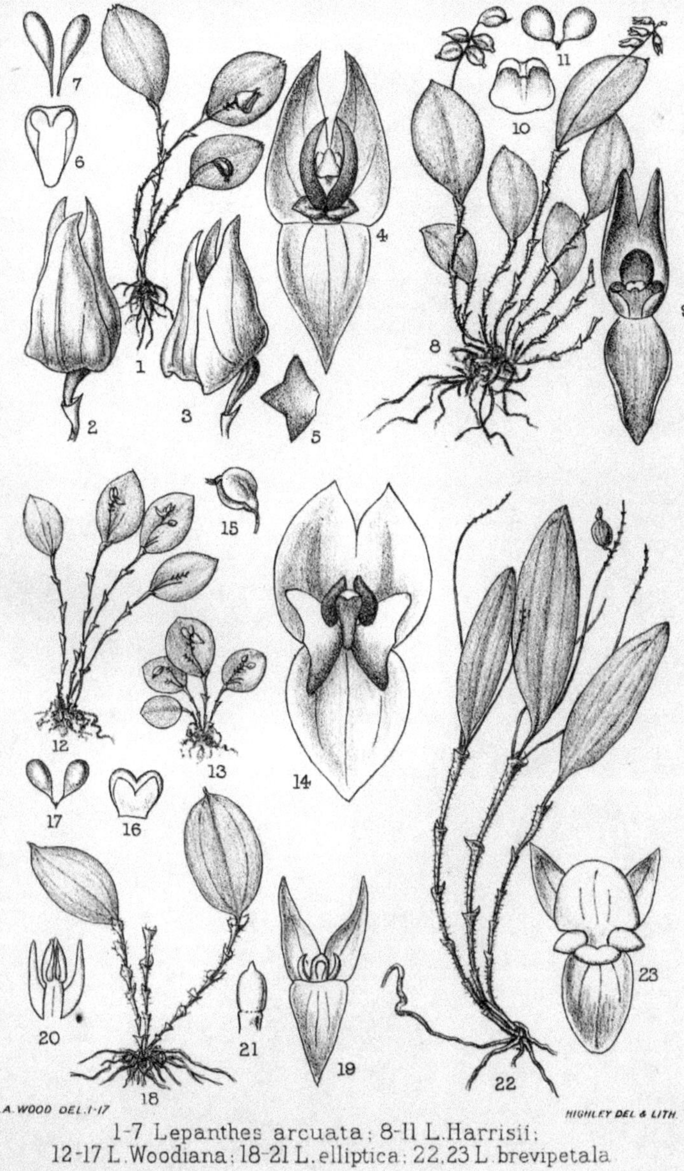

1-7 Lepanthes arcuata; 8-11 L. Harrisii;
12-17 L. Woodiana; 18-21 L. elliptica; 22, 23 L. brevipetala

EXPLANATION OF PLATE 12

Brachionidium Sherringii

1. Plant, natural size.
2. Leaf showing position of sheaths × 2.
3. Flower × 1½, from front.
4. Do. from back.
5. Dorsal sepal × about 2.
6. Lateral sepals × about 2.
7. Petal × about 2.
8. Lip and column × 17, side view.
9. Do. back view.
10. Anther case enlarged.
11. Pollinia enlarged.

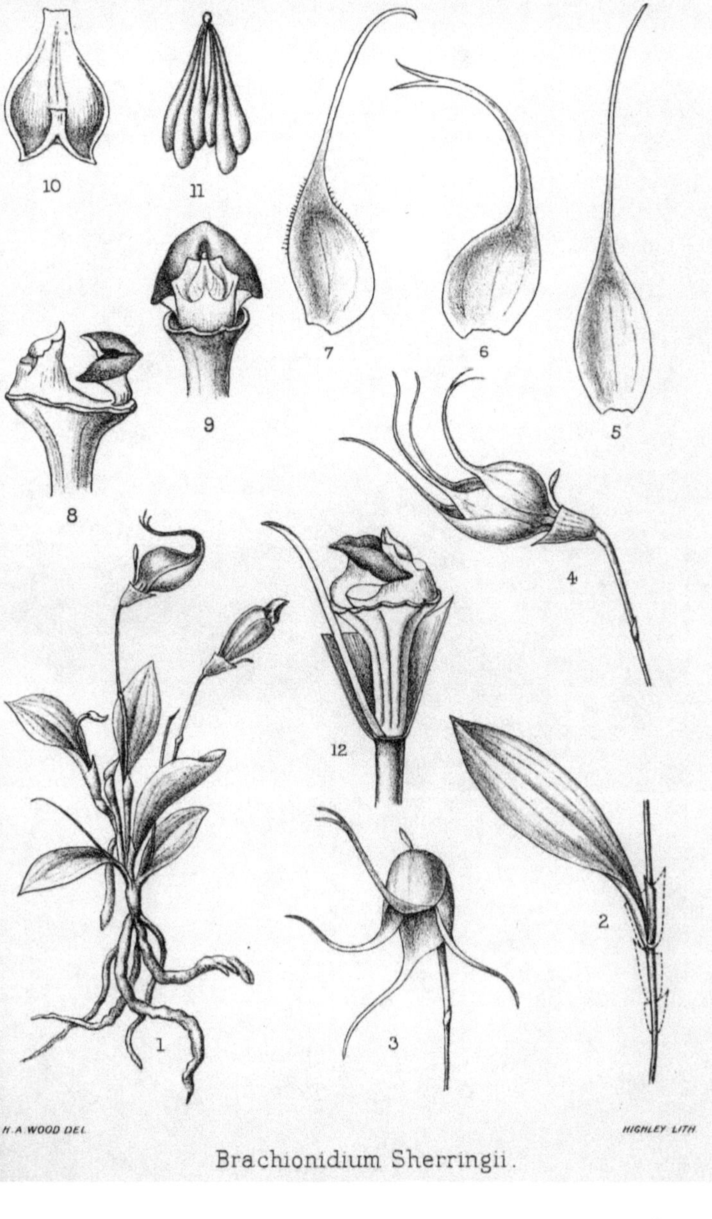

Brachionidium Sherringii.

EXPLANATION OF PLATE 13

Cœlia triptera

1. Plant, half natural size.
2. Flower with dorsal sepal held down × 2.
3. Flower with sepals and petals removed × 2.
4. Do. from one side × 2.
5. Column enlarged.
6. Anther enlarged.
7. Pollinia enlarged.

Isochilus linearis

8. Plant, half nat. size.
9. Flower × nearly 2.
10. Flower, with half the sepals and one petal removed, × 3.
11. Lip × 3.
12. Anther enlarged.
13. Pollinia enlarged.

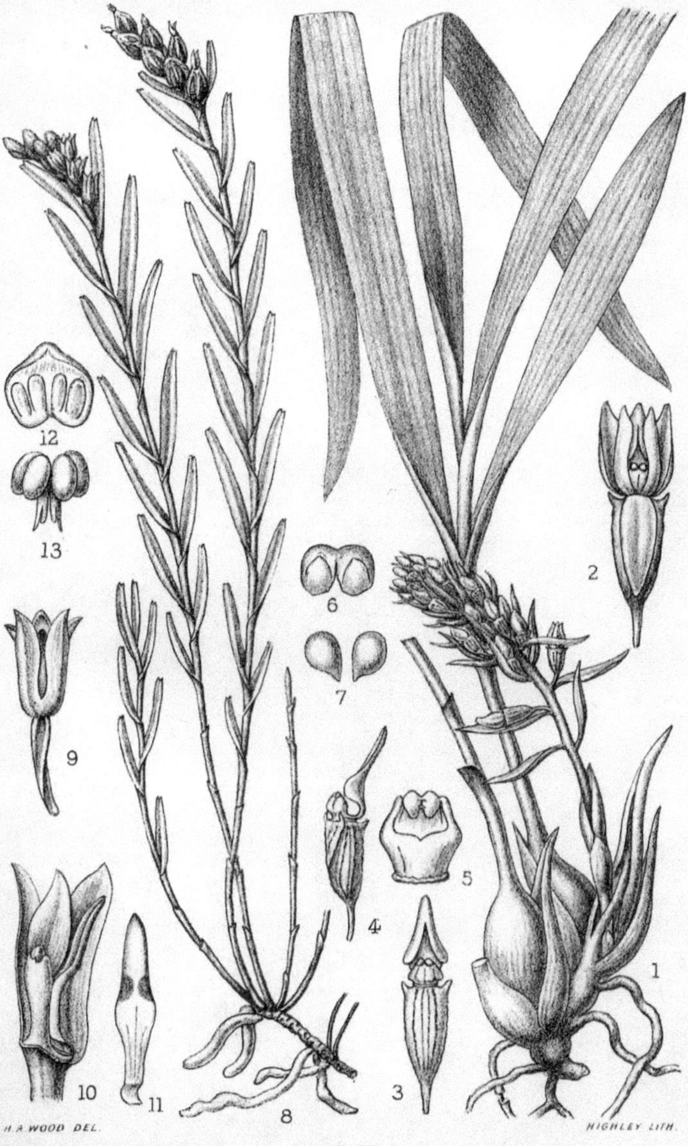

1·7 Cœlia triptera; 8-13 Isochilus linearis.

EXPLANATION OF PLATE 14

Seraphyta diffusa

1. Flower × about 3.
2. Anther enlarged.
3. Pollinia enlarged.

Arpophyllum giganteum

4. Upper portion of spike, natural size.
5. Flower × 2.
6. Lip and column × 2.
7. Do. with lip depressed.
8. Anther enlarged.
9. Pollinia enlarged.

Broughtonia sanguinea

10. Flower, with one sepal and one half and one petal removed, nat. size.
11. Lip slightly enlarged.
12. Column × about 2.
13. Anther enlarged.
14. Pollinia enlarged.

Broughtonia domingensis

15. Upper portion of spike, nat. size.
16. Lip, nat. size.
17. Column × about 2.
18. Anther enlarged.
19. Pollinia enlarged.

1-3 Seraphyta diffusa; 4-9 Arpophyllum giganteum; 10-14 Broughtonia sanguinea; 15-19 B. domingensis.

EXPLANATION OF PLATE 15

Epidendrum angustilobum

1. Flower, natural size.

Epidendrum nocturnum

2. Lip and column, nat. size.

Epidendrum difforme

3. Flower × 2.

Epidendrum serrulatum

4. Flower with lateral sepal and petal removed × 5.

Epidendrum nutans

5. Flower × 1½.

Epidendrum bifarium

6. Flower × 6.

Epidendrum rivulare

7. Flower slightly enlarged.

Epidendrum Ottonis

8. Flower × 4.

Epidendrum imbricatum

9. Flower × 3.

Epidendrum anceps

10. Flower slightly enlarged.

Epidendrum verrucosum

11. Flower slightly enlarged.

Epidendrum ramosum

12. Lip and column × 4.

H.A.WOOD DEL. HIGHLEY LITH.

Epidendrum spp.

EXPLANATION OF PLATE 16

Epidendrum monticolum

1. Plant, natural size.
2. Flower × 1½.
3. Petal × 2½.
4. Lip × 2½.

Epidendrum parvilobum

5. Plant, lower part, with leaf, nat. size.
6. Petal × 3.
7. Lip × 3.

Epidendrum belvederense

8. Petal × 3.
9. Lip × 3.
10. Capsule, nat. size.

H.A. WOOD DEL. 1-4

HIGHLEY DEL & LITH.

1-4 Epidendrum monticolum; 5-7 E. parvilobum; 8-10 E. belvederense.

EXPLANATION OF PLATE 17

Hormidium tripterum

1. Portion of plant, natural size.
2. Flower × 3.
3. Column and lip, anther removed × about 7.
4. Anther enlarged, side view.
5. Do. front view.

Lælia monophylla

6. Stem with roots and leaf, nat. size.
7. Flower, nat. size.
8. Lip and column × 4.
9. Lip spread open × 4.
10. Anther enlarged from back.
11. Do. from front.
12. Pollinia enlarged.

Octadesmia montana

13. Portion of stem, nat. size.
14. Column enlarged front view.
15. Do. back view.
16. Pollinia enlarged.

1-5 Hormidium tripterum; 6-12 Lælia monophylla;
13-16 Octadesmia montana.

EXPLANATION OF PLATE 18

Homalopetalum vomeriforme (*H. jamaicense*)

1. Plant, natural size.
2. Column and lip × 3.
3. Lip flattened × 3.
4. Anther with pollinia, enlarged.
5. Pollinia enlarged.
6. Do. spread out.

Tetramicra parviflora

7. Plant, nat. size.
8. Flower × about 3.
9. Anther with pollinia, enlarged.
10. Pollinia enlarged.

1-6 Homalopetalum jamaicense: 7-10 Tetramicra parviflora.

EXPLANATION OF PLATE 19

Schomburgkia Lyonsii

1. Portion of raceme, natural size.
2. Column × 2, back view.
3. Do. front view.
4. Anther enlarged, front view.
5. Do. back view.
6. Pollinia, as attached to stigma, enlarged.

Brassavola cordata

7. Flower, nat. size.
8. Column × 3.
9. Anther, with pollinia, enlarged.
10. Empty anther case enlarged.
11. Pollinia enlarged.

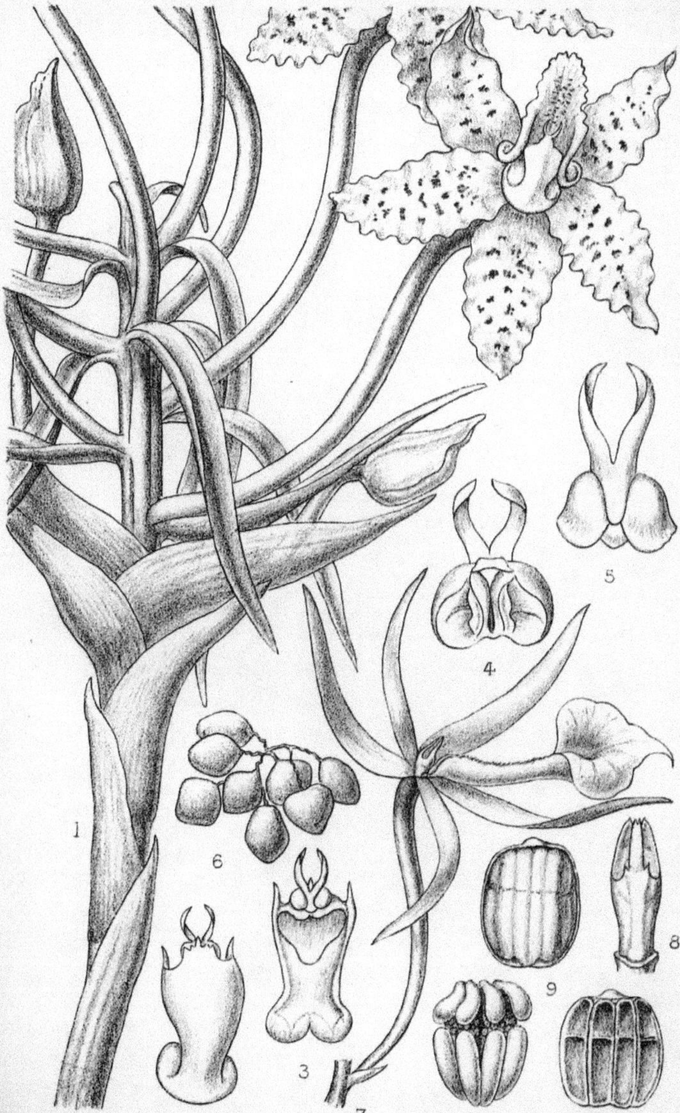

H. A. WOOD DEL. HIGHLEY LITH.

1-6 Schomburgkia Lyonsii; 7-11 Brassavola cordata.

EXPLANATION OF PLATE 20

Elleanthus capitatus

1. Apex of stem with head of flowers, natural size.
2. Flowers, nat. size, from back.
3. Do. from side.
4. Column × about 2, partly front view.
5. Do. from side.
6. Anther enlarged.
7. Pollinia enlarged.

Elleanthus longibracteatus

8. Flower with bract × 2.
9. Column and lip × 2.
10. Do., lip thrown back, × 2½.
11. Lip × 2½.
12. Pollinia enlarged.

Phaius Tancarvilleæ (P. grandifolius)

13. Flower × about ½.
14. Column slightly enlarged from side.
15. Do. from front.
16. Anther enlarged.
17. Pollinia enlarged.

Calanthe mexicana

18. Raceme, nat. size.
19. Flower, with one sepal and one half and one petal removed, × 2½.
20. Lip flattened × 3.
21. Anther enlarged.
22. Pollinia enlarged.

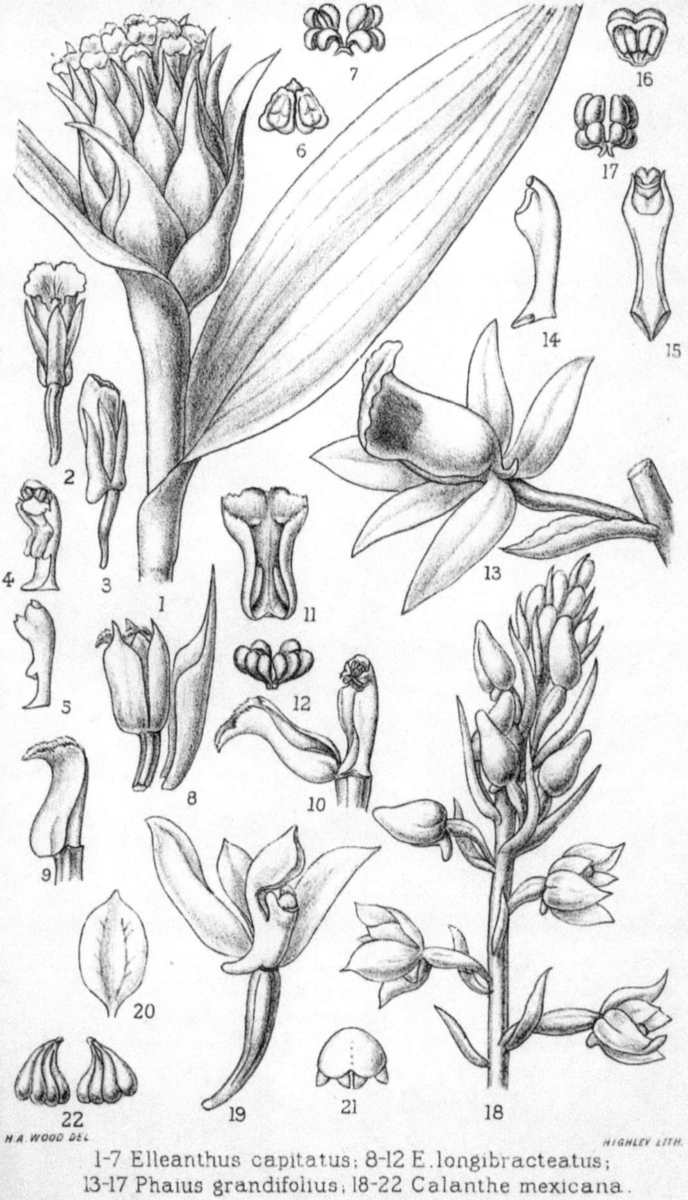

H. A. WOOD DEL. HIGHLEY LITH.

1-7 Elleanthus capitatus; 8-12 E. longibracteatus; 13-17 Phaius grandifolius; 18-22 Calanthe mexicana.

EXPLANATION OF PLATE 21

Bletia purpurea (*B. alta*)

1. Top of raceme, natural size.
2. Flower spread out × 2.
3. Lip and column of bud, nat. size.
4. Lip and column of opening flower × 2.
5. Lip flattened out × 2.
6. Anther enlarged.
7. Pollinia enlarged.

Bletia florida

8. Flower, with one sepal and one half and petal removed, nat. size.
9. Lip and column of bud, nat. size.
10. Lip × nearly 2.
11. Column × 2.

Figs. 3, 9, 10, 11 are copies of Salisbury's original drawings in the Department of Botany.

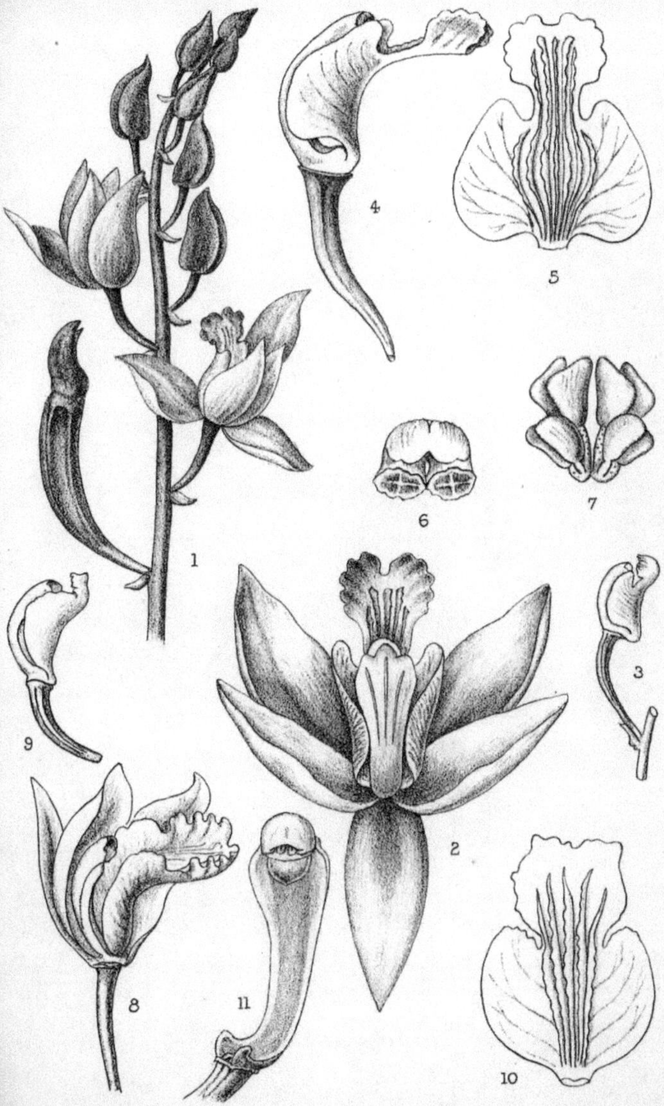

H. A. WOOD DEL. HIGHLEY LITH.

1-7 Bletia alta; 8-11 B. florida.

EXPLANATION OF PLATE 22

Govenia utriculata

1. Lower portion of stem, about natural size.
2. Upper portion of raceme slightly enlarged.
3. Flower slightly enlarged.

Eulophia alta (by error *longifolia*)

4. Raceme, nat. size.
5. Flower with lip removed, nat. size.
6. Lip, nat. size.
7. Anther enlarged.
8. Pollinia enlarged.

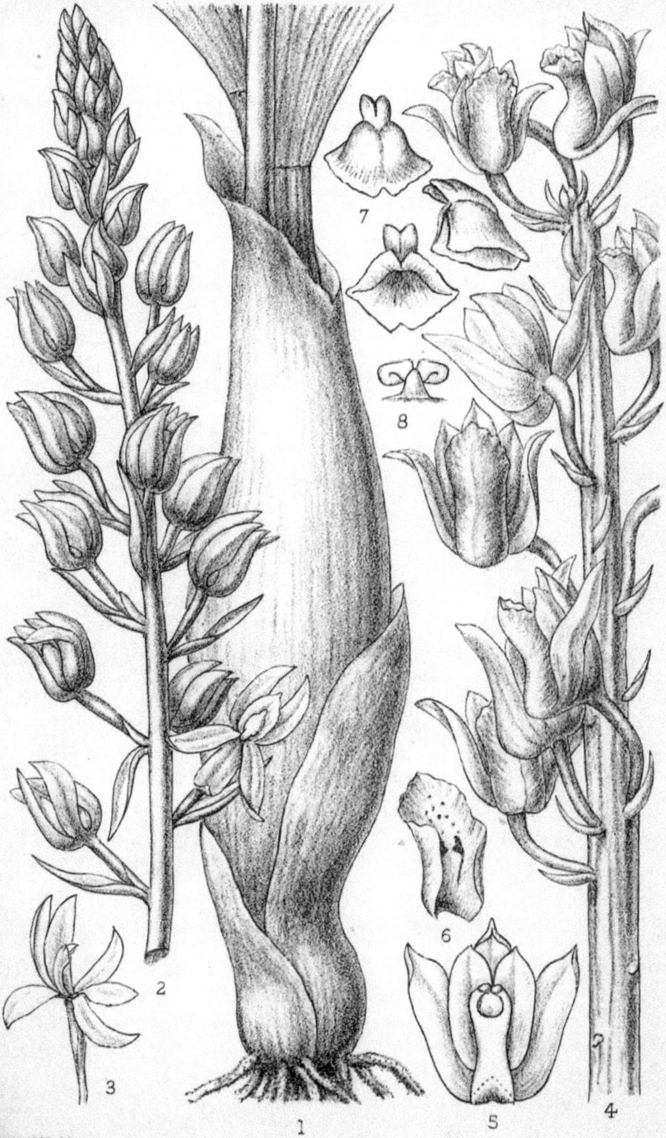

H. A. WOOD DEL. HIGHLEY LITH.

1-3 Govenia utriculata; 4-8 Eulophia longifolia.

EXPLANATION OF PLATE 23

Lycaste Barringtoniæ

1. Scapes and pseudobulb, with sepals and petals removed in one flower, half natural size.
2. Anther enlarged.
3. Pollinia enlarged.

Xylobium palmifolium

4. Scape and pseudobulb, half nat. size.
5. Flower with perianth removed, except one sepal and one petal, slightly enlarged.
6. Anther enlarged.
7. Pollinia enlarged.

Bulbophyllum jamaicense

8. Stem and pseudobulb, nat. size.
9. Flower × 3.
10. Flower, sepals and one petal removed × 6.
11. Column and lip × 6.
12. Anther enlarged, from front.
13. Do. from back.
14. Pollinia enlarged.

Bulbophyllum pachyrrhachis

15. Upper portion of stem, nat. size.
16. Flower, sepals removed, × 7.

Zygopetalum flabelliforme (*Z. cochleare*)

17. Lower portion of plant, nat. size.
18. Anther enlarged.
19. Pollinia enlarged.

H. A. WOOD DEL. HIGHLEY LITH.

1-3 Lycaste Barringtoniæ; 4-7 Xylobium palmifolium; 8-14 Bulbophyllum jamaicense; 15,16 B. pachyrrhachis; 17-19 Zygopetalum cochleare.

EXPLANATION OF PLATE 24

Neo-urbania adendrobium

1. Plant, half natural size, lower part.
2. Do. upper part.
3. Flower and peduncle with bracts × 3.
4. Flower, half the sepals and one petal removed × 7.
5. Lip from side × 7.
6. Lip × 7.
7. Column × about 8.
8. Anther enlarged, from above.
9. Do. from below.
10. Pollinia enlarged.

H. A. WOOD DEL. HIGHLEY LITH.

Neo-urbania adendrobium.

EXPLANATION OF PLATE 25

Maxillaria alba

1. Plant, half natural size.
2. Flower, nat. size.
3. Lip and column slightly enlarged.
4. Lip flattened slightly enlarged.
5. Anther enlarged.
6. Pollinia enlarged.

Maxillaria sessilis

7. Tip of leaf nat. size.
8. Flower nat. size.
9. Lip flattened nat. size.
10. Column enlarged.
11. Anther enlarged.
12. Pollinia enlarged.

Maxillaria rufescens

13. Plant, half nat. size.
14. Lip flattened $\times 1\frac{1}{2}$.
15. Anther enlarged.
16. Pollinia enlarged.

Maxillaria rufescens var minor.

17. Flower, nat. size.
18. Lip flattened slightly enlarged.
19. Column, nat. size.

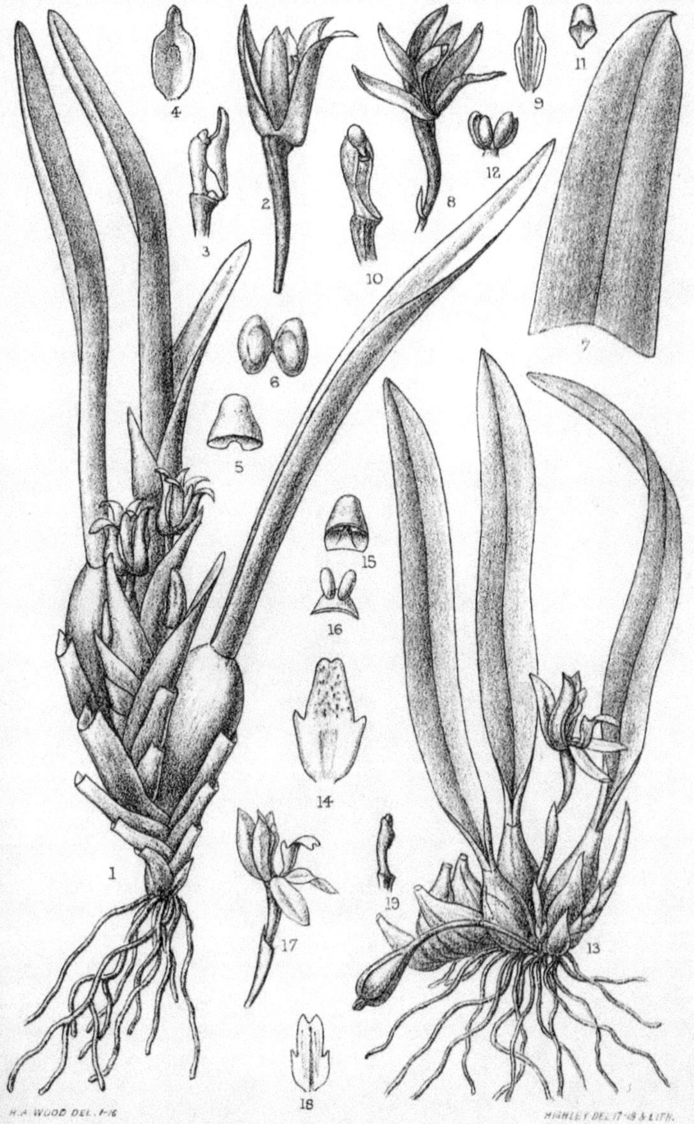

H. A. WOOD DEL. 1-16 HIGHLEY DEL. 17-19 & LITH.

1-6 Maxillaria alba; 7-12 M. sessilis; 13-16 M. rufescens; 17-19 M. rufescens var. minor.

EXPLANATION OF PLATE 26

Ornithidium vestitum

1. Portion of plant, natural size.
2. Flower, with one sepal and one half and one petal removed × 5.
3. Anther enlarged.
4. Pollinia enlarged.

Ornithidium proliferum

5. Portion of plant, nat. size.
6. Flower × about 2.
7. Lip flattened × about 2.

H. A. WOOD DEL. 1, 3, 4. HIGHLEY DEL. & LITH.

1-4 Ornithidium vestitum; 5-7 O. proliferum.

EXPLANATION OF PLATE 27

Ionopsis satyrioides (I. testiculata)

1. Plant, natural size.
2. Flower, with one sepal and one half and one petal removed, × between 3 and 4.
3. Pollinia enlarged.

Ionopsis utricularioides

4. Portion of panicle, nat. size.
5. Flower, with half perianth removed, slightly enlarged.
6. Pollinia enlarged.

Macradenia lutescens

7. Plant, nat. size.
8. Flower, with sepals and petals removed, × about 2.
9. Pollinia enlarged, from side.
10. Do. from back.

1-3 Ionopsis testiculata; 4-6 I. utricularioides; 7-10 Macradenia lutescens.

EXPLANATION OF PLATE 28

Brassia maculata

1. Portion of plant, half natural size.
2. Portion of raceme, half nat. size.
3. Column × 4, from side.
4. Do. from front.
5. Anther enlarged.
6. Pollinia enlarged.

Brassia caudata

7. Flower, with one sepal and one petal removed, nat. size.
8. Anther enlarged.
9. Pollinia enlarged.

Comparettia falcata

10. Plant, half nat. size.
11. Flower, with sepals and one petal removed, slightly enlarged.
12. Sepals and ovary, slightly enlarged.
13. Column × 3.
14. Anther enlarged.
15. Pollinia enlarged.

H. A. WOOD DEL. HIGHLEY LITH.

1-6 Brassia maculata; 7-9 B. caudata; 10-15 Comparettia falcata.

EXPLANATION OF PLATE 29

Trichopilia jamaicensis

1. Plant, natural size.
2. Column × about 3.
3. Pollinia enlarged.

Oncidium pulchellum

4. Flower, nat. size.
5. Column × 6.
6. Anther enlarged.
7. Pollinia enlarged, from front.
8. Do. from side.

Oncidium luridum

9. Flower, nat. size.
10. Anther enlarged.
11. Pollinia enlarged.

Oncidium triquetrum

12. Flower, slightly enlarged.
13. Column × 4 or 5, from front.
14. Do. from back.
15. Anther enlarged.
16. Pollinia enlarged, from front.
17. Do. from back.

Oncidium guttatum (*O. tetrapetalum*)

18. Flower, nat. size.
19. Flower, with one sepal and one petal removed, × $2\frac{1}{2}$.
20. Anther enlarged.
21. Pollinia enlarged.

Oncidium sphacelatum (by error *altissimum*)

22. Flower, nat. size.
23. Anther enlarged.
24. Pollinia enlarged, from front.
25. Do. from side.

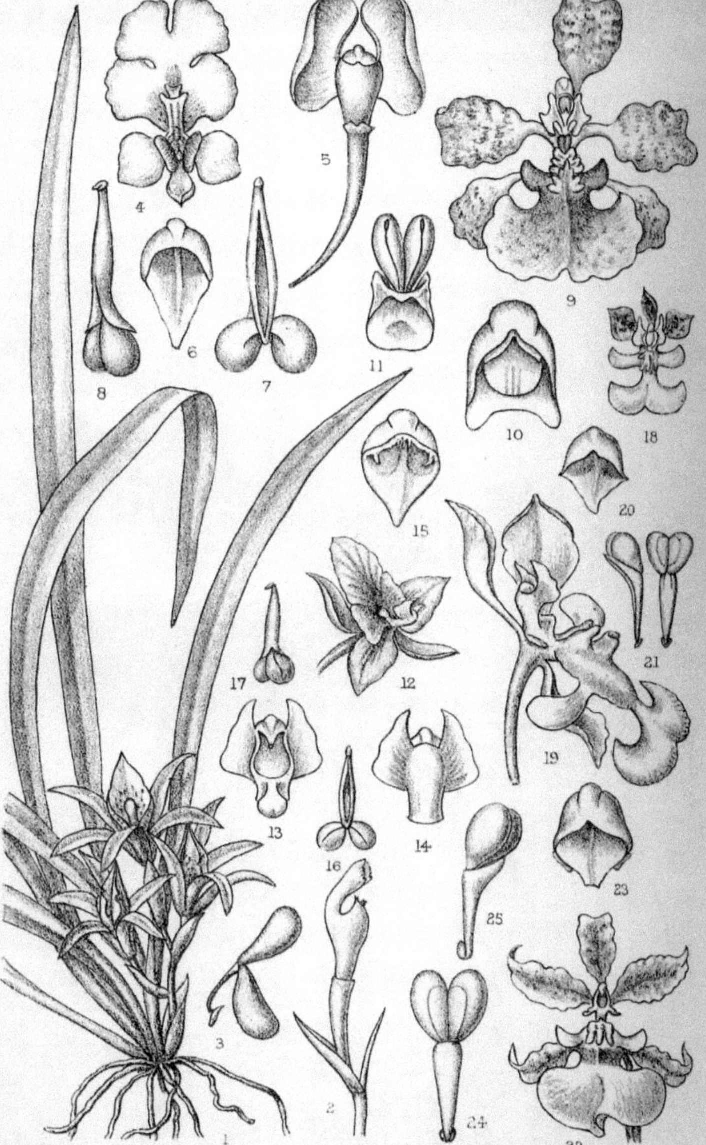

1-3 Trichopilia jamaicensis; 4-8 Oncidium pulchellum; 9-11 O. luridum; 12-17 O. triquetrum; 18-21 O. tetrapetalum; 22-25 O. altissimum.

EXPLANATION OF PLATE 30

Leochilus labiatus
(*L. cochlearis*)

1. Plant, natural size.
2. Flower, with one sepal and one half and one petal removed × about 4.
3. Lateral sepals × about 4.
4. Column × about 9.
5. Pollinia enlarged.

Cryptarrhena lunata

6. Flower × about 4.
7. Lip × about 4.

Dichæa graminoides

8. Plant, half nat. size.
9. Lateral sepal × 3.
10. Median sepal × 3.
11. Petal × 3.
12. Lip × 3.

Dichæa glauca

13. Lateral sepal × 3.
14. Median sepal × 3.
15. Petal × 3.
16. Lip × 3.

Dichæa trichocarpa

17. Lateral sepal × 3.
18. Median sepal × 3.
19. Petal × 3.
20. Lip × 3.
21. Hairs from capsule × 3.

Dichæa muricata

22. Sepal × 3.
23. Petal × 3.
24. Lip × 3.
25. Hairs from capsule × 3.

Dichæa echinocarpa

26. Lateral sepal × 3.
27. Median sepal × 3.
28. Petal × 3.
29. Lip × 3.
30. Hairs from capsule × 3.

Dichæa Morrisii

31. Lateral sepal, nat. size.
32. Median sepal, nat. size.
33. Petal, nat. size.
34. Lip and column × 3.
35. Lip × 3.
36. Hairs from capsule × 3.

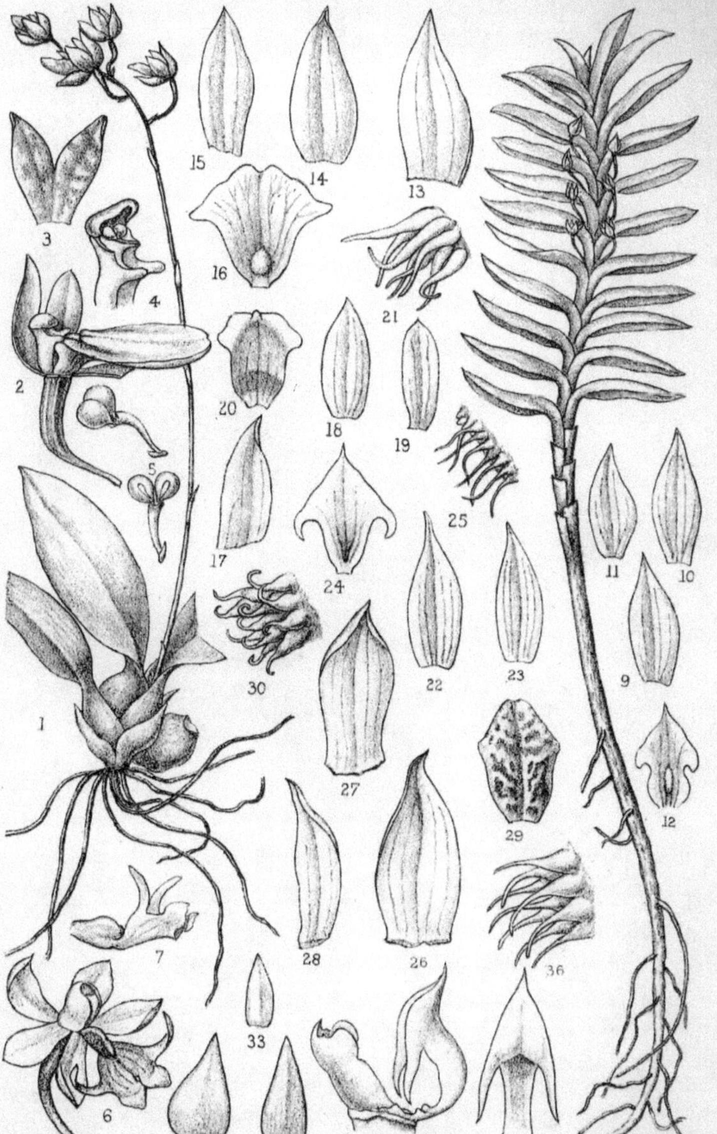

1-5 Leochilus cochlearis; 6, 7 Cryptarrhena lunata; 8-12 Dichæa graminoides; 13-16 D. glauca; 17-21 D. trichocarpa; 22-25 D. muricata; 26-30 D. echinocarpa; 31-35 D. Morrisii.

EXPLANATION OF PLATE 31

Campylocentrum Sullivanii

1. Plant, natural size.
2. Flower × 11 times.
3. Flower, with sepal and petal removed, × 11 times.
4. Lip × 11 times.

Campylocentrum minus

5. Leaf and spike, nat. size.
6. Flower × 10.
7. Flower, lateral sepals removed, and lip held down, × 10.
8. Lip × 10.

Campylocentrum Barrettiæ

9. Plant, nat. size.
10. Flower × 5.
11. Lip, spur, column and ovary × 8.
12. Lip × 8.
13. Anther, enlarged, from side.
14. Do. from front.
15. Pollinia enlarged.

Harrisella porrecta

16. Plant, nat. size.
17. Flower × 8.
18. Flower, with sepals and petals removed, × 8.
19. Anther, enlarged, from front.
20. Do. from back.
21. Pollinia enlarged.

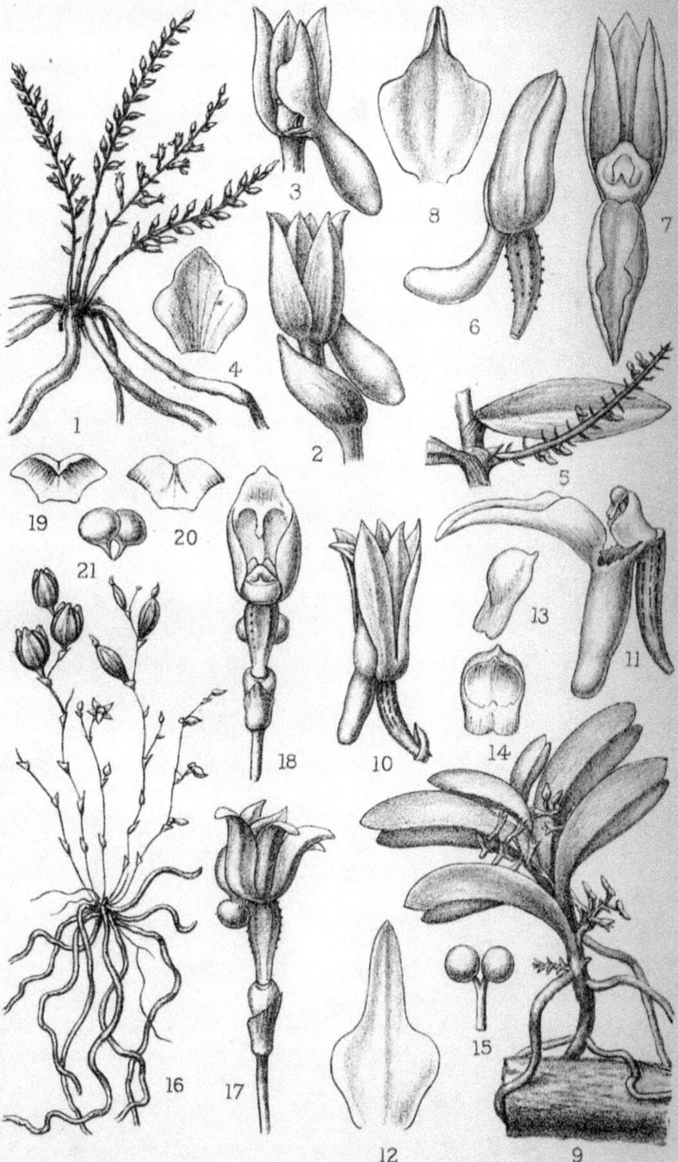

H. A. WOOD DEL. HIGHLEY LITH.

1-4 Campylocentrum Sullivanii; 5-8 C. minus; 9-15 C. Barrettiæ; 16-21 Harrisella porrecta.

EXPLANATION OF PLATE 32

Dendrophylax funalis

1. Plant, natural size.
2. Flower cut to show entrance to spur and lip flattened out, not quite full size.
3. Column with anther raised to show position of pollinia × 3 times.
4. Anther enlarged, from back.
5. Do. from front.
6. Pollinia enlarged, front view.
7. Do. side view.

H. A. WOOD DEL. HIGHLEY LITH.

Dendrophylax funalis.

B.M. (N.H.).
No. 21.

LIST OF THE CURRENT

NATURAL HISTORY PUBLICATIONS OF THE TRUSTEES OF THE BRITISH MUSEUM.

The following publications can be purchased through the Agency of *Messrs.* LONGMANS & Co., 39, *Paternoster Row, London, E.C.*; *Mr.* QUARITCH, 11, *Grafton Street, New Bond Street, W.*; and *Messrs.* DULAU & Co., Ltd., 37, *Soho Square, W.*; or at the NATURAL HISTORY MUSEUM, *Cromwell Road, London, S.W.*

The History of the Collections contained in the Natural History Departments of the British Museum :—

Vol. I. Libraries; Botany; Geology; Minerals. Pp. xvii., 442. 1904, 8vo. 15*s.*

Vol. II. Separate historical accounts of the several collections included in the Department of Zoology. Pp. 782. 1906, 8vo. 1*l.* 10*s.*

Catalogue of the Books, Manuscripts, Maps, and Drawings in the British Museum (Natural History). Compiled by B. B. Woodward, Assistant in charge of the General Library, with some clerical assistance :—

Vol. I. A—D. Pp. viii., 500. 1903, 4to. 1*l.*
Vol. II. E—K. Pp. 501–1038. 1904, 4to. 1*l.*

Catalogue of the Specimens and Drawings of Mammals, Birds, Reptiles, and Fishes of Nepal and Tibet. Presented by B. H. Hodgson, Esq., to the British Museum. 2nd edition. By John Edward Gray. Pp. xii., 90. [With an account of the Collection by Mr. Hodgson.] 1863, 12mo. 2*s.* 3*d.*

(16335r—29.) Wt. 2696—4. 1500. 5/10. D & S. (S.) A

Report on the Zoological Collections made in the Indo-Pacific Ocean during the voyage of H.M.S. "Alert," 1881–2. Pp. xxv., 684. 54 Plates. 1884, 8vo. 1*l.* 10*s.*

Summary of the Voyage... ... ... By Dr. R. W. Coppinger.
Mammalia ... ... „ O. Thomas.
Aves... ... ... „ R. B. Sharpe.
Reptilia, Batrachia, Pisces ... ... „ A. Günther.
Mollusca ... ... „ E. A. Smith.
Echinodermata ... ... „ F. J. Bell.
Crustacea ... ... „ E. J. Miers.
Coleoptera ... ... „ C. O. Waterhouse.
Lepidoptera ... ... „ A. G. Butler.
Alcyonaria and Spongiida ... ... „ S. O. Ridley.

Report on the Collections of Natural History made in the Antarctic Regions during the Voyage of the "Southern Cross." Pp. ix., 344. 53 Plates. 1902, Royal 8vo. 2*l.*

Mammalia ... ... By Capt. G. E. H. Barrett-Hamilton.
Notes on Antarctic Seals ... „ E. A. Wilson, M.B.
Extracts from the "Diary" of the late Nicolai Hanson. —
Aves ... ... „ Dr. R. Bowdler Sharpe.
Pisces... ... ... „ G. A. Boulenger, F.R.S.
Tunicata ... ... „ Prof. W. A. Herdman, F.R.S.
Mollusca ... ... „ E. A. Smith.
Echinoderma ... ... „ Prof. F. J. Bell.
Insecta ... ... „ G. H. Carpenter and Hon. N. C Rothschild.
Arachnida ... ... „ Dr. E. L. Trouessart.
Crustacea ... ... „ T. V. Hodgson.
Polychæta ... ... „ Dr. A. Willey.
Gephyrea ... ... „ A. E. Shipley.
Nematoda ... ... }
Cestoda ... ... } „ Dr. von Linstow.
Polyzoa ... ... }
Porifera ... ... } „ R. Kirkpatrick.
Anthozoa ... ... „ Dr. L. Roule and S. J. Hickson, F.R.S.
Actiniæ ... ... „ J. A. Clubb.
Hydrozoa ... ... „ E. T. Browne.
Cryptogamia ... ... „ A. Gepp, V. H. Blackman, and Miss E. S. Barton.
Rock Specimens ... ... „ Dr. G. T. Prior.

National Antarctic Expedition, 1901–1904. [*Discovery Report.*] Natural History :—

Vol. I. Geology (Field-Geology : Petrography). Pp. xii., 160 : 10 plates, 72 text-figures, 2 maps. [With Index.] 1907, 4to. 1*l.* 10*s.*

Field-Geology ... ... By H. T. Ferrar, M.A., F.G.S.
Rock-Specimens ... ... „ G. T. Prior, M.A., D.Sc., F.G.S.

Vol. II. Zoology (Vertebrata : Mollusca : Crustacea). Pp. xiv., 355 : 33 plates (17 coloured), 146 text-figures, 1 map. 1907, 4to. 3*l.*

Mammalia (pp. 69 : 5 pls. [3 col.], 36 figs.) }
Aves (pp. 121 : 13 pls. col., 46 figs.) ... } By E. A. Wilson, M.B.
Anatomy of Penguins (pp. 28 : 1 pl., 8 figs.) „ W. P. Pycraft.

National Antarctic Expedition, 1901-1904—*continued.*

Vol. II—*continued.*

Fishes (pp. 5; 2 pls.)	By G. A. Boulenger, F.R.S.
Cephalodiscus (pp. 67: 7 pls. [1 col.], 17 figs.)	" W. G. Ridewood, D.Sc.
Mollusca. I.—Cephalopoda (pp. 2: 1 fig.)	" W. E. Hoyle, D.Sc.
" II.—Gastropoda / " III.—Amphineura (pp. 12, 1: 2 pls.)	" E. A. Smith, I.S.O.
" IV.—Nudibranchiata (pp. 28: 1 pl., 30 figs.)	" Sir Chas. Eliot, K.C.M.G., LL.D.
" V.—Lamellibranchiata (pp. 7: 1 pl.) / Brachiopoda (pp. 2: 4 figs.)	" E. A. Smith, I.S.O.
Crustacea I.—Decapoda (pp. 7) ... / " II.—Cumacea (pp. 6: 1 pl., 4 figs.)	" W. T. Calman, D.Sc.

Vol. III. Zoology and Botany (Invertebrata: Marine Algæ, Musci). Pp. vi., 273: 51 plates, 8 text-figures, and 1 chart. 1907, 4to. 2*l.* 10*s.*

On Collecting in Antarctic Seas (pp. 10)	By T. V. Hodgson, F.L.S.
Mollusca. VI.—Pteropoda (pp. 15: 2 pls.)	" Sir Charles Eliot, K.C.M.G., LL.D.
Crustacea. III.—Amphipoda (pp. 39: 13 pls.)	" A. O. Walker, F.L.S.
" IV.—Leptostraca (pp. 2: 2 figs.)	" Dr. J. Thiele.
" V.—Ostracoda (pp. 9: 3 pls.)	" Prof. G. S. Brady, F.R.S.
" VI.—Cirripedia (pp. 4: 1 pl.)	" Prof. A. Gruvel.
Pycnogonida (pp. 72: 10 pls.)	" T. V. Hodgson, F.L.S.
Acari (pp. 6: 1 pl.)	" Dr. E. L. Trouessart.
Chætognatha (pp. 6: 1 chart)	" Dr. G. H. Fowler, F.L.S.
Nematoda (pp. 4: 1 pl.)	" Dr. O. von Linstow.
Cestoda (pp. 6: 1 pl.)	" A. E. Shipley, F.R.S.
Coelentera. I.—Alcyonaria (pp. 15: 2 pls.)	" Prof. S. J. Hickson, F.R.S.
" II.—Hydroid Zoophytes (pp. 34: 4 pls.)	" Prof S. J. Hickson, F.R.S., and F. H. Gravely.
" III.—Tentacles of a Siphonophore (pp. 3: 5 figs.)	" Dr. J. Rennie.
Porifera. I.—Hexactinellida (pp. 25: 7 pls.)	" R. Kirkpatrick.
Marine Algæ. I.—Phæophyceæ and Florideæ (pp. 15: 4 pls.)	" A. Gepp and Mrs. E. S. Gepp.
" " II.—Corallinaceæ (pp. 2: 1 fig.)	" M. Foslie.
Musci (pp. 6: 2 pls.)	" J. Cardot.

Vol. IV. Zoology (various Invertebrata). Pp. iv., 281: 65 plates, 1 text-figure. 1908, 4to. 1*l.* 15*s.*

Mollusca. VII.—Solenogastres (pp. 13: 2 pls.)	By Dr. H. F. Nierstrasz.
Insecta.—Aptera (pp. 5: 1 pl.) ...	" G. H. Carpenter, B.Sc., M.R.I.A.
Crustacea. VII.—Schizopoda (pp. 42: 8 pls.)	" W. M. Tattersall, M.Sc.
" VIII.—Copepoda (pp. 44: 7 pls.)	" R. Norris Wolfenden, M.D.

National Antarctic Expedition, 1901-1904—*continued.*

Vol. IV—*continued.*

Echinoderma (pp. 16 : 5 pls.)	By F. Jeffrey Bell, M.A.
Echinoderm Larvæ (pp. 9 : 1 pl.) ...	„ E. W. McBride, M.A., F.R.S., and J. C. Simpson, B.Sc.
Myzostomidæ (pp. 26 : 1 pl. and 1 figure)	„ Dr. Rudolf Ritter von Stummer-Traunfels.
Sipunculoidea (pp. 6)	„ W. F. Lanchester, M.A.
Coelentera. IV.—Actiniæ (pp. 12 : 3 pls.)	„ J. A. Clubb, M.Sc.
Porifera. II.—Tetraxonida (pp. 56 : 19 pls.)	„ R. Kirkpatrick.
„ III.—Calcarea (pp. 52 : 12 pls.)	„ C. F. Jenkin, B.A.

Vol. V. Zoology and Botany. Pp. ix., 212 : 28 plates, and 19 text-figures. [With a Summary of the Contents of Vols. II.-V.] 1910, 4to. 1*l.* 10*s.*

Vertebrata. IA.—Mammalia (Seal-Embryos) (pp. 21 : 2 pls., 1 text-figure)	By Dr. H. W. Marett Tims.
Tunicata (pp. 26 : 7 pls., 2 text-figures)	„ Prof. W. A. Herdman, D.Sc., F.R.S.
Crustacea. IX.—Isopoda (pp. 77 : 10 pls)	„ T. V. Hodgson, F.L.S.
Nemertinea (pp. 15 : 1 pl., 16 text-figures)	„ Prof. L. Joubin.
Cœlentera. V.—Medusæ (pp. 62 : 7 pls.)	„ Edward T. Browne.
Lichenes (pp. 11 : 1 pl.)...	„ Dr. O. V. Darbishire.

A Monograph of Christmas Island (Indian Ocean) : Physical Features and Geology by C. W. Andrews, B.A., B.Sc., F.G.S., with descriptions of the Fauna and Flora by numerous contributors. Pp. xv., 337: 22 plates (7 coloured), a map, and 27 illustrations in text. [With Index.] 1900, 8vo. 1*l.*

First Report on Economic Zoology. By Fred. V. Theobald, M.A., &c. [With an Introduction, containing a Classification of Animals from the point of view of Economic Zoology, by Prof. E. Ray Lankester, LL.D., F.R.S.] Pp. xxxiv., 192. 18 Woodcuts. 1903, Roy. 8vo. 6*s.*

Second Report on Economic Zoology. By Fred. V. Theobald, M.A., &c. Pp. x., 197. 29 Illustrations. 1904, Roy. 8vo. 6*s.*

MAMMALS.

Catalogue of the Bones of Mammalia in the Collection of the British Museum. By Edward Gerrard. Pp. iv., 296. 1862, 8vo. 5*s.*

Catalogue of Monkeys, Lemurs, and Fruit-eating Bats in the Collection of the British Museum. By Dr. J. E. Gray, F.R.S., &c. Pp. viii., 137. 21 Woodcuts. 1870, 8vo. 4*s.*

Catalogue of Carnivorous, Pachydermatous, and Edentate Mammalia in the British Museum. By John Edward Gray, F.R.S., &c. Pp. vii., 398. 47 Woodcuts. 1869, 8vo. 6*s.* 6*d.*

Catalogue of Seals and Whales in the British Museum. By John Edward Gray, F.R.S., &c. 2nd Edition. Pp. vii., 402. 101 Woodcuts. 1866, 8vo. 8*s*.

——— Supplement. By John Edward Gray, F.R.S., &c., Pp. vi., 103. 11 Woodcuts. 1871, 8vo. 2*s*. 6*d*.

List of the Specimens of Cetacea in the Zoological Department of the British Museum. By William Henry Flower, LL.D., F.R.S., &c. [With Systematic and Alphabetical Indexes.] Pp. iv., 36. 1885, 8vo. 1*s*. 6*d*.

Catalogue of Ruminant Mammalia (*Pecora*, Linnæus) in the British Museum. By John Edward Gray, F.R.S., &c. Pp. viii., 102. 4 Plates. 1872, 8vo. 3*s*. 6*d*.

Catalogue of the Marsupialia and Monotremata in the Collection of the British Museum. By Oldfield Thomas. Pp. xiii., 401. 4 Coloured and 24 plain Plates. [With Systematic and Alphabetical Indexes.] 1888, 8vo. 1*l*. 8*s*.

BIRDS.

Catalogue of the Birds in the British Museum :—

Vol. VII. Catalogue of the Passeriformes, or Perching Birds, in the Collection of the British Museum. *Cichlomorphæ* : Part IV., containing the concluding portion of the family Timeliidæ (Babbling Thrushes). By R. Bowdler Sharpe. Pp. xvi., 698. Woodcuts and 15 coloured Plates. [With Systematic and Alphabetical Indexes.] 1883, 8vo. 1*l*. 6*s*.

Vol. VIII. Catalogue of the Passeriformes, or Perching Birds, in the Collection of the British Museum. *Cichlomorphæ* : Part V., containing the families Paridæ and Laniidæ (Titmice and Shrikes) ; and *Certhiomorphæ* (Creepers and Nuthatches). By Hans Gadow, M.A., Ph.D. Pp. xiii., 386. Woodcuts and 9 coloured Plates. [With Systematic and Alphabetical Indexes.] 1883, 8vo. 17*s*.

Vol. X. Catalogue of the Passeriformes, or Perching Birds, in the Collection of the British Museum. *Fringilliformes* : Part I., containing the families Dicæidæ, Hirundinidæ, Ampelidæ, Mniotiltidæ, and Motacillidæ. By R. Bowdler Sharpe. Pp. xiii., 682. Woodcuts and 12 coloured Plates. [With Systematic and Alphabetical Indexes.] 1885, 8vo. 1*l*. 2*s*.

Vol. XI. Catalogue of the Passeriformes, or Perching Birds, in the Collection of the British Museum. *Fringilliformes* : Part II., containing the families Cœrebidæ, Tanagridæ, and Icteridæ. By Philip Lutley Sclater, M.A., F.R.S. Pp. xvii., 431. Woodcuts and 18 coloured Plates. [With Systematic and Alphabetical Indexes.] 1886, 8vo. 1*l*.

Catalogue of the Birds in the British Museum—*continued.*

Vol. XII. Catalogue of the Passeriformes, or Perching Birds, in the Collection of the British Museum. *Fringilliformes*: Part III., containing the family Fringillidæ. By R. Bowdler Sharpe. Pp. xv., 871. Woodcuts and 16 coloured Plates. [With Systematic and Alphabetical Indexes.] 1888, 8vo. 1*l*. 8*s*.

Vol. XIII. Catalogue of the Passeriformes, or Perching Birds, in the Collection of the British Museum. *Sturniformes*, containing the families Artamidæ, Sturnidæ, Ploceidæ, and Alaudidæ. Also the families Atrichiidæ and Menuridæ. By R. Bowdler Sharpe. Pp. xvi., 701. Woodcuts and 15 coloured Plates. [With Systematic and Alphabetical Indexes.] 1890, 8vo. 1*l*. 8*s*.

Vol. XIV. Catalogue of the Passeriformes, or Perching Birds, in the Collection of the British Museum. *Oligomyodæ*, or the families Tyrannidæ, Oxyrhamphidæ, Pipridæ, Cotingidæ, Phytotomidæ, Philepittidæ, Pittidæ, Xenicidæ, and Eurylæmidæ. By Philip Lutley Sclater, M.A., F.R.S. Pp. xix., 494. Woodcuts and 26 coloured Plates. [With Systematic and Alphabetical Indexes.] 1888, 8vo. 1*l*. 4*s*.

Vol. XV. Catalogue of the Passeriformes, or Perching Birds, in the Collection of the British Museum. *Tracheophonæ*, or the families Dendrocolaptidæ, Formicariidæ, Conopophagidæ, and Pteroptochidæ. By Philip Lutley Sclater, M.A., F.R.S. Pp. xvii., 371. Woodcuts and 20 coloured Plates. [With Systematic and Alphabetical Indexes.] 1890, 8vo. 1*l*.

Vol. XVI. Catalogue of the Picariæ in the Collection of the British Museum. *Upupæ* and *Trochili*, by Osbert Salvin. *Coraciæ*, of the families Cypselidæ, Caprimulgidæ, Podargidæ, and Steatornithidæ, by Ernst Hartert. Pp. xvi., 703. Woodcuts and 14 coloured Plates. [With Systematic and Alphabetical Indexes.] 1892, 8vo. 1*l*. 16*s*.

Vol. XVII. Catalogue of the Picariæ in the Collection of the British Museum. *Coraciæ* (contin.) and *Halcyones*, with the families Leptosomatidæ, Coraciidæ, Meropidæ, Alcedinidæ, Momotidæ, Totidæ and Coliidæ, by R. Bowdler Sharpe. *Bucerotes* and *Trogones*, by W. R. Ogilvie Grant. Pp. xi., 522. Woodcuts and 17 coloured Plates. [With Systematic and Alphabetical Indexes.] 1892, 8vo. 1*l*. 10*s*.

Catalogue of the Birds in the British Museum—*continued.*

Vol. XVIII. Catalogue of the Picariæ in the Collection of the British Museum. *Scansores*, containing the family Picidæ. By Edward Hargitt. Pp. xv., 597. Woodcuts and 15 coloured Plates. [With Systematic and Alphabetical Indexes.] 1890, 8vo. 1*l.* 6*s.*

Vol. XIX. Catalogue of the Picariæ in the Collection of the British Museum. *Scansores* and *Coccyges*: containing the families Rhamphastidæ, Galbulidæ, and Bucconidæ, by P. L. Sclater; and the families Indicatoridæ, Capitonidæ, Cuculidæ, and Musophagidæ, by G. E. Shelley. Pp. xii., 484: 13 coloured Plates. [With Systematic and Alphabetical Indexes.] 1891, 8vo. 1*l.* 5*s.*

Vol. XX. Catalogue of the Psittaci, or Parrots, in the Collection of the British Museum. By T. Salvadori. Pp. xvii., 658. Woodcuts and 18 coloured Plates. [With Systematic and Alphabetical Indexes.] 1891, 8vo. 1*l.* 10*s.*

Vol. XXI. Catalogue of the Columbæ, or Pigeons, in the Collection of the British Museum. By T. Salvadori. Pp. xvii., 676. 15 coloured Plates. [With Systematic and Alphabetical Indexes.] 1893, 8vo. 1*l.* 10*s.*

Vol. XXII. Catalogue of the Game Birds (*Pterocletes, Gallinæ, Opisthocomi, Hemipodii*) in the Collection of the British Museum. By W. R. Ogilvie Grant. Pp. xvi., 585. 8 coloured Plates. [With Systematic and Alphabetical Indexes.] 1893, 8vo. 1*l.* 6*s.*

Vol. XXIII. Catalogue of the Fulicariæ (Rallidæ and Heliornithidæ) and Alectorides (Aramidæ, Eurypygidæ, Mesitidæ, Rhinochetidæ, Gruidæ, Psophiidæ, and Otididæ) in the Collection of the British Museum. By R. Bowdler Sharpe. Pp. xiii., 353. 9 coloured Plates. [With Systematic and Alphabetical Indexes.] 1894, 8vo. 1*l.*

Vol. XXIV. Catalogue of the Limicolæ in the Collection of the British Museum. By R. Bowdler Sharpe. Pp. xii., 794. Woodcuts and 7 coloured Plates. [With Systematic and Alphabetical Indexes.] 1896, 8vo. 1*l.* 5*s.*

Vol. XXV. Catalogue of the Gaviæ and Tubinares in the Collection of the British Museum. Gaviæ (Terns, Gulls, and Skuas), by Howard Saunders. Tubinares (Petrels and Albatrosses), by Osbert Salvin. Pp. xv., 475. Woodcuts and 8 coloured Plates. [With Systematic and Alphabetical Indexes.] 1896, 8vo. 1*l.* 1*s.*

Catalogue of Birds in British Museum—*continued.*

Vol. XXVI. Catalogue of the Plataleæ, Herodiones, Steganopodes, Pygopodes, Alcæ, and Impennes in the Collection of the British Museum. Plataleæ (Ibises and Spoonbills) and Herodiones (Herons and Storks), by R. Bowdler Sharpe. Steganopodes (Cormorants, Gannets, Frigate-birds, Tropic-birds, and Pelicans), Pygopodes (Divers and Grebes), Alcæ (Auks), and Impennes (Penguins), by W. R. Ogilvie-Grant. Pp. xvii., 687. Woodcuts and 14 coloured Plates. [With Systematic and Alphabetical Indexes.] 1898, 8vo. 1*l.* 5*s.*

Vol. XXVII. Catalogue of the Chenomorphæ (Palamedeæ, Phoenicopteri, Anseres), Crypturi, and Ratitæ in the Collection of the British Museum. By T. Salvadori. Pp. xv., 636. 19 coloured Plates. [With Systematic and Alphabetical Indexes.] 1895, 8vo. 1*l.* 12*s.*

A Hand-list of the Genera and Species of Birds. [Nomenclator Avium tum Fossilium tum Viventium.] By R. Bowdler Sharpe, LL.D. :—

Vol. II. Pp. xv., 312. [With Systematic Index, and an Alphabetical Index to Vols. I. and II.] 1900, 8vo. 10*s.*

Vol. III. Pp. xii., 367. [With Systematic and Alphabetical Indexes.] 1901, 8vo. 10*s.*

Vol. IV. Pp. xii., 391. [With Systematic and Alphabetical Indexes.] 1903, 8vo. 10*s.*

Vol. V. Pp. xx., 678. [With Systematic and Alphabetical Indexes.] 1909, 8vo. £1.

List of the Specimens of Birds in the Collection of the British Museum. By George Robert Gray :—

Part III., Sections III. and IV. Capitonidæ and Picidæ. Pp. 137. [With Index.] 1868, 12mo. 1*s.* 6*d.*

Part IV. Columbæ. Pp. 73. [With Index.] 1856, 12mo. 1*s.* 9*d.*

Part V. Gallinæ. Pp. iv., 120. [With an Alphabetical Index.] 1867, 12mo. 1*s.* 6*d.*

Catalogue of the Birds of the Tropical Islands of the Pacific Ocean in the Collection of the British Museum. By George Robert Gray, F.L.S., &c. Pp. 72. [With an Alphabetical Index.] 1859, 8vo. 1*s.* 6*d.*

Catalogue of the Collection of Birds' Eggs in the British Museum (Natural History) :—

Vol. I. Ratitæ. Carinatæ (Tinamiformes—Lariformes). By Eugene W. Oates. Pp. xxiii., 252. 18 Coloured Plates. [With Systematic and Alphabetical Indexes.] 1901, 8vo. 1*l*. 10*s*.

Vol. II. Carinatæ (Charadriiformes—Strigiformes). By Eugene W. Oates. Pp. xx., 400. 15 Coloured Plates. [With Systematic and Alphabetical Indexes.] 1902, 8vo. 1*l*. 10*s*.

Vol. III. Carinatæ (Psittaciformes — Passeriformes). By Eugene W. Oates and Capt. Savile G. Reid. Pp. xxiii., 349. 10 Coloured Plates. [With Systematic and Alphabetical Indexes.] 1903, 8vo. 1*l*. 5*s*.

Vol. IV. Carinatæ (Passeriformes continued). By Eugene W. Oates, assisted by Capt. Savile G. Reid. Pp. xviii., 352. 14 Coloured Plates. [With Systematic and Alphabetical Indexes.] 1905, 8vo. 1*l*. 10*s*.

REPTILES.

Catalogue of the Tortoises, Crocodiles, and Amphisbænians in the Collection of the British Museum. By Dr. J. E. Gray, F.R.S., &c. Pp. viii., 80. [With an Alphabetical Index.] 1844, 12mo. 1*s*.

Catalogue of Shield Reptiles in the Collection of the British Museum. By John Edward Gray, F.R.S., &c.:—

Appendix. Pp. 28. 1872, 4to. 2*s*. 6*d*.

Part II. Emydosaurians, Rhynchocephalia, and Amphisbænians. Pp. vi., 41. 25 Woodcuts. 1872, 4to. 3*s*. 6*d*.

Hand-List of the Specimens of Shield Reptiles in the British Museum. By Dr. J. E. Gray, F.R.S., F.L.S., &c. Pp. iv., 124. [With an Alphabetical Index.] 1873, 8vo. 4*s*.

Catalogue of the Chelonians, Rhynchocephalians, and Crocodiles in the British Museum (Natural History). New Edition. By George Albert Boulenger. Pp. x., 311. 73 Woodcuts and 6 Plates. [With Systematic and Alphabetical Indexes.] 1889, 8vo. 15*s*.

Catalogue of the Lizards in the British Museum (Natural History). Second Edition. By George Albert Boulenger :—

Vol. I. Geckonidæ, Eublepharidæ, Uroplatidæ, Pygopodidæ, Agamidæ. Pp. xii., 436. 32 Plates. [With Systematic and Alphabetical Indexes.] 1885, 8vo. 1*l*.

Vol. II. Iguanidæ, Xenosauridæ, Zonuridæ, Anguidæ, Anniellidæ, Helodermatidæ, Varanidæ, Xantusiidæ, Teiidæ, Amphisbænidæ. Pp. xiii., 497. 24 Plates. [With Systematic and Alphabetical Indexes.] 1885, 8vo. 1*l*.

Vol. III. Lacertidæ, Gerrhosauridæ, Scincidæ, Anelytropidæ, Dibamidæ, Chamæleontidæ. Pp. xii., 575. 40 Plates. [With a Systematic Index and an Alphabetical Index to the three volumes.] 1887, 8vo. 1*l*. 6*s*.

Catalogue of the Snakes in the British Museum (Natural History). By George Albert Boulenger, F.R.S., &c. :—

Vol. I., containing the families Typhlopidæ, Glauconiidæ, Boidæ, Ilysiidæ, Uropeltidæ, Xenopeltidæ, and Colubridæ aglyphæ (part). Pp. xiii., 448 : 26 Woodcuts and 28 Plates. [With Systematic and Alphabetical Indexes.] 1893, 8vo. 1*l*. 1*s*.

Vol. II., containing the conclusion of the Colubridæ aglyphæ. Pp. xi., 382 : 25 Woodcuts and 20 Plates. [With Systematic and Alphabetical Indexes.] 1894, 8vo. 17*s*. 6*d*.

Vol. III., containing the Colubridæ (Opisthoglyphæ and Proteroglyphæ), Amblycephalidæ, and Viperidæ. Pp. xiv., 727 : 37 Woodcuts and 25 Plates. [With Systematic Index, and Alphabetical Index to the 3 volumes.] 1896, 8vo. 1*l*. 6*s*.

Catalogue of Colubrine Snakes in the Collection of the British Museum. By Dr. Albert Günther. Pp. xvi., 281. [With Geographic, Systematic, and Alphabetical Indexes.] 1858, 12mo. 4*s*.

BATRACHIANS.

Catalogue of the Batrachia Salientia in the Collection of the British Museum. By Dr. Albert Günther. Pp. xvi., 160. 12 Plates. [With Systematic, Geographic, and Alphabetical Indexes.] 1858, 8vo. 6*s*.

FISHES.

Catalogue of the Fishes in the British Museum. Second edition. Vol. I. Catalogue of the Perciform Fishes in the British Museum. Vol. I. Containing the Centrarchidæ,

Catalogue of Fishes in the British Museum—*continued.*

Percidæ, and Serranidæ (part). By George Albert Boulenger, F.R.S. Pp. xix., 394. Woodcuts and 15 Plates. [With Systematic and Alphabetical Indexes.] 1895, 8vo. 15*s.*

Catalogue of Fish collected and described by Laurence Theodore Gronow, now in the British Museum. Pp. vii., 196. [With a Systematic Index.] 1854, 12mo. 3*s.* 6*d.*

Catalogue of Lophobranchiate Fish in the Collection of the British Museum. By J. J. Kaup, Ph.D., &c. Pp. iv., 80. 4 Plates. [With an Alphabetical Index.] 1856, 12mo. 2*s.*

Catalogue of the Fresh-water Fishes of Africa in the British Museum (Natural History). Vol. I. By G. A. Boulenger, F.R.S. Pp. xi., 373 : 270 text-figures. [With Systematic and Alphabetical Indexes.] 1909, imp. 8vo. £1 12*s.* 6*d.*

MOLLUSCA.

Guide to the Systematic Distribution of Mollusca in the British Museum. Part I. By John Edward Gray, Ph.D., F.R.S., &c. Pp. xii., 230. 121 Woodcuts. 1857, 8vo. 5*s.*

Catalogue of the Collection of Mazatlan Shells in the British Museum, collected by Frederick Reigen. Described by Philip P. Carpenter. Pp. xvi., 552. 1857, 12mo. 8*s.*

Catalogue of Pulmonata, or Air Breathing Mollusca, in the Collection of the British Museum. Part I. By Dr. Louis Pfeiffer. Pp. iv., 192. Woodcuts. 1855, 12mo. 2*s.* 6*d.*

Catalogue of the Auriculidæ, Proserpinidæ, and Truncatellidæ in the Collection of the British Museum. By Dr. Louis Pfeiffer. Pp. iv., 150. Woodcuts. 1857, 12mo. 1*s.* 9*d.*

List of the Mollusca in the Collection of the British Museum. By John Edward Gray, Ph.D., F.R.S., &c. :—

Part II. Olividæ. Pp. 41. 1865, 12mo. 1*s.*

Catalogue of the Conchifera, or Bivalve Shells, in the Collection of the British Museum. By M. Deshayes :—

Part I. Veneridæ, Cyprinidæ, Glauconomidæ, and Petricoladæ. Pp. iv., 216. 1853, 12mo. 3*s.*

Part II. Petricoladæ (concluded); Corbiculadæ. Pp. 217–292. [With an Alphabetical Index to the two parts.] 1854, 12mo. 6*d.*

BRACHIOPODA.

Catalogue of Brachiopoda Ancylopoda or Lamp Shells in the Collection of the British Museum. [*Issued as* "Catalogue of the Mollusca, Part IV."] Pp. iv., 128. 25 Woodcuts. [With an Alphabetical Index.] 1853, 12mo. 3*s*.

POLYZOA.

Catalogue of Marine Polyzoa in the Collection of the British Museum. Part III. Cyclostomata. By George Busk, F.R.S. Pp. viii., 39. 38 Plates. [With a Systematic Index.] 1875, 8vo. 5*s*.

CRUSTACEA.

Catalogue of the Specimens of Amphipodous Crustacea in the Collection of the British Museum. By C. Spence Bate, F.R.S., &c. Pp. iv., 399. 58 Plates. [With an Alphabetical Index.] 1862, 8vo. 1*l*. 5*s*.

ARACHNIDA.

Descriptive Catalogue of the Spiders of Burma, based upon the Collection made by Eugene W. Oates and preserved in the British Museum. By T. Thorell. Pp. xxxvi., 406. [With Systematic List and Alphabetical Index.] 1895, 8vo. 10*s*. 6*d*.

INSECTS.

Coleopterous Insects.

Nomenclature of Coleopterous Insects in the Collection of the British Museum:—

Part VII. Longicornia, I. By Adam White. Pp. iv., 174. 4 Plates. 1853, 12mo. 2*s*. 6*d*.

Part VIII. Longicornia, II. By Adam White. Pp. 237. 6 Plates. 1855, 12mo. 3*s*. 6*d*.

Illustrations of Typical Specimens of Coleoptera in the Collection of the British Museum. Part I. Lycidæ. By Charles Owen Waterhouse. Pp. x., 83. 18 Coloured Plates. [With Systematic and Alphabetical Indexes.] 1879, 8vo. 16*s*.

Catalogue of the Coleopterous Insects of Madeira in the Collection of the British Museum. By T. Vernon Wollaston, M.A., F.L.S. Pp. xvi., 234: 1 Plate. [With a Topographical Catalogue and an Alphabetical Index.] 1857, 8vo. 3*s*.

Catalogue of the Coleopterous Insects of the Canaries in the Collection of the British Museum. By T. Vernon Wollaston, M.A., F.L.S. Pp. xiii., 648. [With Topographical and Alphabetical Indexes.] 1864, 8vo. 10*s*. 6*d*.

Catalogue of Halticidæ in the Collection of the British Museum. By the Rev. Hamlet Clark, M.A., F.L.S. Physapodes and Œdipodes. Part I. Pp. xii., 301. Frontispiece and 9 Plates. 1860, 8vo. 7*s*.

Catalogue of Hispidæ in the Collection of the British Museum. By Joseph S. Baly, M.E.S., &c. Part I. Pp. x., 172. 9 Plates. [With an Alphabetical Index.] 1858, 8vo. 6*s*.

Hymenopterous Insects.

Catalogue of Hymenopterous Insects in the Collection of the British Museum. By Frederick Smith. 12mo.:—

Part I. Andrenidæ and Apidæ. Pp. 197. 6 Plates. 1853, 2*s*. 6*d*.

Part II. Apidæ. Pp. 199–465. 6 Plates. [With an Alphabetical Index.] 1854. 6*s*.

Part III. Mutillidæ and Pompilidæ. Pp. 206. 6 Plates. 1855. 6*s*

Part IV. Sphegidæ, Larridæ, and Crabronidæ. Pp. 207–497. 6 Plates. [With an Alphabetical Index.] 1856. 6*s*.

Part V. Vespidæ. Pp. 147. 6 Plates. [With an Alphabetical Index.] 1857. 6*s*.

Part VI. Formicidæ. Pp. 216. 14 Plates. [With an Alphabetical Index.] 1858. 6*s*.

Part VII. Dorylidæ and Thynnidæ. Pp. 76. 3 Plates. [With an Alphabetical Index.] 1859. 2*s*.

List of Hymenoptera, with descriptions and figures of the Typical Specimens in the British Museum. Vol. I., Tenthredinidæ and Siricidæ. By W. F. Kirby. Pp. xxviii., 450. 16 Coloured Plates. [With Systematic and Alphabetical Indexes.] 1882, 8vo. 1*l*. 18*s*.

Dipterous Insects.

A Monograph of the Culicidæ, or Mosquitoes. Mainly compiled from the Collections received at the British Museum from various parts of the world in connection with the

Investigation into the cause of Malaria conducted by the Colonial Office and the Royal Society. By Fred. V. Theobald, M.A., &c. :—

Vol. III. Pp. xvii., 359 : 17 plates, 1 diagram, and 193 illustrations in text. 1903, 8vo. 1*l*. 1*s*.

Vol. IV. Pp. xix., 639 : 16 plates and 297 text-figures. [With Index.] 1907, 8vo. 1*l*. 12*s*. 6*d*.

A Monograph of the Tsetse-Flies (Genus Glossina, Westwood), based on the Collection in the British Museum. By Ernest Edward Austen. With a chapter on Mouth-parts by H. J. Hansen, Phil. Doc. Pp. ix., 319 : 9 plates (7 coloured), 16 woodcuts, 1 map. 1903, Roy. 8vo. 15*s*.

Illustrations of African Blood-sucking Flies other than Mosquitoes and Tsetse-Flies. By Ernest Edward Austen, with coloured figures by Grace Edwards. Pp. xv., 221 : 13 coloured plates, 3 text-figures. 1909, roy. 8vo. £1 7*s*. 6*d*.

Lepidopterous Insects.

Catalogue of the Lepidoptera Phalænæ in the British Museum. By Sir George F. Hampson, Bart. :—

Vol. I. Catalogue of the Syntomidæ in the Collection of the British Museum. Pp. xxi., 559 : 285 woodcuts. [With Systematic and Alphabetical Indexes.] 1898, 8vo. 15*s*.

——Atlas of 17 Coloured Plates, 8vo. 15*s*.

Vol. II. Catalogue of the Arctiadæ (Nolinæ, Lithosianæ) in the Collection of the British Museum. Pp. xx., 589 : 411 woodcuts. [With Systematic and Alphabetical Indexes.] 1900, 8vo. 18*s*.

——Atlas of 18 Coloured Plates (xviii.–xxxv.), 8vo. 15*s*.

Vol. III. Catalogue of the Arctiadæ (Arctianæ) and Agaristidæ in the Collection of the British Museum. Pp. xix., 690 : 294 woodcuts. [With Systematic and Alphabetical Indexes.] 1901, 8vo. 15*s*.

——Atlas of 19 Coloured Plates (xxxvi–liv.), 8vo. 16*s*.

Vol. IV. Catalogue of the Noctuidæ [Agrotinæ]. Pp. xx., 689 : 125 woodcuts. [With Systematic and Alphabetical Indexes.] 1903, 8vo. 15*s*.

——Atlas of 23 Coloured Plates (lv.–lxxvii), 8vo. 16*s*.

Vol. V. Catalogue of the Noctuidæ [Hadeninæ]. Pp. xvi., 634 : 172 woodcuts. [With Systematic and Alphabetical Indexes.] 1905, 8vo. 15*s*.

——Atlas of 18 Coloured Plates (lxxviii.–xcv.), 8vo. 15*s*.

Catalogue of Lepidoptera Phalænæ—*continued.*

Vol. VI. Catalogue of the Noctuidæ [Cucullianæ]. Pp. xiv., 532: 172 woodcuts. [With Systematic and Alphabetical Indexes.] 1906, 8vo. 15*s.*

——Atlas of 12 Coloured Plates (xcvi.–cvii.), 8vo. 10*s.*

Vol. VII. Catalogue of the Noctuidæ [Acronyctinæ]. Pp. xv., 709: 184 woodcuts. [With Systematic and Alphabetical Indexes.] 1908, 8vo. 17*s.*

——Atlas of 15 Coloured Plates (cviii.–cxxii.), 8vo. 13*s.*

Vol. VIII. Catalogue of the Noctuidæ [Acronyctinæ, II.]. Pp. xiv., 583: 162 woodcuts. [With table of the Phylogeny of the Acronyctinæ, and Systematic and Alphabetical Indexes.] 1909, 8vo. 15*s.*

——Atlas of 14 Coloured Plates (cxxiii.–cxxxvi.), 8vo. 12*s.*

Vol. IX. Catalogue of the Noctuidæ [Acronyctinæ, III.] Pp. xv., 552. 247 Woodcuts. [With Table of the Phylogeny of the Acronyctinæ, and Systematic and Alphabetical Indexes.] 1910, 8vo. 15*s.*

——Atlas of 11 Coloured Plates (cxxxvii.–cxlvii.), 8vo. 12*s.*

Illustrations of Typical Specimens of Lepidoptera Heterocera in the Collection of the British Museum:—

Part V. By Arthur Gardiner Butler. Pp. xii., 74. 78–100 Coloured Plates. [With a Systematic Index.] 1881, 4to. 2*l.* 10*s.*

Part VI. By Arthur Gardiner Butler. Pp. xv., 89. 101–120 Coloured Plates. [With a Systematic Index.] 1886, 4to. 2*l.* 4*s.*

Part VII. By Arthur Gardiner Butler. Pp. iv., 124. 121–138 Coloured Plates. [With a Systematic List.] 1889, 4to. 2*l.*

Part VIII. The Lepidoptera Heterocera of the Nilgiri District. By George Francis Hampson. Pp. iv., 144. 139–156 Coloured Plates. [With a Systematic List.] 1891, 4to. 2*l.*

Part IX. The Macrolepidoptera Heterocera of Ceylon. By George Francis Hampson. Pp. v., 182. 157–176 Coloured Plates. [With a General Systematic List of Species collected in, or recorded from, Ceylon.] 1893, 4to. 2*l.* 2*s.*

Catalogue of the Collection of Palæarctic Butterflies formed by the late John Henry Leech, and presented to the Trustees of the British Museum by his Mother, Mrs. Eliza Leech. By Richard South, F.E.S. Pp. vi., 228. 2 Coloured Plates. With a Portrait and Biographical Memoir of Mr. Leech. 1902, 4to. 1*l.*

Catalogue of Diurnal Lepidoptera described by Fabricius in the Collection of the British Museum. By Arthur Gardiner Butler, F.L.S., &c. Pp. iv., 303. 3 Plates. 1869, 8vo. 7*s.* 6*d.*

List of the Specimens of Lepidopterous Insects in the Collection of the British Museum. By Francis Walker. 12mo.:—

Part XXIII. Geometrites. Pp. 756–1020. 1861. 3*s.* 6*d.*

Part XXV. ———— Pp. 1281–1477. 1862. 3*s.*

Part XXVI. ———— Pp. 1478–1796. [With an Alphabetical Index to Parts XX.–XXVI.] 1862. 4*s.* 6*d.*

Part XXVII. Crambites and Tortricites. Pp. 1–286. 1863. 4*s.*

Part XXVIII. Tortricites and Tineites. Pp. 287–561. 1863. 4*s.*

Part XXIX. Tineites. Pp. 562–835. 1864. 4*s.*

Part XXX. ———— Pp. 836–1096. [With an Alphabetical Index to Parts XXVII.–XXX.] 1864. 4*s.*

Part XXXI. Supplement. Pp. 1–321. 1864. 5*s.*

Part XXXIII.———— Part 3. Pp. 707–1120. 1865. 6*s.*

Part XXXIV.———— Part 4. Pp. 1121–1533. 1865. 5*s.* 6*d.*

Neuropterous Insects.

Catalogue of the Specimens of Neuropterous Insects in the Collection of the British Museum. By Dr. H. Hagen. Part I. Termitina. Pp. 34. 1858, 12mo. 6*d.*

Orthopterous Insects.

Catalogue of Orthopterous Insects in the Collection of the British Museum. Part I. Phasmidæ. By John Obadiah Westwood, F.L.S., &c. Pp. 195. 48 Plates. [With an Alphabetical Index.] 1859, 4to. 3*l.*

Catalogue of the Specimens of Blattariæ in the Collection of the British Museum. By Francis Walker, F.L.S., &c. Pp. 239. [With an Alphabetical Index.] 1868, 8vo. 5*s.* 6*d.*

Catalogue of the Specimens of Dermaptera Saltatoria in the Collection of the British Museum. By Francis Walker, F.L.S., &c. :—

Part II. Locustidæ (continued). Pp. 225–423. [With an Alphabetical Index.] 1869, 8vo. 4*s*. 6*d*.

Part III. Locustidæ (continued).—Acrididæ. Pp. 425–604. [With an Alphabetical Index.] 1870, 8vo. 4*s*.

Part IV. Acrididæ (continued). Pp. 605–809. [With an Alphabetical Index.] 1870, 8vo. 6*s*.

Part V. Tettigidæ.—Supplement to the Catalogue of Blattariæ.—Supplement to the Catalogue of Dermaptera Saltatoria (with remarks on the Geographical Distribution of Dermaptera). Pp. 811–850; 43; 116. [With Alphabetical Indexes.] 1870, 8vo. 6*s*.

Synonymic Catalogue of Orthoptera. By W. F. Kirby :—

Vol. I. Orthoptera Euplexoptera, Cursoria, et Gressoria. (Forficulidæ, Hemimeridæ, Blattidæ, Mantidæ, Phasmidæ.) Pp. x., 501. [With Index.] 1904, 8vo. 10*s*.

Vol. II. Orthoptera Saltatoria, Part I. (Achetidæ et Phasgonuridæ.) Pp. viii., 562. [With Index.] 1906, 8vo. 15*s*.

Vol. III. Orthoptera Saltatoria, Part II. (Locustidæ vel Acridiidæ.) Pp. vii., 674. [With Index.] 1910, 8vo. £1.

Hemipterous Insects.

Catalogue of the Specimens of Heteropterous Hemiptera in the Collection of the British Museum. By Francis Walker, F.L.S., &c. 8vo. :—

Part VI. Pp. 210. [With Alphabetical Index.] 1873. 5*s*.

Part VII. Pp. 213. [With Alphabetical Index.] 1873. 6*s*.

Part VIII. Pp. 220. [With Alphabetical Index.] 1873. [6*s*. 6*d*.

Homopterous Insects.

A Synonymic Catalogue of Homoptera. Part I. Cicadidæ. By W. L. Distant. Pp. 207. [Index.] 1906, 8vo. 5*s*.

VERMES.

Catalogue of the Species of Entozoa, or Intestinal Worms, contained in the Collection of the British Museum. By Dr. Baird. Pp. iv., 132. 2 Plates. [With an Index of the Animals in which the Entozoa mentioned in the Catalogue are found, and an Index of Genera and Species.] 1853, 12mo. 2*s.*

ANTHOZOA.

Catalogue of Sea-pens or Pennatulariidæ in the Collection of the British Museum. By J. E. Gray, F.R.S., &c. Pp. iv., 40. 2 Woodcuts. 1870, 8vo. 1*s.* 6*d.*

Catalogue of Lithophytes or Stony Corals in the Collection of the British Museum. By J. E. Gray, F.R.S., &c. Pp. iv., 51. 14 Woodcuts. 1870, 8vo. 3*s.*

Catalogue of the Madreporarian Corals in the British Museum (Natural History) :—

Vol. I. The Genus Madrepora. By George Brook. Pp. xi., 212. 35 Collotype Plates. [With Systematic and Alphabetical Indexes, and Explanation of the Plates.] 1893, 4to. 1*l.* 4*s.*

Vol. II. The Genus Turbinaria; the Genus Astræopora. By Henry M. Bernard, M.A. Cantab., F.L.S., F.Z.S. Pp. iv., 106. 30 Collotype and 3 Lithographic Plates. [With Index of Generic and Specific Names, and Explanation of the Plates.] 1896, 4to. 18*s.*

Vol. III. The Genus Montipora; the Genus Anacropora. By Henry M. Bernard, M.A., &c. Pp. vii., 192. 30 Collotype and 4 Lithographic Plates. [With Systematic Index, Index of Generic and Specific Names, and Explanation of the Plates.] 1897, 4to. 1*l.* 4*s.*

Vol. IV. The Family Poritidæ. I.—The Genus Goniopora. By Henry M. Bernard, M.A. Pp. viii., 206. 12 Collotype and 4 Lithographic Plates. [With Index of Generic and Specific Names, and Explanation of the Plates.] 1903, 4to. 1*l.*

Vol. V. The Family Poritidæ. II.—The Genus Porites. Part I.—Porites of the Indo-Pacific Region. By Henry M. Bernard, M.A. Pp. vi., 303. 35 Plates. [With Index of Generic and Specific Names and Explanation of the Plates.] 1905, 4to. £1 15*s.*

Vol. VI. The Family Poritidæ. II.—The Genus Porites. Part II.—Porites of the Atlantic and West Indies, with the European Fossil Forms. The Genus Goniopora, a supplement to Vol. IV. By Henry M. Bernard, M.A. Pp. vi., 173. 16 Collotype and 1 Lithographic Plates. [With Index of Generic and Specific Names, and Explanation of the Plates.] 1906, 4to. £1.

BRITISH ANIMALS.

Catalogue of British Birds in the Collection of the British Museum. By George Robert Gray, F.L.S., F.Z.S., &c. Pp. xii., 248. [With a List of Species.] 1863, 8vo. 3*s*. 6*d*.

Catalogue of British Hymenoptera in the Collection of the British Museum. Second edition. Part I. Andrenidæ and Apidæ. By Frederick Smith, M.E.S. New issue. Pp. xi., 236. 11 Plates. [With Systematic and Alphabetical Index. 1891, 8vo. 6*s*.

Catalogue of British Fossorial Hymenoptera, Formicidæ, and Vespidæ in the Collection of the British Museum. By Frederick Smith, V.P.E.S. Pp. 236. 6 Plates. [With an Alphabetical Index.] 1858, 12mo. 6*s*.

Catalogue of British Hymenoptera of the Family Chalcididæ. By Claude Morley, F.Z.S., F.E.S. Pp. 74. [Index.] 1910, 8vo. 3*s*. 6*d*.

Illustrations of British Blood-sucking Flies, with notes by Ernest Edward Austen, Assistant, Department of Zoology, British Museum (N.H.). Pp. 74. 34 Coloured Plates. 1906, roy. 8vo. £1 5*s*.

A Catalogue of the British Non-parasitical Worms in the Collection of the British Museum. By George Johnston, M.D., Edin., F.R.C.L., Ed., LL.D., Marischal Coll., Aberdeen, &c. Pp. 365. Woodcuts and 24 Plates. [With an Alphabetical Index.] 1865, 8vo. 7*s*.

Catalogue of the British Echinoderms in the British Museum (Natural History). By F. Jeffrey Bell, M.A. Pp. xvii., 202. Woodcuts and 16 Plates (2 Coloured). [With Table of Contents, Tables of Distribution, Alphabetical Index, Description of the Plates, &c.] 1892, 8vo. 12*s*. 6*d*.

List of the Specimens of British Animals in the Collection of the British Museum; with Synonyma and References to figures. 12mo.:—

Part V. Lepidoptera. By J. F. Stephens. 2nd Edition. Revised by H. T. Stainton and E. Shepherd. Pp. iv., 224. 1856. 1*s*. 9*d*.

Part VI. Hymenoptera. By F. Smith. Pp. 134. 1851. 2*s*.

Part VII. Mollusca, Acephala and Brachiopoda. By Dr. J. E. Gray. Pp. iv., 167. 1851. 3*s*. 6*d*.

Part VIII. Fish. By Adam White. Pp. xxiii., 164. (With Index and List of Donors.) 1851. 3*s*. 6*d*.

List of the Specimens of British Animals in the Collection of the British Museum—*continued.*

Part IX. Eggs of British Birds. By George Robert Gray. Pp. 143. 1852. 2*s.* 6*d.*

Part XI. Anoplura, or Parasitic Insects. By H. Denny. Pp. iv., 51. 1852. 1*s.*

Part XII. Lepidoptera (continued). By James F. Stephens. Pp. iv., 54. 1852. 9*d.*

Part XIII. Nomenclature of Hymenoptera. By Frederick Smith. Pp. iv., 74. 1853. 1*s.* 4*d.*

Part XIV. Nomenclature of Neuroptera. By Adam White. Pp. iv., 16. 1853. 6*d.*

Part XV. Nomenclature of Diptera, I. By Adam White. Pp. iv., 42. 1853. 1*s.*

Part XVI. Lepidoptera (completed). By H. T. Stainton. Pp. 199. [With an Index.] 1854. 3*s.*

PLANTS.

Illustrations of Australian Plants collected in 1770 during Captain Cook's Voyage round the World in H.M.S. "Endeavour." By the Right Hon. Sir Joseph Banks, Bart., K.B., P.R.S., and Dr. Daniel Solander, F.R.S. [Being a series of lithographic reproductions of copper-plates engraved after paintings by F. P. Nodder, James Miller, J. F. Miller, and John Cleveley.] With Introduction and Determinations by James Britten, F.L.S., Senior Assistant, Department of Botany, British Museum :—

Part I.—101 Plates, with 31 pages of descriptive text. 1900, fol. £1 5*s.*

Part II.—142 Plates (pls. 101–243), with 41 pages of descriptive text (pp. 35–75). 1901, fol. £1 15*s.*

Part III.—77 Plates (pls. 244–318, 45A, and 122), with 26 pages of descriptive text, including Index to the whole work (pp. 77–102), and 3 maps. 1905, fol. £1 5*s.*

Catalogue of the African Plants collected by Dr. Friedrich Welwitsch in 1853–61 :—

Vol. I. Dicotyledons. By William Philip Hiern, M.A., F.L.S., &c. :—

Part I. [Ranunculaceæ to Rhizophoraceæ.] Pd. xxvi., 336. [With Portrait of Dr. Welwitsch. Introduction, Bibliography, and Index of Genera.] 1896. 8vo. 7*s.* 6*d.*

Catalogue of the African Plants collected by Dr. Friedrich Welwitsch in 1853–61—*continued.*

Vol. I.—*continued.*

Part II. Combretaceæ to Rubiaceæ. Pp. 337–510. [With Index of Genera.] 1898, 8vo. 4*s.*

Part III. Dipsaceæ to Scrophulariaceæ. Pp. 511–784. [With Index of Genera.] 1898, 8vo. 5*s.*

Part IV. Lentibulariaceæ to Ceratophylleæ. Pp. 785–1035. [With Index.] 1900, 8vo. 5*s.*

Vol. II. Monocotyledons, Gymnosperms, and Cryptogams :—

Part I. Monocotyledons and Gymnosperms. By Alfred Barton Rendle, M.A., D.Sc., F.L.S., Assistant, Department of Botany. Pp. 260. [With Index of Genera.] 1899, 8vo. 6*s.*

Part II. Cryptogamia. Pp. 261–566. [With Table of Errata, and General Index to the whole work.] 1901, 8vo. 6*s.*

Vascular Cryptogams ...	By	William Carruthers, F.R.S.
Mosses	,,	Antony Gepp, M.A., F.L.S.
Hepatics	,,	F. Stephani.
Marine Algæ	,,	Ethel S. Barton.
Freshwater Algæ	,,	W. West, F.L.S., and G. S. West, B.A.
Diatomaceæ	,,	Thomas Comber, F.L.S.
Lichenes	,,	E. A. Wainio.
Fungi	,,	Annie Lorrain Smith.
Mycetozoa	,,	Arthur Lister, F.R.S.

Synopsis of the British Basidiomycetes : a Descriptive Catalogue of the Drawings and Specimens in the Department of Botany, British Museum. By Worthington George Smith, F.L.S. Pp. 531. 5 Plates and 145 Figures in Text. [With Index.] 1908, 8vo. 10*s.*

A Monograph of Lichens found in Britain : being a Descriptive Catalogue of the Species in the Herbarium of the British Museum. By the Rev. James M. Crombie, M.A., F.L.S., F.G.S., &c. Part I. Pp. viii., 519 : 74 Woodcuts. [With Glossary, Synopsis, Tabular Conspectus, and Index.] 1894, 8vo. 16*s.*

List of British Diatomaceæ in the Collection of the British Museum. By the Rev. W. Smith, F.L.S., &c. Pp. iv., 55. 1859, 12mo. 1*s.*

FOSSILS.

Catalogue of the Fossil Mammalia in the British Museum (Natural History). By Richard Lydekker, B.A., F.G.S.:—

Part I. Containing the Orders Primates, Chiroptera, Insectivora, Carnivora, and Rodentia. Pp. xxx., 268. 33 Woodcuts. [With Systematic and Alphabetical Indexes.] 1885, 8vo. 5*s.*

Part II. Containing the Order Ungulata, Suborder Artiodactyla. Pp. xxii., 324. 39 Woodcuts. [With Systematic and Alphabetical Indexes.] 1885, 8vo. 6*s.*

Part III. Containing the Order Ungulata, Suborders Perissodactyla, Toxodontia, Condylarthra, and Amblypoda. Pp. xvi., 186. 30 Woodcuts. [With Systematic Index, and Alphabetical Index of Genera and Species, including Synonyms.] 1886, 8vo. 4*s.*

Part IV. Containing the Order Ungulata, Suborder Proboscidea. Pp. xxiv., 235. 33 Woodcuts. [With Systematic Index, and Alphabetical Index of Genera and Species, including Synonyms.] 1886, 8vo. 5*s.*

Part V. Containing the Group Tillodontia, the Orders Sirenia, Cetacea, Edentata, Marsupialia, Monotremata, and Supplement. Pp. xxxv., 345. 55 Woodcuts. [With Systematic Index, and Alphabetical Index of Genera and Species, including Synonyms.] 1887, 8vo. 6*s.*

Catalogue of the Fossil Birds in the British Museum (Natural History). By Richard Lydekker, B.A. Pp. xxvii., 368. 75 Woodcuts. [With Systematic Index, and Alphabetical Index of Genera and Species, including Synonyms.] 1891, 8vo. 10*s.* 6*d.*

Catalogue of the Fossil Reptilia and Amphibia in the British Museum (Natural History). By Richard Lydekker, B.A., F.G.S. :—

Part I. Containing the Orders Ornithosauria, Crocodilia, Dinosauria, Squamata, Rhynchocephalia, and Proterosauria. Pp. xxviii., 309. 69 Woodcuts. [With Systematic Index, and Alphabetical Index of Genera and Species, including Synonyms.] 1888, 8vo. 7*s.* 6*d.*

Part II. Containing the Orders Ichthyopterygia and Sauropterygia. Pp. xxi., 307. 85 Woodcuts. [With Systematic Index, and Alphabetical Index of Genera and Species, including Synonyms.] 1889, 8vo. 7*s.* 6*d.*

Part III. Containing the Order Chelonia. Pp. xviii., 239. 53 Woodcuts. [With Systematic Index, and Alphabetical Index of Genera and Species, including Synonyms.] 1889, 8vo. 7*s.* 6*d.*

Catalogue of the Fossil Reptilia and Amphibia in the British Museum—*continued.*

Part IV. Containing the Orders Anomodontia, Ecaudata, Caudata, and Labyrinthodontia; and Supplement. Pp. xxiii., 295. 66 Woodcuts. [With Systematic Index, Alphabetical Index of Genera and Species, including Synonyms, and Alphabetical Index of Genera and Species to the entire work.] 1890, 8vo. 7*s.* 6*d.*

Catalogue of the Fossil Fishes in the British Museum (Natural History). By Arthur Smith Woodward, LL.D., F.R.S., F.G.S., &c. :—

Part I. Containing the Elasmobranchii. Pp. xlvii., 474. 13 Woodcuts and 17 Plates. [With Alphabetical Index, and Systematic Index of Genera and Species.] 1889, 8vo. 1*l.* 1*s.*

Part II. Containing the Elasmobranchii (Acanthodii), Holocephali, Ichthyodorulites, Ostracodermi, Dipnoi, and Teleostomi (Crossopterygii and Chondrostean Actinopterygii). Pp. xliv., 567. 58 Woodcuts and 16 Plates. [With Alphabetical Index, and Systematic Index of Genera and Species.] 1891, 8vo. 1*l.* 1*s.*

Part III. Containing the Actinopterygian Teleostomi of the Orders Chondrostei (concluded), Protospondyli, Aetheospondyli, and Isospondyli (in part). Pp. xlii., 544. 45 Woodcuts and 18 Plates. [With Alphabetical Index, and Systematic Index of Genera and Species.] 1895, 8vo. 1*l.* 1*s.*

Part IV. Containing the Actinopterygian Teleostomi of the Suborders Isospondyli (in part), Ostariophysi, Apodes, Percesoces, Hemibranchii, Acanthopterygii, and Anacanthini. Pp. xxxix., 636. 22 Woodcuts and 19 Plates. [With Alphabetical Index, and Systematic Index of Genera and Species.] 1901, 8vo. 1*l.* 1*s.*

A descriptive Catalogue of the Tertiary Vertebrata of the Fayûm, Egypt. Based on the Collection of the Egyptian Government in the Geological Museum, Cairo, and on the Collection in the British Museum (Natural History), London. By C. W. Andrews, D.Sc. Pp. xxxvii., 324: 98 Text Figures and 26 Plates. [With Systematic and Alphabetical Indexes.] 1906, 4to. 1*l.* 15*s.*

Systematic List of the Edwards Collection of British Oligocene and Eocene Mollusca in the British Museum (Natural History), with references to the type-specimens from similar horizons contained in other collections belonging to the Geological Department of the Museum. By Richard Bullen Newton, F.G.S. Pp. xxviii., 365. [With table of Families and Genera, Bibliography, Correlation-table, Appendix, and Alphabetical Index.] 1891, 8vo. 6*s*.

Catalogue of Tertiary Mollusca in the Department of Geology, British Museum (Natural History). Part I. The Australasian Tertiary Mollusca. By George F. Harris, F.G.S., &c. Pp. xxvi., 407. 8 Plates. [With Table of Families, Genera, and Sub-Genera, and Index.] 1897, 8vo. 10*s*.

Catalogue of the Fossil Cephalopoda in the British Museum (Natural History) :—

Part I. Containing part of the Suborder Nautiloidea, consisting of the families Orthoceratidæ, Endoceratidæ, Actinoceratidæ, Gomphoceratidæ, Ascoceratidæ, Poterioceratidæ, Cyrtoceratidæ, and Supplement. By Arthur H. Foord, F.G.S. Pp. xxxi., 344. 51 Woodcuts. [With Systematic Index, and Alphabetical Index of Genera and Species, including Synonyms.] 1888, 8vo. 10*s*. 6*d*.

Part II. Containing the remainder of the Suborder Nautiloidea, consisting of the families Lituitidæ, Trochoceratidæ, Nautilidæ, and Supplement. By Arthur H. Foord, F.G.S. Pp. xxviii., 407. 86 Woodcuts. [With Systematic Index, and Alphabetical Index of Genera and Species, including Synonyms.] 1891, 8vo. 15*s*.

Part III. Containing the Bactritidæ, and part of the Suborder Ammonoidea. By Arthur H. Foord, Ph.D., F.G.S., and George Charles Crick, A.R.S.M., F.G.S. Pp. xxxiii., 303. 146 Woodcuts. [With Systematic Index of Genera and Species, and Alphabetical Index.] 1897, 8vo. 12*s*. 6*d*.

List of the Types and Figured Specimens of Fossil Cephalopoda in the British Museum (Natural History). By G. C. Crick, F.G.S. Pp. 103. [With Index.] 1898, 8vo. 2*s*. 6*d*.

A Catalogue of British Fossil Crustacea, with their Synonyms and the Range in Time of each Genus and Order. By Henry Woodward, F.R.S. Pp. xii., 155. [With an Alphabetical Index.] 1877, 8vo. 5*s*.

Catalogue of the Fossil Bryozoa in the Department of Geology, British Museum (Natural History):—

The Jurassic Bryozoa. By J. W. Gregory, D.Sc., F.G.S., F.Z.S. Pp. [viii.,] 239 : 22 Woodcuts and 11 Plates. [With List of Species and Distribution, Bibliography, Index, and Explanation of Plates.] 1896, 8vo. 10*s*.

Catalogue of the Fossil Bryozoa in the British Museum—*cont.*

The Cretaceous Bryozoa. By J. W. Gregory, D.Sc., F.R.S., &c. :—

Vol. I. Pp. xiv., 457 : 64 Woodcuts and 17 Plates. [With Index and Explanation of Plates.] 1899, 8vo. 16*s.*

Vol. II. Pp. xlviii., 346. 75 Woodcuts and 9 Plates. [With List of Localities, Bibliography, Subject and Systematic Indexes, and Explanation of Plates.] 1909, 8vo. 13*s.*

Catalogue of the Blastoidea in the Geological Department of the British Museum (Natural History), with an account of the morphology and systematic position of the group, and a revision of the genera and species. By Robert Etheridge, jun., of the Department of Geology, British Museum (Natural History), and P. Herbert Carpenter, D.Sc., F.R.S., F.L.S. (of Eton College). Pp. xv., 322. 20 Plates. [With Preface by Dr. H. Woodward, Table of Contents, General Index, Explanations of the Plates, &c.] 1886, 4to. 1*l.* 5*s.*

The Genera and Species of Blastoidea, with a List of the Specimens in the British Museum (Natural History). By F. A. Bather, M.A., F.G.S., of the Geological Department. Pp. x., 70. 1 Woodcut. 1899, 8vo. 3*s.*

Catalogue of the Palæozoic Plants in the Department of Geology and Palæontology, British Museum (Natural History). By Robert Kidston, F.G.S. Pp. viii., 288. [With a list of works quoted, and an Index.] 1886, 8vo. 5*s.*

Catalogue of the Mesozoic Plants in the Department of Geology, British Museum (Natural History). By A. C. Seward, M.A., F.R.S., F.G.S., University Lecturer in Botany and Fellow of Emanuel College, Cambridge :—

Part I. The Wealden Flora. Part I. Thallophyta—Pteridophyta. Pp. xxxviii., 179. 17 Woodcuts and 11 Plates. [With Alphabetical Index, Explanations of the Plates, &c.] 1894, 8vo. 10*s.*

Part II. The Wealden Flora. Part II. Gymnospermæ. Pp. viii., 259. 9 Woodcuts and 20 Plates. [With Alphabetical Index, Explanations of the Plates, &c.] 1895, 8vo. 15*s.*

Part III. The Jurassic Flora. Part I. The Yorkshire Coast. Pp. xii., 341. 53 Woodcuts and 21 Plates. [With Alphabetical Index, Explanations of the Plates, &c.] 1900, 8vo. 1*l.*

Catalogue of the Mesozoic Plants in the Department of Geology, British Museum—*continued.*

Part IV. The Jurassic Flora. II.—Liassic and Oolitic Floras of England (excluding the Inferior Oolite Plants of the Yorkshire Coast). Pp. xv., 192. 20 Woodcuts and 13 Plates. [With Alphabetical Index, Explanations of the Plates, &c.] 1904, 8vo. 10*s.*

Catalogue of the Fossil Plants of the Glossopteris Flora in the Department of Geology, British Museum (Natural History). Being a Monograph of the Permo-carboniferous Flora of India and the Southern Hemisphere. By E. A. Newell Arber, M.A., F.L.S., F.G.S. Pp. lxxiv., 255 : 51 Text-Figures and 8 Plates. [With Bibliography and Alphabetical Index.] 1905, 8vo. 12*s.* 6*d.*

GUIDE-BOOKS, ETC.

A General Guide to the British Museum (Natural History), Cromwell Road, London, S.W. Twelfth Edition. With 59 woodcuts, 2 plans, 2 views of the building, and an illustrated cover. Pp. x., 117. 1909, 8vo. 3*d.*

Guide to the Specimens illustrating the Races of Mankind (Anthropology), exhibited in the Department of Zoology, British Museum (Natural History). [By R. Lydekker, F.R.S.] Illustrated by 16 Figures. Pp. 31. 1908, 8vo. 4*d.*

Guide to the Galleries of Mammals (other than Ungulates) in the Department of Zoology of the British Museum (Natural History). Eighth Edition. Pp. 101. 52 Woodcuts and 4 plans. [Index.] 1906, 8vo. 6*d.*

Guide to the Great Game Animals (Ungulata) in the Department of Zoology, British Museum (Natural History). Pp. 93. 53 Text and other figures. With list of Horns, Antlers and Tusks, and Index. 1907, 8vo. 1*s.*

Guide to the Elephants (Recent and Fossil) exhibited in the Department of Geology and Palæontology in the British Museum (Natural History). [By Dr. C. W. Andrews, F.R.S.] Illustrated by 31 text-figures. Pp. 46. 1908, 8vo. 6*d.*

Guide to the Specimens of the Horse Family (Equidæ) exhibited in the Department of Zoology, British Museum (Natural History). [By R. Lydekker, F.R.S.] Pp. 42 26 Figures. 1907, 8vo. 1*s.*

Guide to the Domesticated Animals (other than Horses) exhibited in the Central and North Halls of the British Museum (Natural History). [By R. Lydekker, F.R.S.] Illustrated by 24 Figures. Pp. 55. [With table of Contents, List of Illustrations, and Index.] 1908, 8vo. 6*d.*

Guide to the Whales, Porpoises, and Dolphins (order Cetacea) exhibited in the Department of Zoology, British Museum (Natural History). [By R. Lydekker, F.R.S.] Illustrated by 33 Figures. Pp. 47. [With Index.] 1909, 8vo. 4*d*.

Guide to the Gallery of Birds in the Department of Zoology, British Museum (Natural History). [By W. R. Ogilvie Grant.] Second Edition. Pp. iv., 228. 25 Plates, and 7 Illustrations in text. [With Index.] 1910, 4to. 2*s*. 6*d*.

Guide to the Gallery of Birds in the Department of Zoology, British Museum (Natural History). [By W. R. Ogilvie Grant.] :—

Part I. General Series. Pp. 149. [With Index.] 1905, 4to. 6*d*.

Part II. Nesting Series of British Birds. Second Edition. Pp. 62. 4 Plates. [Index.] 1909, 4to. 4*d*.

Guide to the Gallery of Reptilia and Amphibia in the Department of Zoology of the British Museum (Natural History). [By R. Lydekker, F.R.S.] Illustrated by 76 text and other Figures. Pp. iv., 75. [With Table of Contents.] 1906, 8vo. 6*d*.

Guide to the Gallery of Fishes in the Department of Zoology of the British Museum (Natural History). [By Dr. W. G. Ridewood.] Illustrated by 96 Figures. Pp. v., 209. [With Preface by Sir E. Ray Lankester, Table of Classification, and Index.] 1908, 8vo. 1*s*.

Guide to the British Vertebrates Exhibited in the Department of Zoology, British Museum (Natural History). [By W. P. Pycraft.] Pp. iv., 116. 26 Text-Figures, 1 Plan. 1910, 8vo. 1*s*.

Guide to the exhibited series of Insects in the Department of Zoology, British Museum (Natural History). By C. O. Waterhouse. Second Edition. Pp. 65: 62 text- and full-page Illustrations. [With Table of Contents and Index.] 1909, 8vo. 1*s*.

Guide to the Crustacea, Arachnida, Onychophora and Myriopoda exhibited in the Department of Zoology, British Museum (Natural History). (By W. T. Calman, D.Sc., A. S. Hirst, and F. J. Bell.) Pp. 133: 90 Text-Figures. [With Table of Contents and Index.] 1910, 8vo. 1*s*.

Guide to the Shell and Starfish Galleries (Mollusca, Polyzoa, Brachiopoda, Tunicata, Echinoderma, and Worms). Department of Zoology, British Museum (Natural History). Fifth Edition. Pp. iv., 133. 125 Woodcuts, Plan and Indexes. 1908, 8vo. 6*d*.

Guide to the Coral Gallery (Protozoa, Porifera or Sponges, Hydrozoa, and Anthozoa) in the Department of Zoology, British Museum (Natural History). Second Edition. Pp. [iv., 8] 73. 90 Illustrations, Plan and Index. 1907, 8vo. 1*s*.

A Guide to the Fossil Mammals and Birds in the Department of Geology and Palæontology in the British Museum (Natural History). Ninth Edition. [By A. S. Woodward, LL.D., F.R.S.] Pp. xvi., 100. 6 Plates, 88 Woodcuts. [With List of Illustrations, Table of Stratified Rocks, and Index.] 1909, 8vo. 6*d.*

A Guide to the Fossil Reptiles, Amphibians, and Fishes in the Department of Geology and Palæontology in the British Museum (Natural History). Eighth Edition. [By A. S. Woodward, LL.D., F.R.S.] Pp. xviii., 110. 8 Plates and 116 Text-Figures. [With List of Illustrations, Geological Time-Scale and Index.] 1905, 8vo. 6*d.*

A Guide to the Fossil Invertebrate Animals in the Department of Geology and Palæontology in the British Museum (Natural History). [By F. A. Bather, D.Sc.] Pp. ix., 182. 7 Plates and 96 Text-Figures. [With List of Illustrations, Geological Time-scale, and Index.] 1907, 8vo. 1*s.*

List of British Seed-plants and Ferns exhibited in the Department of Botany, British Museum (Natural History). [By A. B. Rendle, D.Sc., F.R.S., and J. Britten, F.L.S.] With table of Sequence of Orders, and Index of Genera Pp. 44. 1907, 8vo. 4*d.*

Guide to Sowerby's Models of British Fungi in the Department of Botany, British Museum (Natural History). Second Edition, revised. By Worthington G. Smith, F.L.S. Pp. 85. 91 Woodcuts. With Table of Diagnostic Characters, Glossary, and Index. 1908, 8vo. 4*d.*

Guide to the British Mycetozoa exhibited in the Department of Botany, British Museum (Natural History). [By Arthur Lister, F.R.S.] Third Edition, revised. Pp. 49. 46 Woodcuts. Index. 1909, 8vo. 3*d.*

A Guide to the Mineral Gallery of the British Museum (Natural History). Tenth Edition. [By L. Fletcher, M.A., F.R.S.] Pp. 32. Plan. 1908, 8vo. 1*d.*

The Student's Index to the Collection of Minerals, British Museum (Natural History). Twenty-third Edition. [By L. Fletcher, M.A., F.R.S.] Pp. 36. With a Plan of the Mineral Gallery. 1908, 8vo. 2*d.*

An Introduction to the Study of Minerals, with a Guide to the Mineral Gallery of the British Museum (Natural History). By L. Fletcher, M.A., F.R.S. Thirteenth Edition. Pp. 123. 41 Woodcuts. With Plan of the Mineral Gallery and Index. 1910, 8vo. 6*d.*

An Introduction to the Study of Rocks and Guide to the Museum Collection. Fourth Edition. By L. Fletcher, M.A., F.R.S. Pp. 155. [With Plan of the Mineral Gallery, Table of Contents, and Index.] 1909, 8vo. 1*s*.

An Introduction to the Study of Meteorites, with a List of the Meteorites represented in the Collection. By L. Fletcher, M.A., F.R.S., &c. Tenth Edition. Pp. 120. [With a Plan of the Mineral Gallery, and an Index to the Meteorites represented in the Collection.] 1908, 8vo. 6*d*.

SPECIAL GUIDES.

No. 2.—Books and Portraits illustrating the History of Plant Classification exhibited in the Department of Botany. Second Edition. [By A. B. Rendle, M.A., D.Sc., F.R.S.] Pp. 19. 4 Plates. 1909, 8vo. 4*d*.

No. 3.—Memorials of Linnæus: a collection of Portraits, Manuscripts, Specimens, and Books exhibited to commemorate the Bicentenary of his Birth. [By A. B. Rendle, M.A., D.Sc.] Pp. 16. 2 Plates. 1907, 8vo. 3*d*.

No. 4.—Memorials of Charles Darwin: a Collection of Manuscripts, Portraits, Medals, Books, and Natural History Specimens to commemorate the Centenary of his Birth and the Fiftieth Anniversary of the Publication of "The Origin of Species." (Second Edition.) [By W. G. Ridewood, D.Sc.] Pp. vi., 50. 2 Plates. 1910, 8vo. 6*d*.

INSTRUCTIONS FOR COLLECTORS.

Handbook of Instructions for Collectors, issued by the British Museum (Natural History). With Illustrations. Third Edition. Pp. 144. Index. 1906, 8vo. 1*s*. 6*d*.

Instructions for Collectors:—

No. 1.—Mammals. Third Edition. Pp. 12. Text illust. 1905, 8vo. 3*d*.

No. 2.—Birds. Fourth Edition. Pp. 10. 5 figures in text. 1908, 8vo. 3*d*.

No. 3.—Reptiles, Batrachians, and Fishes. [Third Edition.] Pp. 12. 1903, 8vo. 4*d*.

No. 4.—Insects. Fourth Edition. Pp. 11. Text illust. 1907, 8vo. 3*d*.

No. 5.—Diptera (Two-winged Flies). Third Edition. Pp. 16. Text illust. 1908, 8vo. 3*d*.

No. 6.—Mosquitoes (Culicidæ). [Third Edition.] Pp. 8. 1 Plate, 1 figure in text. 1904, 8vo. 3*d.*

No. 7.—Blood-sucking Flies, Ticks, &c. By E. E. Austen. Third Edition. Pp. 24 : 13 figures in text. 1907, 8vo. 3*d.*

No. 8.—Spiders, Centipedes, &c. Second Edition. Pp. 4. 1906, 8vo. 3*d.*

No. 9.—Soft-bodied and other Invertebrate Animals ; Shells of Molluscs. Third Edition. Pp. 18. 1909, 8vo. 3*d.*

No. 10.—Plants. Fourth Edition. Pp. 10 : 3 figures in text. 1909, 8vo. 3*d.*

No. 11.—Fossils and Minerals. Third Edition. Pp. 8. 1906, 8vo. 3*d.*

British Museum (Natural History),
Cromwell Road,
London, S.W.
May, 1910.

Made in United States
Troutdale, OR
11/03/2026

56844491R00186